Aerodynamics of Wings and Bodies

HOLT ASHLEY
*Professor, Aeronautics/Astronautics
and Mechanical Engineering, Stanford University*

MARTEN LANDAHL
*Department of Aeronautics and Astronautics
Massachusetts Institute of Technology
    and
Department of Mechanics
The Royal Institute of Technology, Stockholm*

DOVER PUBLICATIONS, INC., NEW YORK

Published in Canada by General Publishing Company, Ltd., 30 Lesmill Road, Don Mills, Toronto, Ontario.
Published in the United Kingdom by Constable and Company, Ltd., 10 Orange Street, London WC2H 7EG.

This Dover edition, first published in 1985, is an unabridged and unaltered republication of the work first published by the Addison-Wesley Publishing Company, Inc., Reading, Mass., in 1965.

Manufactured in the United States of America
Dover Publications, Inc., 31 East 2nd Street, Mineola, N.Y. 11501

Library of Congress Cataloging in Publication Data

Ashley, Holt.
　　Aerodynamics of wings and bodies.

　　Reprint. Originally published: Reading, Mass. : Addison-Wesley, c1965.
　　Bibliography: p.
　　Includes index.
　　1. Aerodynamics.　　I. Landahl, Marten.　　II. Title.
TL570.A76　　1985　　　　629.132′3　　　　　85-6992
ISBN 0-486-64899-0

Preface

The manuscript of this book has gradually evolved from lecture notes for a two-term course presented by the authors to graduate students in the M.I.T. Department of Aeronautics and Astronautics. We shared with some colleagues a concern lest the essential content of classical aerodynamic theory—still useful background for the practice of our profession and the foundation for currently vital research problems—be squeezed out of the aeronautical curriculum by competition from such lively topics as hypersonic fluid mechanics, heat transfer, nonequilibrium phenomena, and magnetogasdynamics. We sought efficiency, a challenge to the enthusiasm of modern students with their orientation toward scientific rigor, and the comprehensiveness that can accompany an advanced point of view. Our initial fear that certain mathematical complexities might submerge physical understanding, or obscure the utility of most of these techniques for engineering applications, now seems unwarranted. The course has been well-received for three years, and it has made a noticeable impact on graduate research in our department.

We are able to offer a textbook that has successfully met its preliminary tests. We have tried to keep it short. Problems and extended numerical demonstrations are omitted in the interest of brevity. We believe that the instructor who essays the pattern of presentation suggested here may find it stimulating to devise his own examples. It is also our hope that working engineers, with a need for the results of modern aerodynamic research and a willingness to accept new analytical tools, will derive something of value.

There is no shortage of books on aerodynamic theory, so let us point to two threads which make this one certainly distinct and possibly an improvement. The first is the realization that the method of matched asymptotic expansions, developed primarily by Kaplun and Lagerstrom, provides a unifying framework for introducing the boundary-value problems of external flow over thin wings and bodies. Not always an unavoidable necessity nor the clearest introduction to a new idea, this method nevertheless rewards the student's patience with a power and open-ended versatility that are startling. For instance, as apparently first realized by Friedrichs, it furnishes a systematic, rigorous explanation of lifting-line theory—the only approach of its kind which we know. In principle, every approximate development carried out along this avenue implies the possibility of improvements to include the higher-order effects of thickness, angle of attack, or any other small parameter characterizing the problem.

Our second innovation is to embrace the important role of the high-speed computer in aerodynamics. Analytical closed-form solutions for simple flow models are certainly invaluable for gaining an understanding of the physical and mathematical structure of a problem. However, rarely are these directly applicable to practical aerodynamic configurations. For such, one usually has to resort to some approximate or numerical scheme. Theoretical predictions are becoming increasingly important as a practical tool supplementing wind-tunnel measurements, in particular for the purpose of aerodynamic optimization taking advantage of the gains to be realized, for example, by employing unusual plan-

iii

form shapes or camber distributions, from imaginative use of wing-body inter-ference, adoption of complicated multiplane lifting surfaces, etc. We have, accordingly, tried to give adequate prominence to techniques for loading pre-diction which go far beyond the more elegant solutions at the price of voluminous routine numerical computation.

The plan of presentation begins with a short review of fundamentals, followed by a larger chapter on inviscid, constant-density flow, which recognizes the special character of that single division of our subject where the exact field differential equation is linear. Chapter 3 introduces the matched asymptotic expansions, as applied to one situation where physical understanding is attained with a minimum of subtlety. Chapter 4 warns the reader about the limitations and penalties for neglecting viscosity, and also illustrates the Kaplun-Lagerstrom method in its most powerful application. Thin airfoils in two dimensions and slender bodies are then analyzed. Three-dimensional lifting surfaces are treated *in extenso*, proceeding from low to high flight speeds and from single, planar configurations to general interfering systems. Chapter 12 attacks the especially difficult topic of steady transonic flow, and the book ends with a brief review on unsteady motion of wings. Only the surface is scratched, however, in these last two discussions.

Although we closely collaborated on all parts of the book, the responsibility for the initial preparation fell on Marten Landahl for Chapters 3, 4, 5, 6, 8, 9, 10, and 12, and on Holt Ashley for Chapters 1, 2, 7, 11, and 13. The first three groups of students in departmental courses 16.071 and 16.072 furnished in-valuable feedback as to the quality of the writing and accuracy of the develop-ment, but any flaws which remain are attributable solely to us.

We recognize the generous assistance of numerous friends and colleagues in bringing this project to fruition. Professor Milton VanDyke of Stanford Uni-versity and Arnold Kuethe of the University of Michigan most helpfully went over the penultimate draft. There were numerous discussions with others in our department, among them Professors Shatswell Ober, Erik Mollö-Christensen, Judson Baron, Saul Abarbanel, and Garabed Zartarian. Drs. Richard Kaplan and Sheila Widnall assisted with preliminary manuscript preparation and class-room examples. Invaluable help was given in the final stages by Dr. Jerzy Kacprzynski with the proofreading and various suggestions for improvements. The art staff of Addison-Wesley Publishing Company drew an excellent set of final figures. Typing and reproduction of the manuscript were skillfully handled by Mrs. Theodate Cline, Mrs. Katherine Cassidy, Mrs. Linda Furcht, Miss Robin Leadbetter, and Miss Ruth Aldrich; we are particularly appreciative of Miss Aldrich's devoted efforts toward completing the final version under trying circumstances. Finally, we wish to acknowledge the support of the Ford Founda-tion. The preparation of the notes on which this book is based was supported in part by a grant made to the Massachusetts Institute of Technology by the Ford Foundation for the purpose of aiding in the improvement of engineering education.

Cambridge, Massachusetts   H.A.,
May 1965   M.T.L.

Table of Contents

To Margit and Frannie,
Two Beautiful Ladies With
More Patience Than We Deserve

Review of
Fundamentals of Fluid Mechanics

1-1 General Assumptions and Basic Differential Equations

Four general assumptions regarding the properties of the liquids and gases that form the subject of this book are made and retained throughout except in one or two special developments:

(1) the fluid is a continuum;
(2) it is inviscid and adiabatic;
(3) it is either a perfect gas or a constant-density fluid;
(4) discontinuities, such as shocks, compression and expansion waves, or vortex sheets, may be present but will normally be treated as *separate and serve as boundaries* for continuous portions of the flow field.

The laws of motion of the fluid will be found derived in any fundamental text on hydrodynamics or gas dynamics. Lamb (1945), Milne-Thompson (1960), or Shapiro (1953) are good examples. The differential equations which apply the basic laws of physics to this situation are the following.*

1. Continuity Equation or Law of Conservation of Mass

$$\frac{\partial \rho}{\partial t} + \nabla \cdot (\rho \mathbf{Q}) \equiv \frac{D\rho}{Dt} + \rho \nabla \cdot \mathbf{Q} = 0, \qquad (1-1)$$

where p, ρ, and T are static pressure, density, and absolute temperature.

$$\mathbf{Q} \equiv U\mathbf{i} + V\mathbf{j} + W\mathbf{k} \qquad (1-2)$$

is the velocity vector of fluid particles. Here $\mathbf{i}$, $\mathbf{j}$, and $\mathbf{k}$ are unit vectors in the x-, y-, and z-directions of Cartesian coordinates. Naturally, components of any vector may be taken in the directions of whatever set of coordinates is most convenient for the problem at hand.

---

* See Notation List for meanings of symbols which are not defined locally in the text. In the following, D/Dt is the substantial derivative or rate of change following a fluid particle.

2. Newton's Second Law of Motion or the Law of Conservation of Momentum

$$\frac{D\mathbf{Q}}{Dt} = \mathbf{F} - \frac{\nabla p}{\rho}, \tag{1-3}$$

where $\mathbf{F}$ is the distant-acting or body force per unit mass. Often we can write

$$\mathbf{F} = \nabla\Omega, \tag{1-4}$$

where Ω is the potential of the force field. For a gravity field near the surface of a locally plane planet with the z-coordinate taken upward, we have

$$\begin{aligned} \mathbf{F} &= -g\mathbf{k} \\ \Omega &= -gz, \end{aligned} \tag{1-5}$$

g being the gravitational acceleration constant.

3. Law of Conservation of Thermodynamic Energy (Adiabatic Fluid)

$$\frac{D}{Dt}\left[e + \frac{Q^2}{2}\right] = -\frac{\nabla \cdot (p\mathbf{Q})}{\rho} + \mathbf{F} \cdot \mathbf{Q}. \tag{1-6}$$

Here e is the internal energy per unit mass, and Q represents the absolute magnitude of the velocity vector $\mathbf{Q}$, a symbolism which will be adopted uniformly in what follows. By introducing the law of continuity and the definition of enthalpy, $h = e + p/\rho$, we can modify (1–6) to read

$$\rho\frac{D}{Dt}\left[h + \frac{Q^2}{2}\right] = \frac{\partial p}{\partial t} + \rho\mathbf{F} \cdot \mathbf{Q}. \tag{1-7}$$

Newton's law can be used in combination with the second law of thermodynamics to reduce the conservation of energy to the very simple form

$$\frac{Ds}{Dt} = 0, \tag{1-8}$$

where s is the entropy per unit mass. It must be emphasized that none of the foregoing equations, (1–8) in particular, can be applied through a finite discontinuity in the flow field, such as a shock. It is an additional consequence of the second law that through an adiabatic shock s can only increase.

4. Equations of State

For a perfect gas,

$$\begin{aligned} p &= R\rho T, & \text{thermally perfect gas} \\ c_p, c_v &= \text{constants}, & \text{calorically perfect gas.} \end{aligned} \tag{1-9}$$

For a constant-density fluid, or incompressible liquid,

$$\rho = \text{constant.} \tag{1–10}$$

In (1–9), c_p and c_v are, of course, the specific heats at constant pressure and constant volume, respectively; in most classical gas-dynamic theory, they appear only in terms of their ratio $\gamma = c_p/c_v$. The constant-density assumption is used in two distinct contexts. First, for flow of liquids, it is well-known to be an excellent approximation under any circumstances of practical importance, in the absence of cavitation. There are, moreover, many situations in a compressible gas where no serious errors result, such as at low subsonic flight speeds for the external flow over aircraft, in the high-density shock layer ahead of a blunt body in hypersonic flight, and in the crossflow field past a slender body performing longitudinal or lateral motions in a subsonic, transonic, or low supersonic airstream.

1–2 Conservation Laws for a Barotropic Fluid in a Conservative Body Force Field

Under the limitations of the present section, it is easily seen that the law of conservation of momentum, (1–3), can be written

$$\mathbf{a} \equiv \frac{D\mathbf{Q}}{Dt} = \nabla \left[\Omega - \int \frac{dp}{\rho} \right]. \tag{1–11}$$

The term "barotropic" implies a unique pressure-density relation throughout the entire flow field; adiabatic-reversible or isentropic flow is the most important special case. As we shall see, (1–11) can often be integrated to yield a useful relation among the quantities pressure, velocity, density, etc., that holds throughout the entire flow.

Another consequence of barotropy is a simplification of Kelvin's theorem of the rate of change of circulation around a path C always composed of the same set of fluid particles. As shown in elementary textbooks, it is a consequence of the equations of motion for inviscid fluid in a conservative body force field that

$$\frac{D\Gamma}{Dt} = -\oint_C \frac{dp}{\rho} = \oint_C T \, ds, \tag{1–12}$$

where

$$\Gamma \equiv \oint_C \mathbf{Q} \cdot d\mathbf{s} \tag{1–13}$$

is the circulation or closed line integral of the tangential component of the velocity vector. Under the present limitations, we see that the middle member of (1–12) is the integral of a single-valued perfect differential and therefore must vanish. Hence we have the result $D\Gamma/Dt = 0$ for all such fluid paths, which means that the circulation is preserved. In particular, if the circulation around a path is initially zero, it will always remain so. The same result holds in a constant-density fluid where the quantity ρ in the denominator can be taken outside, leaving once more a perfect differential; this is true regardless of what assumptions are made about the thermodynamic behavior of the fluid.

1–3 Some Geometric or Kinematic Properties of the Velocity Field

Next we specialize for the velocity vector $\mathbf{Q}$ two integral theorems which are valid for any suitably continuous and differentiable vector field.

1. Gauss' Divergence Theorem. Consider any volume V entirely within the field enclosed by a single closed bounding surface S as in Fig. 1–1.

$$\oiint_S \mathbf{Q} \cdot \mathbf{n} \, dS = \iiint_V (\nabla \cdot \mathbf{Q}) \, dV. \tag{1–14}$$

Here $\mathbf{n}$ is the outward-directed unit normal from any differential element of area dS. This result is derived and discussed, for instance, in Sections 2.60 and 2.61 of Milne-Thompson (1960). Several alternative forms and some interesting deductions from Gauss' theorem are listed in Section 2–61. The theorem relates the tendency of the field lines to diverge, or spread out within the volume V, to the net efflux of these lines from the boundary of V. It might therefore be described as an equation of continuity of field lines.

2. Stokes' Theorem on Rotation. Now we consider a closed curve C of the sort employed in (1–12) and (1–13), except that the present result is instantaneous so that there is no question of a moving path composed of the same particles. Let S be any open surface which has the curve C as its boundary, as illustrated in Fig. 1–2. The theorem refers to the circulation around the curve C and reads

$$\Gamma \equiv \oint_C \mathbf{Q} \cdot d\mathbf{s} = \iint_S \mathbf{n} \cdot (\nabla \times \mathbf{Q}) \, dS. \tag{1–15}$$

Here

$$\nabla \times \mathbf{Q} \equiv \boldsymbol{\zeta} \tag{1–16}$$

is called the vorticity and can be shown to be equal to twice the angular velocity of a fluid particle about an axis through its own centroid. The theorem connects the spinning tendency of the particles lying in surface S with the associated inclination of the fluid at the boundary of S to circulate in one direction or the other. Sections 2.50 and 2.51 of Milne-Thompson (1960) provide a derivation and a number of alternative forms.

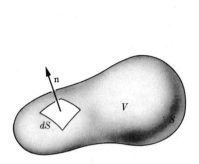

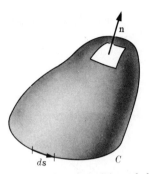

FIG. 1–1.   Finite control volume V surrounded by closed surface S in flow field.

FIG. 1–2.   Open surface S bounded by closed curve C. (Positive relationship of $\mathbf{n}$- and $d\mathbf{s}$-directions as indicated.)

1–4 The Independence of Scale in Inviscid Flows

Consider two bodies of identical shape, but different scales, characterized by the representative lengths l_1 and l_2, moving with the same velocity $\mathbf{Q}_b$ through two unbounded masses of the same inviscid fluid. See Fig. 1–3. This motion is governed by the differential equations developed in Section 1–1 plus the following boundary conditions: (1) disturbances die out at infinity, (2) $Q_n = \mathbf{Q}_b \cdot \mathbf{n}$ at *corresponding* points on the two surfaces, $\mathbf{n}$ being the normal directed into the fluid.

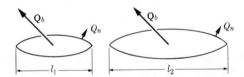

FIG. 1–3. Two bodies of different sizes but identical shapes and motions.

We first treat the case of a steady flow which has gone on for a long period of time so that all derivatives with respect to t vanish. We can then apply the Newtonian transformation, giving the fluid at infinity a uniform motion minus $\mathbf{Q}_b$ and simultaneously bringing the body to rest. Thus the boundary condition No. 2 becomes $Q_n = 0$ all over the surfaces. We take the differential equations governing these two similar problems and make the following changes of variable: in the first case,

$$x_1 = x/l_1, \qquad y_1 = y/l_1, \qquad z_1 = z/l_1; \qquad (1\text{–}17)$$

in the second case,

$$x_2 = x/l_2, \qquad y_2 = y/l_2, \qquad z_2 = z/l_2. \qquad (1\text{–}18)$$

In the dimensionless coordinate systems the two bodies are congruent with each other. Moreover, the differential equations become identical. That is, for instance, continuity reads, in the first case,

$$\frac{\partial(\rho U)}{\partial x_1} + \frac{\partial(\rho V)}{\partial y_1} + \frac{\partial(\rho W)}{\partial z_1} = 0; \tag{1-19}$$

and in the second case,

$$\frac{\partial(\rho U)}{\partial x_2} + \frac{\partial(\rho V)}{\partial y_2} + \frac{\partial(\rho W)}{\partial z_2} = 0. \tag{1-20}$$

We are led to conclude, assuming only uniqueness, that the two flows are identical except for scale. Velocities, gas properties, and all other dependent variables are equal at corresponding points in the two physical fields. This is a result which certainly seems reasonable on grounds of experience.

The same sort of reasoning can be extended to unsteady flows by also scaling time in proportion to length:

$$x_1 = x/l_1, \qquad y_1 = y/l_1, \qquad z_1 = z/l_1, \qquad t_1 = t/l_1, \tag{1-21}$$

$$x_2 = x/l_2, \qquad y_2 = y/l_2, \qquad z_2 = z/l_2, \qquad t_2 = t/l_2, \tag{1-22}$$

in the two cases. With regard to boundary conditions, it must also be specified that over all time $Q_b(t_1)$ in the first case and $Q_b(t_2)$ are identical functions.

Evidently, it makes no difference what the linear scale of an ideal fluid flow is. In what follows, therefore, we shall move back and forth occasionally from dimensional to dimensionless space and time coordinates, even using the same symbols for both. A related simplification, which is often encountered in the literature, consists of stating that the wing chord, body length, etc., will be taken as unity throughout a theoretical development; the time scale is then established by equating the free-stream to unity as well.

One important warning: the introduction of viscosity and heat conduction causes terms containing second derivatives, the viscosity coefficient, and coefficient of heat conductivity to appear in some of the governing differential equations. When dimensionless variables are introduced, the Reynolds number and possibly the Prandtl number* appear as additional

---

* It should be noted that more parameters are introduced when one takes account of the dependence on state of viscosity coefficient, conductivity, and specific heats.

parameters in these equations, spoiling the above-described similarity. The two flows will no longer scale because the Reynolds numbers are different in the two cases. Even in inviscid flow, the appearance of relaxation or of finite reaction rates introduces an additional length scale that destroys the similarity. It is satisfactory, however, for real gas which remains either in chemical equilibrium or in the "frozen" condition thermodynamically.

If shock waves and/or vortex wakes are present in the field, it can be reasoned without difficulty that they also scale in the same manner as a flow without discontinuities. An excellent discussion of this whole subject of invariance to scale changes will be found in Section 2.2 of Hayes and Probstein (1959). Finally, it might be observed that differences between the two flows in the ambient state of the fluid at infinity, such as the pressure, density, and temperature there, may also be included in the scaling by referring the appropriate state variables to these reference values. The Mach numbers in the two flows must then be the same.

1–5 Vortex Theorems for the Ideal Fluid

In connection with the study of wing wakes, separation, and related phenomena, it is of value to study the properties of the field vorticity vector ζ, (1–16). The reader is assumed to be familiar with ways of describing the field of the velocity vector Q and with the concept of an instantaneous pattern of streamlines, drawn at a given time, everywhere tangent to this vector. A related idea is the "stream tube," defined to be a bundle of streamlines sufficiently small that property variations across a normal section are negligible by comparison with variations along the length of the tube. Similar concepts can be defined for any other vector field, in particular the field of ζ. Thus one is led to the idea of a vortex line and a vortex tube, the arrows along such lines and tubes being directed according to the right-hand rule of spin of fluid particles.

Because ζ is the curl of another vector, the field of vortex lines has certain properties that not all vector fields possess. Two of these are identified by the first two vortex theorems of Helmholtz. Although these theorems will be stated for the vorticity field, they are purely geometrical in nature and are unrelated in any way to the physics or dynamics of the fluid, or even to the requirement of continuity of mass.

1. First Vortex Theorem. The circulation around a given vortex tube ("strength" of the vortex) is the same everywhere along its length.

This result can be proved in a variety of ways, one simple approach being to apply Stokes' theorem to a closed path in the surface of the vortex tube constructed as indicated in Fig. 1–4.

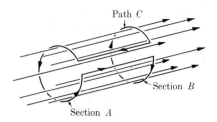

FIG. 1–4. Two cross sections of a vortex tube.

We turn to (1–15) and choose for S the cylindrical surface lying in a wall of the tube. Obviously, no vortex lines cross S, so that

$$\mathbf{n} \cdot \boldsymbol{\zeta} = \mathbf{n} \cdot (\nabla \times \mathbf{Q}) = 0. \qquad (1\text{–}23)$$

Hence the circulation Γ around the whole of the curve C vanishes. By examining C, it is clear that

$$0 = \Gamma = \Gamma_B - \Gamma_A + \text{(two pieces which cancel each other).} \qquad (1\text{–}24)$$

Hence $\Gamma_A = \Gamma_B$. Sections A and B can be chosen arbitrarily, however, so the circulation around the vortex is the same at all sections.

Incidentally, the circulation around the tube always equals $\iint \mathbf{n} \cdot \boldsymbol{\zeta} \, dS$, where the integral is taken over any surface which cuts through the tube but does not intersect any other vortex lines. It can be concluded that this integral has the same value regardless of the orientation of the area used to cut through the tube. A physical interpretation is that the number of vortex lines which go to make up the tube, or bundle, is everywhere the same.

2. Second Vortex Theorem. A vortex tube can never end in the fluid, but must close onto itself, end at a boundary, or go to infinity.

Examples of the three kinds of behavior mentioned in this theorem are a smoke ring, a vortex bound to a two-dimensional airfoil spanning across from one wall to the other in a two-dimensional wind tunnel, and the downstream ends of horseshoe vortices representing the loading on a three-dimensional wing. This second theorem can be quite easily deduced from the continuity of circulation asserted by the first theorem; one simply notes that assuming an end for a vortex tube leads to a situation where the circulation is changing from one section to another along its length.

The first two vortex theorems are closely connected to the fact that the field of $\boldsymbol{\zeta}$ is *solenoidal*, that is,

$$\nabla \cdot \boldsymbol{\zeta} = 0. \qquad (1\text{–}25)$$

When this result is inserted into Gauss' theorem, (1–14), we see that just as many vortex lines must enter any closed surface as leave it.

There is a useful mathematical analogy between the ζ-field and that of the magnetic induction vector $\mathbf{B}$. The latter satisfies one of the basic Maxwell equations,

$$\nabla \cdot \mathbf{B} = 0. \tag{1-26}$$

Although no such analogy generally exists with the fluid velocity vector field, it does so when the density is constant, which simplifies the continuity equation to

$$\nabla \cdot \mathbf{Q} = 0. \tag{1-27}$$

Hence, flow streamlines cannot end, and the volume flux through any section is the same as that through any other section at a given instant of time. One may examine in the same light the field of tubes of the vector $\rho\mathbf{Q}$ in a *steady* compressible flow.

3. Third Vortex Theorem. We now proceed to derive the third vortex theorem, which is connected with the dynamical properties of the fluid. Following Milne-Thompson (1960, Section 3.53), we start from the vector identity

$$\mathbf{a} \equiv \frac{D\mathbf{Q}}{Dt} = \frac{\partial \mathbf{Q}}{\partial t} + (\mathbf{Q} \cdot \nabla)\mathbf{Q} = \frac{\partial \mathbf{Q}}{\partial t} + \nabla \frac{Q^2}{2} - \mathbf{Q} \times \zeta. \tag{1-28}$$

The second step is to take the curl of (1–28), noting that the curl of a gradient vanishes,

$$\nabla \times \mathbf{a} = \nabla \times \frac{\partial \mathbf{Q}}{\partial t} + 0 - \nabla \times (\mathbf{Q} \times \zeta). \tag{1-29}$$

The operations $\nabla\times$ and $\partial/\partial t$ can be interchanged, so that the first term on the right becomes $\partial \zeta/\partial t$. For any two vectors $\mathbf{A}$ and $\mathbf{B}$,

$$\nabla \times (\mathbf{A} \times \mathbf{B}) = (\mathbf{B} \cdot \nabla)\mathbf{A} - (\mathbf{A} \cdot \nabla)\mathbf{B} - \mathbf{B}(\nabla \cdot \mathbf{A}) + \mathbf{A}(\nabla \cdot \mathbf{B}). \tag{1-30}$$

Hence

$$\nabla \times (\mathbf{Q} \times \zeta) = (\zeta \cdot \nabla)\mathbf{Q} - (\mathbf{Q} \cdot \nabla)\zeta - \zeta(\nabla \cdot \mathbf{Q}) + 0. \tag{1-31}$$

Substituting into (1–29), we have

$$\nabla \times \mathbf{a} = \frac{D\zeta}{Dt} - (\zeta \cdot \nabla)\mathbf{Q} + \zeta(\nabla \cdot \mathbf{Q}). \tag{1-32}$$

So far, our results are purely kinematical. We next introduce the conservation of mass, (1–1), second line:

$$\zeta(\nabla \cdot \mathbf{Q}) = -\frac{\zeta}{\rho} \frac{D\rho}{Dt} = \rho\zeta \frac{D}{Dt}\left(\frac{1}{\rho}\right), \tag{1-33}$$

whence, after a little manipulation, (1–32) may be made to read

$$\frac{D}{Dt}\left(\frac{\zeta}{\rho}\right) = \left(\frac{\zeta}{\rho} \cdot \nabla\right)\mathbf{Q} + \frac{\nabla \times \mathbf{a}}{\rho}. \tag{1-34}$$

Under the special conditions behind (1–11), **a** is the gradient of another vector and its curl vanishes. Thus for inviscid, barotropic fluid in a conservative body force field, the foregoing result reduces to

$$\frac{D}{Dt}\left(\frac{\zeta}{\rho}\right) = \left(\frac{\zeta}{\rho} \cdot \nabla\right) Q. \tag{1–35}$$

This last is what is usually known as the third vortex theorem of Helmholtz. In the continuum sense, it is an equation of conservation of angular momentum. If the specific entropy s is not uniform throughout the fluid, one can determine from a combination of dynamical and thermodynamic considerations that

$$\nabla \times \mathbf{a} = \nabla \times (T\nabla s). \tag{1–36}$$

When inserted into (1–34), this demonstrates the role of entropy gradients in generating angular momentum, a result which is often associated with the name of Crocco.

To examine the implications of the third vortex theorem, we shall look at three special cases, in increasing order of complexity. First consider an initially irrotational flow, supposing that at all times previous to some given instant $\zeta = 0$ for all fluid particles. In the absence of singularities or discontinuities, it is possible to write for this initial instant, using (1–35),

$$\frac{D}{Dt}\left(\frac{\zeta}{\rho}\right) = \frac{D^2}{Dt^2}\left(\frac{\zeta}{\rho}\right) = \cdots = \frac{D^n}{Dt^n}\left(\frac{\zeta}{\rho}\right) = \cdots = 0. \tag{1–37}$$

Since the quantity ζ/ρ is an analytic function of space and time, Taylor's theorem shows that it vanishes at all subsequent instants of time. Hence, the vorticity vector itself is zero. We can state that an initially irrotational, inviscid, barotropic flow with a body force potential will remain irrotational. This result can also be proved by a combination of Kelvin's and Stokes' theorems, (1–12) and (1–15).

Next examine a rotational but two-dimensional flow. Here the vorticity vector points in a direction normal to the planes of flow, but derivatives of the velocity **Q** in this direction must vanish. We therefore obtain

$$\frac{D}{Dt}\left(\frac{\zeta}{\rho}\right) = 0. \tag{1–38}$$

Once more, by Taylor's theorem, ζ/ρ remains constant. This is equivalent to the statement that the angular momentum of a fluid particle of fixed mass about an axis through its own center of gravity remains independent of time. In incompressible liquid it reduces to the invariance of vorticity itself, following the fluid.

For our third example we turn to three-dimensional rotational flow. Let us consider an infinitesimal line element ds which moves with the fluid and which at some instant of time is parallel to the vector $\boldsymbol{\zeta}/\rho$. That is,

$$d\mathbf{s} = \epsilon \frac{\boldsymbol{\zeta}}{\rho}, \tag{1-39}$$

where ϵ is a small scalar factor. Since the line element is attached to the fluid particles, the motion of one end relative to the other is determined by the difference in $\mathbf{Q}$ between these ends. Taking the x-component, for instance,

$$\frac{D}{Dt}(dx) = dx\frac{\partial U}{\partial x} + dy\frac{\partial U}{\partial y} + dz\frac{\partial U}{\partial z} = (d\mathbf{s} \cdot \nabla)U. \tag{1-40}$$

In general,

$$\frac{D}{Dt}(d\mathbf{s}) = (d\mathbf{s} \cdot \nabla)\mathbf{Q} = \epsilon\left(\frac{\boldsymbol{\zeta}}{\rho} \cdot \nabla\right)\mathbf{Q}. \tag{1-41}$$

Comparing this last result with the third vortex theorem, (1–35), we are led to

$$\frac{D\epsilon}{Dt} = 0, \qquad \text{or} \qquad \epsilon = \text{const.} \tag{1-42}$$

It follows that (1–39) holds for all subsequent instants of time, and the vector $\boldsymbol{\zeta}/\rho$ moves in the same way as the fluid particles do in three dimensions.

The proportionality of the length of a small fluid element to the quantity $\boldsymbol{\zeta}/\rho$ can be interpreted in terms of conservation of angular momentum in the following way. As implied by (1–39), this length is directed along the axis of spin. Hence, if the length increases, the element itself will shrink in its lateral dimension, and its rate of spin must increase in order to conserve angular momentum. To be precise, the quantity $\boldsymbol{\zeta}/\rho$ is that which increases, because angular momentum is not directly proportional to angular velocity for a variable density particle, but will decrease as the density increases, the density being a measure of how the mass of the fluid particle clusters about the spin axis.

For a constant-density fluid, of course, the vorticity itself is found to be proportional to the length of the fluid particle. In general, we can conclude that vortices are preserved as time passes, and that they cannot decay or disappear except through the action of viscosity or some other dissipative mechanism. Their persistence is revealed by many phenomena in the atmosphere. For example, one often sees the vortex wake, visualized through the mechanism of condensation trails, remaining for many miles after an airplane has passed.

As a final remark, we point out that in flows where only small disturbances from a fixed uniform stream condition occur, it can be proved that the vorticity ζ is preserved (to first order in the small perturbation) in the same way that it is in a constant-density fluid, since the effect of variations in density on the quantity ζ/ρ is of higher order.

1–6 Integral Conservation Theorems for Inviscid Fluid

For later use in connection with the calculation of forces and moments on wings and bodies, we wish to be able to express these quantities in terms of the fluxes of linear and angular momentum through arbitrary control surfaces S. This approach will often be found to have a special convenience, because singularities which occur in velocities and pressures at the surface of a vehicle may not persist at great distances in the flow field around it, so that the integrations which must be carried out are facilitated. The basic tools for carrying out this task are integrated forms of Newton's law of motion known as momentum theorems.

In connection with the presentation of the momentum theorems, we take the opportunity to discuss the question of conservation of other flow properties, as expressed in integral form.

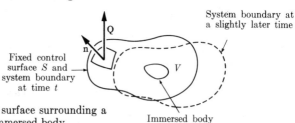

Fixed control surface S and system boundary at time t

System boundary at a slightly later time

Immersed body surface σ

Fig. 1–5. Control surface surrounding a fluid volume and immersed body.

Consider any quantity E which is characteristic of the fluid particles contained within a fixed control volume V. Let V be bounded on the inside by one or more impermeable bodies, whose collective surfaces are denoted by the symbol σ, and bounded on the outside by a larger fixed surface S. See Fig. 1–5. The closed system under examination is that fixed mass of fluid that happens to be contained within V at a certain instant of time. To find the rate of change of the total quantity for this system at the instant it coincides with V, we observe that this change is made up of the sum of all the local changes at points within V plus changes which occur as a result of the motion of the system boundary. Supposing that E is referred to unit volume of the fluid, the former rate of change can be written

$$\frac{\partial}{\partial t} \iiint_V E \, dV = \iiint_V \frac{\partial E}{\partial t} \, dV, \qquad (1\text{--}43)$$

where the interchange of the operations of differentiation and integration is permissible in view of the constancy of the volume V. There is also an increment to the total amount of the quantity E as a result of the fact that the fluid is moving across the bounding surfaces S and σ with a normal velocity component $(\mathbf{Q} \cdot \mathbf{n})$. At points where this scalar product is positive, the fluid adjacent to an area element dS of the boundary takes up new positions outside this boundary, the volume per unit time passing outside the boundary being given by $(\mathbf{Q} \cdot \mathbf{n})\, dS$. Thus the rate of change of E for the system due to passage of fluid across the boundary is given by

$$\oiint_{S+\sigma} E(\mathbf{Q} \cdot \mathbf{n})\, dS.$$

Combining these last two results, we find for the total rate of change of this generalized property for the system,

$$\frac{DE_{\text{tot}}}{Dt} = \iiint_V \frac{\partial E}{\partial t}\, dV + \oiint_{S+\sigma} E(\mathbf{Q} \cdot \mathbf{n})\, dS. \tag{1–44}$$

This general result is now specialized to several cases of interest.

1. Conservation of Mass or Continuity. To derive an integral continuity equation we replace E by the mass per unit volume ρ and observe that, in the absence of sources and sinks, the total mass of the system must remain constant. Thus we are led to

$$\iiint_V \frac{\partial \rho}{\partial t}\, dV + \oiint_{S+\sigma} \rho(\mathbf{Q} \cdot \mathbf{n})\, dS = 0. \tag{1–45}$$

For steady flow around an impermeable body of fixed position, of course, the first integral in (1–45) vanishes, and the contribution to the second integral from the inner surface must be zero because the quantity in parentheses vanishes.

We note incidentally how from Gauss' theorem, (1–14),

$$\oiint_{S+\sigma} \rho(\mathbf{Q} \cdot \mathbf{n})\, dS = \iiint_V \nabla \cdot (\rho \mathbf{Q})\, dV. \tag{1–46}$$

Substituting into (1–45), we obtain

$$\iiint_V \left[\frac{\partial \rho}{\partial t} + \nabla \cdot (\rho \mathbf{Q}) \right] dV = 0. \tag{1–47}$$

Since the volume V is arbitrary, the only way that this integral can vanish is for its integrand to be everywhere zero. Thus, the differential form of the continuity equation, (1–1), is confirmed.

2. Linear Momentum. Let $\sum_i \mathbf{F}_i$ represent the vector sum of all forces applied by the surroundings to the system. According to Newton's second law, this sum is equal to the time rate of change of linear momentum of the system, which corresponds to replacing E with the quantity $\rho \mathbf{Q}$ in (1–44). Consequently, we obtain the following generalized version of the law of conservation of momentum:

$$\sum_i \mathbf{F}_i = \iiint_V \frac{\partial}{\partial t}(\rho \mathbf{Q})\, dV + \oiint_{S+\sigma} \rho \mathbf{Q}(\mathbf{Q} \cdot \mathbf{n})\, dS. \qquad (1\text{–}48)$$

We now examine the various contributions that might appear to the force system in (1–48). If there is a conservative body force field, the left-hand side will include a quantity

$$\iiint_V \rho \mathbf{F}\, dV = \iiint_V \rho \nabla \Omega\, dV. \qquad (1\text{–}49)$$

This will be omitted from what follows because of its relative unimportance in aeronautical applications.

The remaining external force will then be broken into two parts: the reaction $(-\mathbf{F}_{\text{body}})$ to the force exerted by the fluid on the body, and the force exerted across the outer boundary S by the surroundings. Recalling that $\mathbf{n}$ is the outward-directed normal, this latter might take the form

$$\oiint_S [-p\mathbf{n} + \tau]\, dS,$$

where p is the pressure across S, and τ is the sum of shear stress and deviatoric normal stress exerted by the surroundings, if these are significant. We may write τ as the dot product of a dyadic or tensor of deviatoric stress by the unit normal $\mathbf{n}$. Since we are dealing generally with a non-viscous fluid, however, the question will not be elaborated here. Leaving out effects of shear stress, (1–48) can be modified to read

$$\mathbf{F}_{\text{body}} = -\oiint_S p\mathbf{n}\, dS - \oiint_{S+\sigma} \rho \mathbf{Q}(\mathbf{Q} \cdot \mathbf{n})\, dS - \iiint_V \frac{\partial}{\partial t}(\rho \mathbf{Q})\, dV. \qquad (1\text{–}50)$$

Again we remark that if the body is fixed in our coordinate system, the contribution to the second integral on the right from σ will vanish. Also in steady flow the last term on the right is zero, leaving

$$\mathbf{F}_{\text{body}} = -\oiint_S [p\mathbf{n} + \rho \mathbf{Q}(\mathbf{Q} \cdot \mathbf{n})]\, dS. \qquad (1\text{–}51)$$

Equation (1–51) is actually the most useful form for practical applications. The specialized versions of (1–51) which occur when the flow involves small perturbations will be discussed in a later chapter. It usually proves convenient to use an integrated form of the equations of motion (Bernoulli's equation) to replace the pressure in terms of the velocity field.

3. Angular Momentum. Let $\mathbf{r}$ be a vector of position measured from the origin about which moments are to be taken. Then it is an easy matter to derive the following counterpart of the first form of the linear momentum theorem:

$$\sum_i \mathbf{r}_i \times \mathbf{F}_i = \iiint_V \frac{\partial}{\partial t} (\rho \mathbf{r} \times \mathbf{Q}) \, dV + \oiint_{S+\sigma} \rho(\mathbf{r} \times \mathbf{Q})(\mathbf{Q} \cdot \mathbf{n}) \, dS. \quad (1\text{--}52)$$

The summation of moments on the left here can once more be broken up into a body-force term, a reaction to the moment exerted by the fluid through σ and a pressure or shear moment exerted on the system over the outer boundary. Substitutions of this sort, neglecting the deviatoric stress, lead to the working form of the theorem of angular momentum:

$$M_{\text{body}} = - \oiint_S p(\mathbf{r} \times \mathbf{n}) \, dS$$

$$- \oiint_{S+\sigma} \rho(\mathbf{r} \times \mathbf{Q})(\mathbf{Q} \cdot \mathbf{n}) \, dS - \iiint_V \frac{\partial}{\partial t} (\rho \mathbf{r} \times \mathbf{Q}) \, dV. \quad (1\text{--}53)$$

The steady-flow simplification involves dropping the integral over V and over the inner boundary σ.

4. Thermodynamic Energy. Integral forms of the laws of thermodynamics will be found developed in detail in Chapter 2 of Shapiro (1953). Since these will have little direct usefulness in later applications and since many new definitions are involved, none of these results are reproduced here. Shapiro's equation (2.20), for instance, provides an excellent working form of the first law. It is of interest that, when the pressure work exerted on the boundaries is included, the quantity E in the second or boundary term on the right of (1–44) is found to be

$$E = \rho \left[h + \frac{Q^2}{2} + gz \right]. \quad (1\text{--}54)$$

Here z is the distance vertically upward in a parallel gravity field, and h is the enthalpy per unit mass, which proves to be the effective thermodynamic energy in steady flow.

As a sidelight on the question of energy conservation we note that, in a constant-density fluid without body forces, the only way that energy can be stored is in kinetic form. Hence, a very convenient procedure for calculating drag, or fluid resistance, is to find the rate of addition of kinetic energy to the fluid per unit time and to equate this to the work done by the drag. This represents a balance of mechanical rather than thermodynamic energy. When the fluid is compressible and there are still no dissipative mechanisms present, energy can be radiated away by compression waves in an acoustic fashion. Therefore, the problem of computing drag from energy balance becomes a good deal more complicated.

1–7 Irrotational Flow

Enough has been said about the subject of vorticity, its conservation and generation, that it should be obvious that an initially irrotational, uniform, inviscid flow will remain irrotational in the absence of heat transfer and of strong curved shocks. One important consequence of permanent irrotationality is the existence of a velocity potential. That is, the equation

$$\zeta \equiv \nabla \times \mathbf{Q} = 0 \tag{1–55}$$

is a necessary and sufficient condition for the existence of a potential Φ such that

$$\mathbf{Q} = \nabla\Phi, \tag{1–56}$$

where $\Phi(x, y, z, t)$ or $\Phi(\mathbf{r}, t)$ is the potential for the velocity in the entire flow. Its existence permits the replacement of a three-component vector by a single scalar as the principle dependent variable or unknown in theoretical investigations.

Given the existence of Φ, we proceed to derive two important consequences, which will be used repeatedly throughout the work which follows.

1. The Bernoulli Equation for Irrotational Flow (Kelvin's Equation). This integral of the equations of fluid motion is derived by combining (1–3) and (1–28), and assuming a distant acting force potential:

$$\frac{D\mathbf{Q}}{Dt} = \frac{\partial \mathbf{Q}}{\partial t} + \nabla\left(\frac{Q^2}{2}\right) - \mathbf{Q} \times \zeta$$

$$= \frac{\partial \mathbf{Q}}{\partial t} + \nabla\left(\frac{Q^2}{2}\right) = \nabla\Omega - \nabla\int\frac{dp}{\rho}. \tag{1–57}$$

Under our present assumptions,

$$\frac{\partial \mathbf{Q}}{\partial t} = \frac{\partial}{\partial t}(\nabla\Phi) = \nabla\left(\frac{\partial\Phi}{\partial t}\right), \tag{1–58}$$

so that (1–57) can be rearranged into

$$\nabla\left[\frac{\partial\Phi}{\partial t} + \frac{Q^2}{2} + \int\frac{dp}{\rho} - \Omega\right] = 0. \tag{1–59}$$

The vanishing of the gradient implies that, at most, the quantity involved will be a function of time throughout the entire field. Hence the least restricted form of this Bernoulli equation is

$$\frac{\partial \Phi}{\partial t} + \frac{Q^2}{2} + \int \frac{dp}{\rho} - \Omega = F(t). \tag{1-60}$$

In all generality, the undetermined time function here can be eliminated by replacing Φ with

$$\Phi' = \Phi - \int F(t)\, dt. \tag{1-61}$$

This artificiality is usually unnecessary, however, because conditions are commonly known for all time at some reference point in the flow. For instance, suppose there is a uniform stream U_∞ at remote points. There Φ will be constant and the pressure may be set equal to p_∞ and the force potential to Ω_∞ at some reference level.

$$F(t) = \frac{U_\infty^2}{2} + \int^{p_\infty} \frac{dp}{\rho} - \Omega_\infty = \text{const.} \tag{1-62}$$

The simplified version of (1-60) reads

$$\frac{\partial \Phi}{\partial t} + \tfrac{1}{2}[Q^2 - U_\infty^2] + \int_{p_\infty}^{p} \frac{dp}{\rho} + [\Omega_\infty - \Omega] = 0. \tag{1-63}$$

In isentropic flow with constant specific heat ratio γ, (1-63) is easily reorganized into a formula for the local pressure coefficient

$$C_p \equiv \frac{p - p_\infty}{\tfrac{1}{2}\rho_\infty U_\infty^2}$$
$$= \frac{2}{\gamma M^2} \left\{ \left[1 - \frac{\gamma - 1}{a_\infty^2} \left(\frac{\partial \Phi}{\partial t} + \frac{Q^2 - U_\infty^2}{2} + \Omega_\infty - \Omega \right) \right]^{\gamma/(\gamma-1)} - 1 \right\}. \tag{1-64}$$

Here a is the speed of sound and $M \equiv U_\infty/a_\infty$ is Mach number. For certain other purposes, it is convenient to recognize that

$$\int_{p_\infty}^{p} \frac{dp}{\rho} = \int_{a_\infty^2}^{a^2} \frac{d(a^2)}{\gamma - 1} = \frac{1}{\gamma - 1} [a^2 - a_\infty^2], \tag{1-65}$$

where

$$a^2 \equiv \left(\frac{\partial p}{\partial \rho} \right)_s = \frac{dp}{d\rho} = \gamma R T = \frac{\gamma p}{\rho}. \tag{1-66}$$

(Note that, here and below, the particular form chosen for the barotropic relation is isentropic. Under this restriction, $dp/d\rho$, $(\partial p/\partial \rho)_s$, and a^2 all have the same meaning.)

This substitution in (1–63) provides a convenient means of computing the local value of a or of the absolute temperature T,

$$a^2 - a_\infty^2 = -(\gamma - 1)\left\{\frac{\partial\Phi}{\partial t} + \tfrac{1}{2}(Q^2 - U_\infty^2) + (\Omega_\infty - \Omega)\right\}. \quad (1\text{–}67)$$

Finally, we remind the reader that the term containing the body-force potential is usually negligible in aeronautics.

2. The Partial Differential Equation for Φ. By substituting for ρ and Q in the equation of continuity, the differential equation satisfied by the velocity potential can be developed:

$$\frac{1}{\rho}\frac{D\rho}{Dt} + \nabla \cdot \mathbf{Q} = 0. \quad (1\text{–}68)$$

The second term here is written directly in terms of Φ, as follows:

$$\nabla \cdot \mathbf{Q} = \nabla \cdot (\nabla\Phi) \equiv \nabla^2\Phi,$$

which we identify as the familiar Laplacian operator,

$$\left(\nabla^2\Phi = \frac{\partial^2\Phi}{\partial x^2} + \frac{\partial^2\Phi}{\partial y^2} + \frac{\partial^2\Phi}{\partial z^2}, \quad \text{in Cartesians}\right). \quad (1\text{–}69)$$

To modify the first term of (1–68), take the form of Bernoulli's equation appropriate to uniform conditions at infinity, for which, of course, a special case would be that of fluid at rest, $U_\infty = 0$. The body-force term is dropped for convenience, leaving

$$\int_{p_\infty}^{p}\frac{dp}{\rho} = -\frac{\partial\Phi}{\partial t} - \tfrac{1}{2}(Q^2 - U_\infty^2). \quad (1\text{–}70)$$

By the Leibnitz rule for differentiation of a definite integral,

$$\frac{d}{dp}\int_{p_\infty}^{p}\frac{dp}{\rho} = \frac{1}{\rho}. \quad (1\text{–}71)$$

We then apply the substantial derivative operator to (1–71) and make use of the first three members of (1–66),

$$\frac{D}{Dt}\int_{p_\infty}^{p}\frac{dp}{\rho} = \left[\frac{d}{dp}\int_{p_\infty}^{p}\frac{dp}{\rho}\right]\frac{Dp}{Dt}$$

$$= \frac{1}{\rho}\frac{dp}{d\rho}\frac{D\rho}{Dt} = \frac{a^2}{\rho}\frac{D\rho}{Dt}. \quad (1\text{–}72)$$

Dividing by a^2 and introducing the velocity potential, we obtain

$$\frac{1}{\rho} \frac{D\rho}{Dt} = -\frac{1}{a^2} \frac{D}{Dt}\left(\frac{\partial \Phi}{\partial t} + \frac{Q^2}{2}\right)$$

$$= -\frac{1}{a^2}\left(\frac{\partial}{\partial t} + \mathbf{Q} \cdot \nabla\right)\left(\frac{\partial \Phi}{\partial t} + \frac{Q^2}{2}\right)$$

$$= -\frac{1}{a^2}\left[\frac{\partial^2 \Phi}{\partial t^2} + 2\mathbf{Q} \cdot \frac{\partial \mathbf{Q}}{\partial t} + \mathbf{Q} \cdot \nabla\left(\frac{Q^2}{2}\right)\right]. \quad (1\text{-}73)$$

Finally, the insertion of (1–69) and (1–73) into (1–68) produces

$$\nabla^2 \Phi - \frac{1}{a^2}\left[\frac{\partial^2 \Phi}{\partial t^2} + \frac{\partial}{\partial t}\left(Q^2\right) + \mathbf{Q} \cdot \nabla\left(\frac{Q^2}{2}\right)\right] = 0. \quad (1\text{-}74)$$

In view of (1–67) and of the simple relationship between the velocity vector and Φ, this is essentially the desired differential equation. If it is multiplied through by a^2, one sees that it is of third degree in the unknown dependent variable and its derivatives. It reduces to an ordinary wave equation in a situation where the speed of sound does not vary significantly from its ambient values, and where the squares of the velocity components can be neglected by comparison with a^2.

It is of interest that Garrick (1957) has pointed out that (1–74) can be reorganized into

$$\nabla^2 \Phi = \frac{1}{a^2}\left(\frac{\partial}{\partial t} + \mathbf{Q} \cdot \nabla\right)\left(\frac{\partial \Phi}{\partial t} + \mathbf{Q}_c \cdot \nabla \Phi\right) = \frac{1}{a^2} \frac{D_c^2}{Dt^2} \Phi, \quad (1\text{-}75)$$

where the subscript c on $\mathbf{Q}_c$ and on the substantial derivative is intended to indicate that this velocity is treated as a constant during the second application of the operators $\partial/\partial t$ and $(\mathbf{Q} \cdot \nabla)$. Equation (1–75) is just a wave equation (with the propagation speed equal to the local value of a) when the process is observed relative to a coordinate system moving at the local fluid velocity $\mathbf{Q}$.

The question of boundary conditions and the specialization of (1–74) and (1–75) for small-disturbance flows will be deferred to the point where the subject of linearized theory is first taken up.

1–8 The Acceleration Potential

It is of interest that when the equations of fluid motion can be simplified to the form

$$\mathbf{a} \equiv \frac{D\mathbf{Q}}{Dt} = \nabla\left[\Omega - \int \frac{dp}{\rho}\right], \quad (1\text{-}76)$$

it follows immediately that

$$\nabla \times \mathbf{a} = 0. \quad (1\text{-}77)$$

In a manner paralleling the treatment of irrotational flow, we can conclude that

$$\mathbf{a} = \nabla \Psi, \tag{1–78}$$

where $\Psi\,(\mathbf{r}, t)$ is a scalar function called the acceleration potential. Clearly,

$$\Psi = \Omega - \int \frac{dp}{\rho} + G(t), \tag{1–79}$$

$G(t)$ being a function of time that is usually nonessential.

The acceleration potential becomes practically useful when disturbances are small, so that

$$\int_{p_\infty}^{p} \frac{dp}{\rho} \cong \frac{p - p_\infty}{\rho_\infty}. \tag{1–80}$$

In the absence of significant body forces, we then have

$$\Psi \cong \frac{p_\infty - p}{\rho_\infty}. \tag{1–81}$$

This differs only by a constant from the local pressure, and doublets of Ψ prove a very useful tool for representing lifting surfaces. The authors have not been able to construct a suitable partial differential equation for the acceleration potential in the general case, but it satisfies the same equation as the disturbance velocity potential in linearized theory.

$$\mathbf{A} = \Phi\nabla\Phi'. \qquad (2\text{--}3)$$

2

Constant-Density
Inviscid Flow

2-1 Introduction

For the present chapter we adopt all the limitations listed in Section 1–1, plus the following:

(1) $\rho = $ constant everywhere; and
(2) The fluid was initially irrotational.

The former assumption implies essentially an infinite speed of sound, while the latter guarantees the existence of a velocity potential. Turning to (1–74), we see that the flow field is now governed simply by Laplace's equation

$$\nabla^2\Phi = 0. \tag{2-1}$$

Associated with this differential equation, the boundary conditions prescribe the values of the velocity potential or its normal derivative over the surfaces of a series of inner and outer boundaries. These conditions may be given in one of the following forms:

(1) The Neumann problem, in which $v_n = \partial\Phi/\partial n$ is given.
(2) The Dirichlet problem, in which the value of Φ itself is given.
(3) The mixed (Poincaré) problem, in which Φ is given over certain portions of the boundary and $\partial\Phi/\partial n$ is given over the remainder.

A great deal is known about the solution of classical boundary-value problems of this type and more particularly about fluid dynamic applications. Innumerable examples can be found in books like Lamb (1945) and Milne-Thompson (1960). The subject is by no means closed, however, as will become apparent in the light of some of the applications presented in later sections and chapters. A useful recent book, which combines many results of viscous flow theory with old and new developments on the constant-density inviscid problem, is the one edited by Thwaites (1960).

Among the very many concepts and practical solutions that might be considered worthy of presentation, we single out here a few which are especially fundamental and which will prove useful for subsequent work.

2-2 The Three-Dimensional Rigid Solid Moving Through a Liquid

Let us consider a single, finite solid S moving through a large mass of constant-density fluid with an outer boundary Σ, which may for many purposes be regarded as displaced indefinitely toward infinity. For much of what follows, S may consist of several three-dimensional solids rather than a single one. The direction of the normal vector **n** will now be regarded as from the boundary, either the inner boundary S or the outer boundary Σ, toward the fluid volume.

In the early sections of Chapter 3, Lamb (1945) proves the following important but relatively straightforward results for a noncirculatory flow, which are stated here without complete demonstrations:

(1) The flow pattern is determined uniquely at any instant if the boundary values of Φ or $\partial\Phi/\partial n$ are given at all points of S and Σ. One important special case is that of the fluid at rest remote from S; then Φ can be equated to zero at infinity.

(2) The value of Φ cannot have a maximum or minimum at any interior point but only on the boundaries. To be more specific, the mean value of Φ over any spherical surface containing only fluid is equal to the value at the center of this sphere. This result is connected with the interpretation of the Laplacian operator itself, which may be regarded as a measure of the "lumpiness" of the scalar field; Laplace's equation simply states that this "lumpiness" has the smallest possible value in any region.

(3) The magnitude of the velocity vector $Q \equiv |\mathbf{Q}|$ cannot have a maximum in the interior of the flow field but only on the boundary. It can have a minimum value zero at an interior stagnation point.

(4) If $\Phi = 0$ or $\partial\Phi/\partial n = 0$ over all of S and Σ, the fluid will be at rest everywhere. That is, no boundary motion corresponds to no motion in the interior.

Next we proceed to derive some less straightforward results.

1. Green's Theorem. For the moment, let the outer boundary of the flow field remain at a finite distance. We proceed from Gauss' theorem for any vector field **A**, (1–14),

$$\iint\limits_{S+\Sigma} \mathbf{A} \cdot \mathbf{n}\, dS = -\iiint\limits_{V} \nabla \cdot \mathbf{A}\, dV. \tag{2-2}$$

(The minus sign here results from the reversal of direction of the normal vector.) Let Φ and Φ' be two continuous functions with finite, single-valued first and second derivatives throughout the volume V. We do not yet specify that these functions represent velocity potentials of a fluid flow. Let

$$\mathbf{A} = \Phi\nabla\Phi'. \tag{2-3}$$

The integrands in the two sides of (2–2) may be expressed as follows:

$$\mathbf{A} \cdot \mathbf{n} = \Phi[\nabla\Phi' \cdot \mathbf{n}] = \Phi \frac{\partial \Phi'}{\partial n} \qquad (2\text{–}4)$$

$$\nabla \cdot \mathbf{A} = \nabla \cdot (\Phi\nabla\Phi') = \Phi\nabla^2\Phi' + \nabla\Phi \cdot \nabla\Phi'. \qquad (2\text{–}5)$$

Equations (2–4) and (2–5) are now substituted into Gauss' theorem. After writing the result, we interchange the functions Φ and Φ', obtaining two alternative forms of the theorem:

$$\oiint_{S+\Sigma} \Phi \frac{\partial \Phi'}{\partial n} \, dS = -\iiint_V [\nabla\Phi \cdot \nabla\Phi' + \Phi\nabla^2\Phi'] \, dV, \qquad (2\text{–}6)$$

$$\oiint_{S+\Sigma} \Phi' \frac{\partial \Phi}{\partial n} \, dS = -\iiint_V [\nabla\Phi' \cdot \nabla\Phi + \Phi'\nabla^2\Phi] \, dV. \qquad (2\text{–}7)$$

2. Kinetic Energy. As a first illustration of the application of Green's theorem, let Φ in (2–6) be the velocity potential of some flow at a certain instant of time and let $\Phi' = \Phi$. Of course, it follows that

$$\nabla^2\Phi = \nabla^2\Phi' = 0. \qquad (2\text{–}8)$$

We thus obtain a formula for the integral of the square of the fluid particle speed throughout the field

$$\oiint_{S+\Sigma} \Phi \frac{\partial \Phi}{\partial n} \, dS = -\iiint_V |\nabla\Phi|^2 \, dV = -\iiint_V Q^2 \, dV. \qquad (2\text{–}9)$$

Moreover, if we multiply the last member of (2–9) by one-half the fluid density ρ and change its sign, we recognize the total kinetic energy T of the fluid within V. There results

$$T = -\frac{\rho}{2} \oiint_{S+\Sigma} \Phi \frac{\partial \Phi}{\partial n} \, dS. \qquad (2\text{–}10)$$

Such an integral over the boundary is often much easier to evaluate than a triple integral throughout the interior of the volume. In particular, $\partial\Phi/\partial n$ is usually known from the boundary conditions. If the solid S is moving through an unlimited mass of fluid, with $Q = 0$ at infinity, it is a simple matter to prove that the integral over Σ vanishes. It follows that (2–10) need be integrated over only the inner boundary at the surface of the solid itself,

$$T = -\frac{\rho}{2} \oiint_S \Phi \frac{\partial \Phi}{\partial n} \, dS. \qquad (2\text{–}11)$$

.*3. A Reciprocal Theorem.* Another interesting consequence of Green's theorem is obtained by letting Φ and Φ' be the velocity potentials of two different constant-density flows having the same inner and outer bounding surfaces. Then, of course, the two Laplacians in (2–6) and (2–7) vanish, and the right-hand sides of these two relations are found to be equal. Equating the left-hand sides, we deduce

$$\oiint_{S+\Sigma} \Phi \frac{\partial \Phi'}{\partial n} \, dS = \oiint_{S+\Sigma} \Phi' \frac{\partial \Phi}{\partial n} \, dS. \tag{2–12}$$

4. The Physical Interpretation of Φ. To assist in understanding the significance of the last two results and to give a meaning to the velocity potential itself, we next demonstrate an artificial but nevertheless meaningful interpretation of Φ. We begin with Bernoulli's equation in the form (1–63), assuming the fluid at infinity to be at rest and evaluating the pressure integral in consequence of the constancy of ρ,

$$\rho \frac{\partial \Phi}{\partial t} = -\left[(p - p_\infty) + \frac{\rho}{2} Q^2 + \rho(\Omega_\infty - \Omega) \right]. \tag{2–13}$$

Imagine a process in which a system of very large impulsive pressures

$$\mathbf{P} = \int_{(t-\delta t)}^{t} p \, d\tau \tag{2–14}$$

is applied to the fluid, starting from rest, to produce the actual motion existing at a certain time t. In (2–14), τ is a dummy variable of integration. We can make the interval of application of this impulse arbitrarily short, and integrate Bernoulli's equation over it. [Incidentally, the same result is obtainable from the basic equations of fluid motion, (1–3) and (1–4), by a similar integration over the interval δt.] In the limit, the integrals of the pressure p_∞, $\rho Q^2/2$, and $\rho(\Omega_\infty - \Omega)$ become negligible relative to that of the very large p, and we derive

$$\rho \int_{(t-\delta t)}^{t} \frac{\partial \Phi}{\partial \tau} \, d\tau = \rho \Phi(t) = -\mathbf{P}. \tag{2–15}$$

Hence, $\mathbf{P} = -\rho\Phi$ constitutes precisely the system of impulsive pressures required to generate the actual motion swiftly from rest. This process might be carried out, for example, by applying impulsive force and torque to the solid body and simultaneously a suitable distribution of impulsive pressures over the outer boundary Σ. The total impulse thus applied will equal the total momentum in the instantaneous flow described by Φ. Unfortunately, both this momentum and the impulse applied at the outer boundary become indeterminate as Σ spreads outward toward infinity, so that there are certain problems of physical interpretation when dealing with an externally unbounded mass of liquid.

In the light of this interpretation of the velocity potential, we reexamine the kinetic energy in (2–11), rewriting this result

$$T = \frac{1}{2} \oiint_{S+\Sigma} (-\rho\Phi) \frac{\partial\Phi}{\partial n} \, dS = \oiint_{S+\Sigma} \mathbf{P} \frac{v_n}{2} \, dS. \tag{2–16}$$

The work done by an impulse acting on a system which starts from rest is known to be the integral over the boundary of the product of the impulse by one-half the final normal velocity at the boundary. Thus, the starting impulses do a total amount of work given by exactly the last member of (2–16). Since the system is a conservative one, this integral would be expected to equal the change of kinetic energy which, of course, is the final kinetic energy T in the present case. The difficulty in connection with carrying Σ to infinity disappears here, since the work contribution at the outer boundary can be shown to approach zero uniformly. Hence, the kinetic energy of an unbounded mass of constant-density fluid without circulation can be determined entirely from conditions at the inner boundary, and it will always be finite if the fluid is at rest at infinity.

The reciprocal theorem, (2–12), can be manipulated, by multiplication with the density, into the form

$$\oiint_S \mathbf{P} v_n' \, dS = \oiint_S \mathbf{P}' v_n \, dS. \tag{2–17}$$

As such, it becomes a special case of a fairly familiar theorem of dynamics which states that, for any two possible motions of the same system, the sum over all the degrees of freedom of the impulse required to generate one motion multiplied by the velocity in the second motion equals the same summed product taken with the impulses and velocities interchanged.

One final important result is stated without proof: it can be shown that for a given set of boundary conditions the kinetic energy T of a liquid in a finite or infinite region is a minimum when the flow is acyclical and irrotational, relative to all other possible motions.

2–3 The Representation of Φ in Terms of Boundary Values

Let us consider a motion of the type treated in the foregoing sections and examine the question of finding Φ_P at a certain point (x, y, z) arbitrarily located in the flow field. For this purpose we turn to the reciprocal theorem (2–12), making Φ the velocity potential for the actual flow and

$$\Phi' = \frac{1}{r} \equiv \frac{1}{\sqrt{(x - x_1)^2 + (y - y_1)^2 + (z - z_1)^2}}. \tag{2–18}$$

Here (x_1, y_1, z_1) is any other point, one on the boundary for instance. See Fig. 2–1. By direct substitution in Laplace's equation, it is easy to prove that

$$\nabla^2 \Phi' = \nabla^2 \left(\frac{1}{r}\right) = 0 \qquad (2\text{–}19)$$

everywhere except in the immediate vicinity of $r = 0$, where the Laplacian has an impulsive behavior corresponding to the local violation of the requirement of continuity.

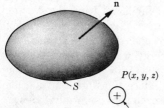

$P(x, y, z)$

Small spherical
surface σ

Fig. 2–1. Arbitrary point P in liquid flow produced by general motion of an inner boundary S.

If we are to use the reciprocal theorem, which requires that the Laplacians of both members vanish, we must exclude the point P from the volume V. This we do by centering a small spherical surface σ around the point, and we obtain

$$\oiint_{S+\Sigma} \Phi \frac{\partial \Phi'}{\partial n} \, dS + \oiint_{\sigma} \Phi \frac{\partial \Phi'}{\partial n} \, d\sigma = \oiint_{S+\Sigma} \Phi' \frac{\partial \Phi}{\partial n} \, dS + \oiint_{\sigma} \Phi' \frac{\partial \Phi}{\partial n} \, d\sigma. \qquad (2\text{–}20)$$

Inserting the value of Φ', we rewrite this

$$\oiint_{\sigma} \Phi \frac{\partial}{\partial n}\left(\frac{1}{r}\right) d\sigma = \oiint_{S+\Sigma} \frac{\partial \Phi}{\partial n} \frac{1}{r} \, dS$$

$$- \oiint_{S+\Sigma} \Phi \frac{\partial}{\partial n}\left(\frac{1}{r}\right) dS + \oiint_{\sigma} \frac{\partial \Phi}{\partial n} \frac{1}{r} \, d\sigma. \qquad (2\text{–}21)$$

If we let σ become a very small sphere, then

$$d\sigma = r^2 \, d\Omega, \qquad (2\text{–}22)$$

where $d\Omega$ is an element of solid angle such that

$$\oiint_{\sigma} d\Omega = 4\pi. \qquad (2\text{–}23)$$

Over the surface of σ,

$$\frac{\partial}{\partial n}\left(\frac{1}{r}\right) = \frac{\partial}{\partial r}\left(\frac{1}{r}\right) = -\frac{1}{r^2}. \tag{2-24}$$

By the mean value theorem of integration, it is possible to replace the finite, continuous Φ by Φ_P in the vicinity of the point P to an acceptable degree of approximation. These considerations lead to

$$\lim_{\sigma \to 0} \iint_\sigma \Phi \frac{\partial}{\partial n}\left(\frac{1}{r}\right) d\sigma = -\Phi_p \lim_{r \to 0} \iint_\sigma \frac{r^2\, d\Omega}{r^2} = -4\pi\Phi_P. \tag{2-25}$$

Moreover, although $\partial\Phi/\partial n$ varies rapidly over σ, it will always be possible to find a bounded average value $(\partial\Phi/\partial n)_{\text{mean}}$ such that

$$\lim_{\sigma \to 0} \iint_\sigma \frac{\partial\Phi}{\partial n} \frac{1}{r}\, d\sigma = \left(\frac{\partial\Phi}{\partial n}\right)_{\text{mean}} \lim_{r \to 0} \iint_\sigma \frac{r^2\, d\Omega}{r} = 0. \tag{2-26}$$

Hence

$$\Phi_P = -\iint_{S+\Sigma} \frac{\partial\Phi}{\partial n} \frac{1}{4\pi r}\, dS + \iint_{S+\Sigma} \Phi \frac{\partial}{\partial n}\left(\frac{1}{4\pi r}\right) dS. \tag{2-27}$$

In the limit as the outer boundary goes to infinity, with the liquid at rest there, one can show that the integrals over Σ vanish, leaving

$$\Phi_P \equiv \Phi(x, y, z;\, t) = -\iint_S \frac{\partial\Phi}{\partial n} \frac{1}{4\pi r}\, dS + \iint_S \Phi \frac{\partial}{\partial n}\left(\frac{1}{4\pi r}\right) dS. \tag{2-28}$$

All the manipulations in (2–27) and (2–28) on the right-hand sides should be carried out using the variables x_1, y_1, z_1. In particular, the normal derivative is expressed in terms of these dummy variables.

The reader will be rewarded by a careful examination of some of the deductions from these results that appear in Sections 57 and 58 of Lamb (1945). For instance, by imagining an artificial fluid motion which goes on in the interior of the bounding surface S, it is possible to reexpress (2–28) entirely in terms of either the boundary values of Φ or its normal derivative. Thus, the determinate nature of the problem when one or the other of these quantities is given from the boundary conditions becomes evident.

The quantities $-1/4\pi r$ and $\partial/\partial n(1/4\pi r)$, which appear in the integrands of (2–27) and (2–28), are fundamental solutions of Laplace's equation that play the role of Green's functions in the representation of the velocity potential. Their names and physical significances are as follows.

1. The Point Source

$$\Phi^S = -\frac{1}{4\pi r}, \qquad (2\text{--}29)$$

where r may be regarded as the radial coordinate in a set of spherical coordinates having the origin at the center of the source. The radial velocity component is

$$U_r = \frac{\partial \Phi^S}{\partial r} = \frac{1}{4\pi r^2}, \qquad (2\text{--}30)$$

whereas the other velocity components vanish. Evidently we have a spherically symmetric outflow with radial streamlines. The volume efflux from the center is easily shown to equal unity. The equipotential surfaces are concentric spheres.

A negative source is referred to as a point sink and has symmetric inflow. When the strength or efflux of the source is different from unity, a strength factor H [dimensions (length)^3 (time)$^{-1}$] is normally applied in (2–29) and (2–30).

2. The Doublet.

The derivative of a source in any arbitrary direction s is called a doublet,

$$\Phi^D = \frac{\partial}{\partial s}\left(\frac{1}{4\pi r}\right). \qquad (2\text{--}31)$$

In particular, a doublet centered at the point (x_1, y_1, z_1) and oriented in the z-direction would have the velocity potential

$$\Phi^D = \frac{\partial}{\partial z}\left(\frac{1}{4\pi r}\right) = -\frac{z - z_1}{4\pi r^3}. \qquad (2\text{--}32)$$

By examining the physical interpretation of the directional derivative, we see that a doublet may be regarded as a source-sink pair of equal strengths, with the line between them oriented in the direction of the doublet's axis, and carried to the limit of infinitesimal separation between them. As this limit is taken, the individual strengths of the source and sink must be allowed to increase in inverse proportion to the separation.

Sources and doublets have familiar two-dimensional counterparts, whose potentials can be constructed by appropriate superposition of the three-dimensional singular solutions. There also exist more complicated and more highly singular solutions of Laplace's equation, which are obtained by taking additional directional derivatives. One involving two differentiations is known as a quadrupole, one with three differentiations an octupole, and so forth.

2-4 Further Examination of the Rigid, Impermeable Solid Moving Through a Constant-Density Fluid Without Circulation

This section reviews some interesting results of incompressible flow theory which highlight the similarities between the dynamics of a rigid body and constant-density flow. They are particularly useful when calculating the resultant forces and moments exerted by the fluid on one or more bodies moving through it. Consider a particular solid body S of the type discussed above. Let there be a set of portable axes attached to the body with origin at O'; $\mathbf{r}$ is the position vector measured instantaneously from O'. Let $\mathbf{u}(t)$ and $\boldsymbol{\omega}(t)$ denote the instantaneous *absolute* linear and angular velocity vectors of the solid relative to the fluid at rest at infinity. See Fig. 2-2. If $\mathbf{n}$ is the outward normal to the surface of S, the boundary condition on the velocity potential is given in terms of the motion by

$$\frac{\partial \Phi}{\partial n} = [\mathbf{u} + \boldsymbol{\omega} \times \mathbf{r}] \cdot \mathbf{n} = \mathbf{u} \cdot \mathbf{n} + \boldsymbol{\omega} \cdot (\mathbf{r} \times \mathbf{n}) \quad \text{on} \quad S = 0. \quad (2\text{-}33)$$

In the second line here the order of multiplication of the triple product has been interchanged in standard fashion.

Now let Φ be written as

$$\Phi = \mathbf{u} \cdot \boldsymbol{\phi} + \boldsymbol{\omega} \cdot \boldsymbol{\chi}. \quad (2\text{-}34)$$

Assuming the components of the vectors $\boldsymbol{\phi}$ and $\boldsymbol{\chi}$ to be solutions of Laplace's equation dying out at infinity, the flow problem will be solved if these coefficients of the linear and angular velocity vectors are made to satisfy the following boundary conditions on S:

$$\frac{\partial \boldsymbol{\phi}}{\partial n} = \mathbf{n}, \quad (2\text{-}35)$$

$$\frac{\partial \boldsymbol{\chi}}{\partial n} = \mathbf{r} \times \mathbf{n}. \quad (2\text{-}36)$$

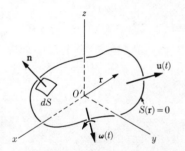

FIG. 2-2. Rigid solid in motion, showing attached coordinate system and linear and angular velocities relative to the liquid at rest.

Evidently $\boldsymbol{\phi}$ and $\boldsymbol{\chi}$ are dependent on the shape and orientation of S, but not on the instantaneous magnitudes of the linear and angular velocities themselves. Thus we see that the total potential will be linearly dependent on the components of $\mathbf{u}$ and $\boldsymbol{\omega}$. Adopting an obvious notation for the various vector components, we write

$$\Phi = u\phi_1 + v\phi_2 + w\phi_3 + p\chi_1 + q\chi_2 + r\chi_3. \quad (2\text{-}37)$$

Turning to (2–11), we determine that the fluid kinetic energy can be expressed as

$$T = -\frac{\rho}{2} \oiint_S \Phi \frac{\partial \Phi}{\partial n} \, dS = -\frac{\rho}{2} \oiint_S [\mathbf{u} \cdot \boldsymbol{\phi} + \boldsymbol{\omega} \cdot \boldsymbol{\chi}][\mathbf{n} \cdot (\mathbf{u} + \boldsymbol{\omega} \times \mathbf{r})] \, dS$$

$$= \tfrac{1}{2}\{Au^2 + Bv^2 + Cw^2 + 2A'vw + 2B'wu + 2C'uv$$

$$+ Pp^2 + Qq^2 + Rr^2 + 2P'qr + 2Q'rp + 2R'pq$$

$$+ 2p[Fu + Gv + Hw] + 2q[F'u + G'v + H'w]$$

$$+ 2r[F''u + G''v + H''w]\}. \tag{2–38}$$

Careful study of (2–37) and the boundary conditions reveals that A, B, C, ... are 21 inertia coefficients directly proportional to density ρ and dependent on the body shape. They are, however, unaffected by the instantaneous motion. Therefore T is a homogeneous, quadratic function of u, v, w, p, q, and r.

By using (2–35) and (2–36) and the reciprocal theorem (2–12), the following examples can be worked out without difficulty:

$$A = -\rho \oiint_S \phi_1 \frac{\partial \phi_1}{\partial n} \, dS = -\rho \oiint_S \phi_1 \gamma_1 \, dS, \tag{2–39}$$

$$A' = -\frac{\rho}{2} \oiint_S \left(\phi_2 \frac{\partial \phi_3}{\partial n} + \phi_3 \frac{\partial \phi_2}{\partial n} \right) dS$$

$$= -\rho \oiint_S \phi_2 \frac{\partial \phi_3}{\partial n} \, dS = -\rho \oiint_S \phi_3 \frac{\partial \phi_2}{\partial n} \, dS$$

$$= -\rho \oiint_S \phi_3 \gamma_2 \, dS. \tag{2–40}$$

Here γ_1 and γ_2 are the direction cosines of the unit normal with respect to the x- and y-axes, respectively.

Some general remarks are in order about the kinetic energy. First, we observe that the introduction of more advanced notation permits a systematization of (2–38). Thus, in terms of matrices,

$$T = \tfrac{1}{2}[u, v, w, p, q, r][m] \begin{Bmatrix} u \\ v \\ w \\ p \\ q \\ r \end{Bmatrix}. \tag{2–41}$$

The first and last factors here are row and column matrices, respectively, while the central one is a 6×6 symmetrical square matrix of inertia coefficients, whose construction is evident from (2–38). In the dyadic or tensor formalism, we can express T

$$T = \tfrac{1}{2}\mathbf{u} \cdot \mathbf{M} \cdot \mathbf{u} + \mathbf{u} \cdot \mathbf{S} \cdot \boldsymbol{\omega} + \tfrac{1}{2}\boldsymbol{\omega} \cdot \mathbf{I} \cdot \boldsymbol{\omega}. \qquad (2\text{–}42)$$

Here **M** and **I** are symmetric tensors of "inertias" and "moments of inertia," while **S** is a nonsymmetrical tensor made up of the inertia coefficients which couple the linear and angular velocity components. For instance, the first of these reads

$$\mathbf{M} = A\mathbf{ii} + B\mathbf{jj} + C\mathbf{kk} + A'[\mathbf{jk} + \mathbf{kj}] + B'[\mathbf{ki} + \mathbf{ik}] + C'[\mathbf{ij} + \mathbf{ji}].$$
$$(2\text{–}43)$$

(Of course, the tensor summation notation might be used in place of dyadics, but the differences are trivial when we are working in Cartesian frames of reference.)

It is suggestive to compare (2–38) with the corresponding formula for the kinetic energy of a rigid body,

$$\begin{aligned}
T_1 = {} & \tfrac{1}{2}m[u^2 + v^2 + w^2] + \tfrac{1}{2}[p^2 I_{xx} + q^2 I_{yy} + r^2 I_{zz}] \\
& - [I_{yz}qr + I_{zx}rp + I_{xy}pq] \\
& + m\{\bar{x}(vr - wq) + \bar{y}(wp - ur) + \bar{z}(uq - vp)\}. \qquad (2\text{–}44)
\end{aligned}$$

Here standard symbols are used for the total mass, moments of inertia, and products of inertia, whereas $\bar{x}$, $\bar{y}$, and $\bar{z}$ are the coordinates of the center of gravity relative to the portable axes with origin at O'. For many purposes, it is convenient to follow Kirchhoff's scheme of analyzing the motion of a combined system consisting of the fluid plus the solid body. It turns out to be possible to derive Lagrangian equations of motion for this combined system, which are little more complex than those for the solid alone.

Another important point concerns the existence of principal axes. It is well known that the number of inertia coefficients for the rigid body can be reduced from ten to four by working with principal axes having their origin at the center of gravity. By analogy with this result, or by reference to the tensor character of the arrays of inertia coefficients, one can show that an appropriate rotation of axes causes the off-diagonal terms of **M** to vanish, while a translation of the origin O' relative to S converts **I** to a diagonal, thus reducing the 21 to 15 inertia coefficients. Sections 124 to 126 of Lamb (1945) furnish the details.

For present purposes, we focus on how the foregoing results can be used to determine the forces and moments exerted by the fluid on the solid.

It is a well-known theorem of dynamics, for a system without potential energy, that the instantaneous external force $\mathbf{F}$ and couple $\mathbf{M}$ are related to the instantaneous linear momentum $\boldsymbol{\xi}$ and angular momentum $\boldsymbol{\lambda}$ by

$$\mathbf{F} = \frac{d\boldsymbol{\xi}}{dt}, \tag{2-45}$$

$$\mathbf{M} = \frac{d\boldsymbol{\lambda}}{dt}. \tag{2-46}$$

Furthermore, these momenta are derivable from the kinetic energy by equations which may be abbreviated

$$\boldsymbol{\xi} = \frac{\partial T_1}{\partial \mathbf{u}}, \tag{2-47}$$

$$\boldsymbol{\lambda} = \frac{\partial T_1}{\partial \boldsymbol{\omega}}. \tag{2-48}$$

The notation for the derivatives is meaningful if the kinetic energy is written in tensor form, (2–42). To assist in understanding, we write out the component forms of (2–47),

$$\xi_1 = \partial T_1/\partial u, \ \ \xi_2 = \partial T_1/\partial v, \ \ \xi_3 = \partial T_1/\partial w. \tag{2-49a,b,c}$$

Lagrange's equations of motion in vector notation are derived by combining (2–45) through (2–48), as follows:

$$\mathbf{F} = \frac{d}{dt}\left(\frac{\partial T_1}{\partial \mathbf{u}}\right), \tag{2-50}$$

$$\mathbf{M} = \frac{d}{dt}\left(\frac{\partial T_1}{\partial \boldsymbol{\omega}}\right). \tag{2-51}$$

The time derivatives here are taken with respect to a nonrotating, non-translating system of inertial coordinates. Some authors prefer these results in terms of a time rate of change of the momenta as seen by an observer rotating with a system of portable axes. If this is done, and if we recognize that there can be a rate of change of angular momentum due to a translation of the linear momentum vector parallel to itself, we can replace (2–45) and (2–46) with

$$\mathbf{F} = \left(\frac{d\boldsymbol{\xi}}{dt}\right)_p + \boldsymbol{\omega} \times \boldsymbol{\xi}, \tag{2-52}$$

$$\mathbf{M} = \left(\frac{d\boldsymbol{\lambda}}{dt}\right)_p + \boldsymbol{\omega} \times \boldsymbol{\lambda} + \mathbf{u} \times \boldsymbol{\xi}. \tag{2-53}$$

The subscript p here refers to the aforementioned differentiation in portable axes. Corresponding corrections to the Lagrange equations are evident.

Since the foregoing constitute a result of rigid-body mechanics, we attempt to see how these ideas can be extended to the surface S moving through an infinite mass of constant-density fluid. For this purpose, we associate with $\boldsymbol{\xi}$ and $\boldsymbol{\lambda}$ the impulsive force and impulsive torque which would have to be exerted over the surface of the solid to produce the motion instantaneously from rest. (Such a combination is referred to as a "wrench.") In view of the relationship between impulsive pressure and velocity potential, the force and torque can be written

$$\boldsymbol{\xi} = -\iint_S \rho \Phi \mathbf{n} \, dS, \tag{2–54}$$

$$\boldsymbol{\lambda} = -\iint_S \mathbf{r} \times (\rho \Phi \mathbf{n}) \, dS. \tag{2–55}$$

These so-called "Kelvin impulses" are no longer equal to the total fluid momenta; the latter are known to be indeterminate in view of the non-vanishing impulses applied across the outer boundary in the limit as it is taken to infinity. Nevertheless, we shall show that the instantaneous force and moment exerted by the body on the liquid in the actual situation are determined from the time rates of change of $\boldsymbol{\xi}$ and $\boldsymbol{\lambda}$.

To derive the required relationship, we resort to a partially physical reasoning that follows Chapter 6 of Lamb (1945). Let us take the linear force and linear impulse for illustration purposes and afterwards deduce the result for the angular quantities by analogy.

Consider any flow with an inner boundary S and an outer boundary Σ, which will later be carried to infinity, and let the velocity potential be Φ. Looking at a short interval between a time t_0 and a time t_1, let us imagine that just prior to t_0 the flow was brought up from rest by a system of impulsive forces $(-\rho \Phi_0)$ and that, just after t_1, it was stopped by $-(-\rho \Phi_1)$. The total time integral of the pressure at any point can be broken up into two impulsive pieces plus a continuous integral over the time interval.

$$\int_{t_0^-}^{t_1^+} p \, dt = (-\rho \Phi_0) - (-\rho \Phi_1) + \int_{t_0}^{t_1} p \, dt. \tag{2–56}$$

Turning to Bernoulli's equation, (2–13), and neglecting the body forces as nonessential, we can also write this integral as follows:

$$\int_{t_0^-}^{t_1^+} p \, dt = -\rho \int_{t_0^-}^{t_1^+} \frac{\partial \Phi}{\partial t} \, dt - \rho \int_{t_0^-}^{t_1^+} \frac{Q^2}{2} \, dt + \text{const} \, [t_1 - t_0]$$

$$= 0 - \rho \int_{t_0}^{t_1} \frac{Q^2}{2} \, dt + \text{const} \, [t_1 - t_0]. \tag{2–57}$$

The first integral on the right here vanishes because the fluid is at rest prior to the starting impulse and after the final one. Also these impulses make no appreciable contribution to the integral of $Q^2/2$.

Suppose now that we integrate these last two equations over the inner and outer boundaries of the flow field, simultaneously applying the unit normal vector $\mathbf{n}$ so as to get the following impulses of resultant force at these boundaries:

$$\iint_S \mathbf{n} \int_{t_0-}^{t_1+} p \, dt \, dS; \qquad \iint_\Sigma \mathbf{n} \int_{t_0-}^{t_1+} p \, dt \, d\Sigma.$$

We assert that each of these impulses must separately be equal to zero if the outer boundary Σ is permitted to pass to infinity. This fact is obvious if we do the integration on the last member of (2–57),

$$\iint_\Sigma \mathbf{n} \int_{t_0-}^{t_1+} p \, dt \, d\Sigma = -\rho \int_{t_0}^{t_1} \mathbf{n} \iint_\Sigma \frac{Q^2}{2} \, d\Sigma \, dt + 0 \to 0 \quad \text{as} \quad \Sigma \to \infty \tag{2–58}$$

The integral of the term containing the constant in (2–57) vanishes since a uniform pressure over a closed surface exerts no resultant force. Furthermore, the integral of $Q^2/2$ vanishes in the limit because $\mathbf{Q}$ can readily be shown to drop off at least as rapidly as the inverse square of the distance from the origin. The overall process starts from a condition of rest and ends with a condition of rest, so that the resultant impulsive force exerted over the inner boundary and the outer boundary must be zero. Equation (2–58) shows that this impulsive force vanishes separately at the outer boundary, so we are led to the result

$$\iint_S \mathbf{n} \int_{t_0-}^{t_1+} p \, dt \, dS = 0. \tag{2–59}$$

In view of (2–59), the right-hand member of (2–56) can then be integrated over S and equated to zero, leading to

$$\iint_S \mathbf{n} \int_{t_0}^{t_1} p \, dt \, dS = \int_{t_0}^{t_1} \mathbf{F} \, dt = \iint_S [-\rho\Phi_1 - (-\rho\Phi_0)]\mathbf{n} \, dS = \boldsymbol{\xi}_1 - \boldsymbol{\xi}_0. \tag{2–60}$$

Finally, we hold t_0 constant and differentiate the second and fourth members of (2–60) with respect to t_1. Replacing t_1 with t, since it may be regarded as representing any given instant, we obtain

$$\mathbf{F} = \frac{d\boldsymbol{\xi}}{dt}. \tag{2–61}$$

This is the desired result. It is a simple matter to include moment arms in the foregoing development, thus working with angular rather than linear momentum, and derive

$$\mathbf{M} = \frac{d\boldsymbol{\lambda}}{dt}, \tag{2-62}$$

where $\boldsymbol{\lambda}$ is the quantity defined by (2–55). The instantaneous force and instantaneous moment about an axis through the origin O' exerted by the body on the fluid are, respectively, $\mathbf{F}$ and $\mathbf{M}$.

If we combine with the foregoing the considerations which led to (2–52) and (2–53), we can reexpress these last relationships in terms of rates of change of the Kelvin linear and angular impulses seen by an observer moving with the portable axes,

$$\mathbf{F} = \left(\frac{d\boldsymbol{\xi}}{dt}\right)_p + \boldsymbol{\omega} \times \boldsymbol{\xi}, \tag{2-63}$$

$$\mathbf{M} = \left(\frac{d\boldsymbol{\lambda}}{dt}\right)_p + \boldsymbol{\omega} \times \boldsymbol{\lambda} + \mathbf{u} \times \boldsymbol{\xi}. \tag{2-64}$$

As might be expected by comparison with rigid body mechanics, $\boldsymbol{\xi}$ and $\boldsymbol{\lambda}$ can be obtained from the fluid kinetic energy T. The rather artificial and tedious development in Lamb (1945) can be bypassed by carefully examining (2–38). The integrals there representing the various inertia coefficients are taken over the inner bounding surface only, and it should be quite apparent, for example, that the quantity obtained by partial differentiation with respect to u is precisely the x-component of the linear impulse defined by (2–54); thus one finds

$$\xi_1, \xi_2, \xi_3 = \frac{\partial T}{\partial u}, \frac{\partial T}{\partial v}, \frac{\partial T}{\partial w} \tag{2-65}$$

and

$$\lambda_1, \lambda_2, \lambda_3 = \frac{\partial T}{\partial p}, \frac{\partial T}{\partial q}, \frac{\partial T}{\partial r}. \tag{2-66}$$

The dyadic contraction of these last six relations is, of course, similar to (2–47) and (2–48).

Suitable combinations between (2–65)–(2–66) and (2–61)–(2–64) can be regarded as Lagrange equations of motion for the infinite fluid medium bounded by the moving solid. For instance, adopting the rates of change as seen by the moving observer, we obtain

$$\mathbf{F} = -\mathbf{F}_{\text{body}} = \left[\frac{d}{dt}\left(\frac{\partial T}{\partial \mathbf{u}}\right)\right]_p + \boldsymbol{\omega} \times \frac{\partial T}{\partial \mathbf{u}}, \tag{2-67}$$

$$\mathbf{M} = -\mathbf{M}_{\text{body}} = \left[\frac{d}{dt}\left(\frac{\partial T}{\partial \boldsymbol{\omega}}\right)\right]_p + \boldsymbol{\omega} \times \frac{\partial T}{\partial \boldsymbol{\omega}} + \mathbf{u} \times \frac{\partial T}{\partial \mathbf{u}}. \tag{2-68}$$

In the above equations, $\mathbf{F}_{\text{body}}$ and $\mathbf{M}_{\text{body}}$ are the force and moment exerted by the fluid on the solid, which are usually the quantities of interest in a practical investigation.

2–5 Some Deductions from Lagrange's Equations of Motion in Particular Cases

In special cases we can deduce a number of interesting results by examining (2–67)–(2–68) together with the general form of the kinetic energy, (2–38).

1. Uniform Rectilinear Motion. Suppose a body moves at constant velocity $\mathbf{u}$ and without rotation, so that $\boldsymbol{\omega} = 0$. The time derivative terms in (2–67) and (2–68) vanish, and they may be reduced to

$$\mathbf{F}_{\text{body}} = 0, \tag{2–69}$$

$$\mathbf{M}_{\text{body}} = -\mathbf{u} \times \frac{\partial T}{\partial \mathbf{u}}. \tag{2–70}$$

The fact that there is no force, either drag or lift, on an arbitrary body moving steadily without circulation is known as d'Alembert's paradox. A pure couple is found to be experienced when the linear Kelvin impulse vector and the velocity are not parallel. These quantities are obviously parallel in many cases of symmetry, and it can also be proved that there are in general three orthogonal directions of motion for which they are parallel, in which cases the entire force-couple system vanishes. Similar results can be deduced for a pure rotation with $\mathbf{u} = 0$.

2. Rectilinear Acceleration. Suppose that the velocity is changing so that $\mathbf{u} = \mathbf{u}(t)$, but still $\boldsymbol{\omega} = 0$. Then a glance at the general formula for kinetic energy shows that

$$\frac{\partial T}{\partial \mathbf{u}} \equiv \mathbf{i}\frac{\partial T}{\partial u} + \mathbf{j}\frac{\partial T}{\partial v} + \mathbf{k}\frac{\partial T}{\partial w}$$

$$= \mathbf{i}[Au + C'v + B'w] + \mathbf{j}[Bv + A'w + C'u]$$
$$+ \mathbf{k}[Cw + A'v + B'u] = \mathbf{M} \cdot \mathbf{u}. \tag{2–71}$$

This derivative is a linear function of the velocity components. Therefore,

$$\mathbf{F}_{\text{body}} = -\mathbf{M} \cdot \frac{d\mathbf{u}}{dt}. \tag{2–72}$$

For instance, if the acceleration occurs entirely parallel to the x-axis, the force is still found to have components in all three coordinate directions:

$$\mathbf{F}_{\text{body}} = -\mathbf{i}A\frac{du}{dt} - \mathbf{j}C'\frac{du}{dt} - \mathbf{k}B'\frac{du}{dt}. \tag{2–73}$$

It is clear from these results why the quantities $A, B, \ldots,$ are referred to as virtual masses or apparent masses. In view of the facts that the virtual masses for translation in different directions are not equal, and that there are also crossed-virtual masses relating velocity components in different directions, the similarity with linear acceleration of a rigid body is only qualitative.

3. Hydrokinetic Symmetries. Many examples of reduction of the system of inertia coefficients for a body moving through a constant-density fluid will be found in Chapter 6 of Lamb (1945). One interesting specialization is that of a solid with three mutually perpendicular axes of symmetry, such as a general ellipsoid. If the coordinate directions are aligned with these axes and the origin is taken at the center of symmetry, we are led to

$$T = \tfrac{1}{2}[Au^2 + Bv^2 + Cw^2 + Pp^2 + Qq^2 + Rr^2]. \qquad (2\text{–}74)$$

The correctness of (2–74) can be reasoned physically because the kinetic energy must be independent of a reversal in the direction of any linear or angular velocity component, provided that its magnitude remains the same. It follows that cross-product terms between any of these components are disallowed. Here we can see that a linear or angular acceleration along any one of the axes will be resisted by an inertia force or couple only in a sense opposite to the acceleration itself. This is a result which also can be obtained by examination of the physical system itself.

2–6 Examples of Two- and Three-Dimensional Flows Without Circulation

We now look at some simple illustrations of constant-density flows without circulation, observe how the mathematical solutions are obtained, how virtual masses are determined, and how other information of physical interest is developed.

1. The Circular Cylinder. It is well known that a circular cylinder of radius c in a uniform stream U_∞ parallel to the negative x-direction is represented by the velocity potential

$$\Phi_{\text{steady}} = -U_\infty\left[r + \frac{c^2}{r}\right]\cos\theta. \qquad (2\text{–}75)$$

Here r and θ are conventional plane polar coordinates, and the cylinder is, of course, centered at their origin.

If we remove from this steady flow the free-stream potential

$$[-U_\infty r \cos\theta \equiv -U_\infty x],$$

we are left with

$$\Phi = -\frac{U_\infty c^2}{r}\cos\theta. \qquad (2\text{–}76)$$

This is just a two-dimensional doublet with its axis pointing in the positive x-direction, and it represents a fluid motion, at rest at infinity, having around the cylinder $r = c$ a velocity

$$\frac{\partial \Phi}{\partial n} = \frac{\partial \Phi}{\partial r} = U_\infty \cos \theta. \tag{2–77}$$

This result can be regarded as describing the motion for all time in a set of portable axes that move with the cylinder's center. Moreover, the velocity U_∞ may be a function of time.

Per unit distance normal to the flow, the kinetic energy of the fluid is

$$T = -\frac{\rho}{2} \oiint_S \Phi \frac{\partial \Phi}{\partial n} \, dS = -\frac{\rho}{2} \int_0^{2\pi} \left[\Phi \frac{\partial \Phi}{\partial n} \right]_{r=c} c \, d\theta$$

$$= -\frac{\rho}{2} (-U_\infty c)(U_\infty c) \int_0^{2\pi} \cos^2 \theta \, d\theta = \tfrac{1}{2}(\rho \pi c^2) U_\infty^2. \tag{2–78}$$

Evidently,

$$A = \rho \pi c^2, \tag{2–79}$$

whereas the inertia coefficient B would have the same value and all other coefficients in the r, θ-plane must vanish. We reach the interesting, if accidental, conclusion that the virtual mass of the circular cylinder is precisely the mass of fluid that would be carried within its interior if it were hollowed out. This mass is the factor of proportionality which would relate an x-acceleration to the inertial force of resistance by the fluid to this acceleration.

These same results might be obtained more efficiently using the complex variable representation of two-dimensional constant-density flow, as discussed in Sections 2–9 ff.

2. Sphere of Radius R Moving in the Positive x-Direction (Fig. 2–3). With the fluid at rest at infinity, the sphere passing the origin of coordinates at time $t = 0$ and proceeding with constant velocity U_∞ has the instantaneous equation, expressed in coordinates fixed to the fluid at infinity,

$$B(x, y, z, t) = (x - U_\infty t)^2 + y^2 + z^2 - R^2 = 0. \tag{2–80}$$

The boundary condition may be expressed by the requirement that B is a constant for any fluid particle in contact with the surface. That is,

$$\frac{DB}{Dt} = 0 \quad \text{on} \quad B = 0. \tag{2–81}$$

Working this out with the use of (2–80),

$$\frac{DB}{Dt} \equiv \frac{\partial B}{\partial t} + \nabla \Phi \cdot \nabla B = -2U_\infty(x - U_\infty t)$$

$$+ 2U(x - U_\infty t) + 2Vy + 2Wz = 0. \tag{2–82}$$

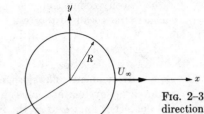

FIG. 2–3.   Sphere moving in the x-direction through a mass of liquid at rest at infinity.

In this example we proceed by trial, attempting to satisfy this condition at $t = 0$ by means of a doublet centered at the origin with its axis in the positive x-direction,

$$\Phi = \frac{\partial}{\partial x}\left(\frac{H}{4\pi r}\right) = -\frac{Hx}{4\pi r^3},$$ (2–83)

where

$$r = \sqrt{x^2 + y^2 + z^2}.$$ (2–84)

We calculate the velocity components

$$U \equiv \frac{\partial \Phi}{\partial x} = \frac{3Hx^2}{4\pi r^5} - \frac{H}{4\pi r^3},$$ (2–85a)

$$V = \frac{3Hxy}{4\pi r^5},$$ (2–85b)

$$W = \frac{3Hxz}{4\pi r^5}.$$ (2–85c)

If (2–85a), (b), and (c) are inserted into (2–82), and r and t are set equal to R and 0, respectively, we are led after some algebra to a formula for the strength H of the doublet:

$$H = 2\pi R^3 U_\infty.$$ (2–86)

The uniqueness theorem, asserted earlier in the chapter, assures that we have the correct answer.

$$\Phi = -\frac{2\pi R^3 U_\infty x}{4\pi r^3} = -\frac{U_\infty x}{2}\frac{R^3}{r^3}.$$ (2–87)

As in the case of the circular cylinder, all kinds of information can be readily obtained once the velocity potential is available. By adding the potential of a uniform stream U_∞ in the negative x-direction one constructs the steady flow and can then obtain streamlines and velocity and pressure patterns around the sphere. The only inertia coefficients are direct virtual masses. It is also of interest that the unsteady potential dies out at least as fast as the inverse square of distance from the origin when one proceeds toward infinity, and the velocity components die out at least as the inverse cube. General considerations lead to the result that the disturbance created in a constant-density fluid by any nonlifting body will resemble that of a single doublet at remote points.

The foregoing example of the sphere is just a special case of a more general technique for constructing axially symmetric flows around bodies of revolution by means of an equilibrating system of sources and sinks along the axis of symmetry. (The doublet is known to be the limit of a single source-sink pair.) This procedure was originally investigated by Rankine, and such figures are referred to as Rankine ovoids. Among many other investigators, von Kármán (1927) has adapted the method to airship hulls.

For blunt shapes like the sphere, the results of potential theory do not agree well with what is measured because of the presence of a large separated wake to the rear, a problem which we discuss further in Chapter 3. It was this sort of discrepancy that dropped theoretical aerodynamics into considerable disrepute during the nineteenth century. The blunt-body flows are described here principally because they provide simple illustrations. When it comes to elongated streamlined shapes, a great deal of useful and accurate information *can* be found without resort to the nonlinear theory of viscous flow, however, and such applications constitute the ultimate objective of this book.

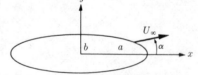

Fig. 2–4. Prolate ellipsoid of revolution moving through constant-density fluid at an angle of attack α in the x, y-plane.

3. Ellipsoids. A complete account of the solution of more complicated boundary-value problems on spheres, prolate or oblate ellipsoids of revolution, and general ellipsoids will be found in the cited references. Space considerations prevent the reproduction of such results here, with the exception that a few formulas will be given for elongated ellipsoids of revolution translating in a direction inclined to the major axis. These have considerable importance relative to effects of angle of attack on fuselages, submarine hulls, and certain missile and booster configurations.

For the case shown in Fig. 2–4, the velocity

$$\mathbf{u} = \mathbf{i}U_\infty \cos \alpha + \mathbf{j}U_\infty \sin \alpha \qquad (2\text{–}88)$$

gives rise to a Kelvin impulse

$$\boldsymbol{\xi} = \mathbf{i}AU_\infty \cos \alpha + \mathbf{j}BU_\infty \sin \alpha$$
$$= \rho[\text{volume}]\left[\mathbf{i}\left(\frac{\alpha_0}{2 - \alpha_0}\right) U_\infty \cos \alpha + \mathbf{j}\left(\frac{\beta_0}{2 - \beta_0}\right) U_\infty \sin \alpha\right]. \qquad (2\text{–}89)$$

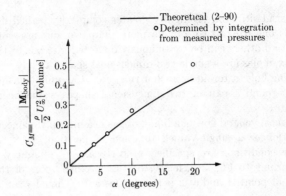

FIG. 2-5. Comparison between predicted and measured moments about a centroidal axis on a prolate spheroid of fineness ratio 4:1. [Adapted from data reported by R. Jones (1925) for a Reynolds number of about 500,000 based on body length.]

The ellipsoid's volume is, of course, $\frac{4}{3}\pi a b^2$, whereas the dimensionless coefficients

$$[\alpha_0/(2 - \alpha_0)], \qquad [\beta_0/(2 - \beta_0)]$$

are functions of eccentricity tabulated in Section 115 of Lamb (1945).

With constant speed and incidence the force is zero, but the vectors $\mathbf{u}$ and $\boldsymbol{\xi}$ are not quite parallel and an overturning couple is generated,

$$\mathbf{M}_{\text{body}} = -\mathbf{u} \times \boldsymbol{\xi}$$
$$= -\mathbf{k}\frac{\rho}{2} U_\infty^2 \,[\text{volume}] \left\{ \frac{\beta_0}{2 - \beta_0} - \frac{\alpha_0}{2 - \alpha_0} \right\} \sin 2\alpha; \quad (2\text{-}90)$$

the inertia factor here in braces is always positive, vanishing for the sphere. At fineness ratios above ten it approaches within a few percent of unity. Thus the familiar unstable contribution of the fuselage to static stability is predicted theoretically. The calculations are in fairly satisfactory accord with measurements when a/b exceeds four or five. Figure 2-5 presents a comparison with pitching-moment data.

2-7 Circulation and the Topology of Flow Regimes

Most of the theorems and other results stated or derived in preceding portions of the chapter refer to finite bodies moving through a finite or infinite mass of liquid. If these bodies have no holes through them, such a field is simply connected in the sense that any closed circuit can be shrunk to a point or continuously distorted into any other closed circuit without ever passing outside the field.

A physical situation of slightly greater complexity, called doubly connected, is one where two, but no more than two, circuits can be found such that all others can be continuously distorted into one or the other of them. Examples are a single two-dimensional shape, a body with a single penetrating hole, a toroid or anchor ring, etc. Carrying this idea further, the flow around a pair of two-dimensional shapes would be triply connected, etc.

In the most general flow of a liquid or gas in a multiply connected region, Φ is no longer a single-valued function of position, even though the boundary conditions are specified properly as in the simply connected case. Turning to Fig. 2–6, let Φ_A be the specified value of the velocity potential at point A and at a particular instant. Then Φ_B can be written either as

$$\Phi_{B_1} = \Phi_A + \int_{ACB} \mathbf{Q} \cdot d\mathbf{s} \qquad (2\text{–}91)$$

or as

$$\Phi_{B_2} = \Phi_A + \int_{ADB} \mathbf{Q} \cdot d\mathbf{s}. \qquad (2\text{–}92)$$

These two results will not necessarily be equal, since we cannot prove the identity of the two line integrals when the region between them is not entirely occupied by fluid. As a matter of fact, the circulation around the closed path is exactly

$$\Gamma = \oint_{ADBCA} \mathbf{Q} \cdot d\mathbf{s} = \Phi_{B_2} - \Phi_{B_1}. \qquad (2\text{–}93)$$

It is an obvious result of Stokes' theorem that Γ is the same for any path completely surrounding just the body illustrated. Hence the difference in the values of Φ taken between paths on one side or the other always turns out to be exactly Γ, wherever the two points A and B are chosen.

Until quite lately it was believed that, to render constant-density fluid motion unique in a multiply connected region, a number of circulations must be prescribed which is one less than the degree of connectivity. A forthcoming book by Hayes shows, however, that more refined topological concepts must be employed to settle this question. He finds that the indeterminacy is associated with a topological property of the region known as the Betti number. Since the mathematical level of these ideas exceeds what is being required of our readers, we confine ourselves to citing the reference and asserting that it confirms the correctness of the simple examples discussed here and in the following section.

Consider the motion of a given two-dimensional figure S with $\Gamma = 0$. A certain set of values of $\partial\Phi/\partial n$ on S can be satisfied by a velocity potential representing the noncirculatory flow around the body. To this basic flow it is possible to add a simple vortex of arbitrary strength Γ for which one

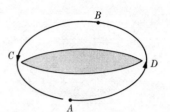

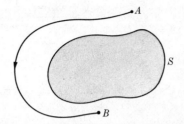

FIG. 2–6. Circuit drawn around a two-dimensional shape surrounded by fluid.   FIG. 2–7. An illustration of the circuit implied in Eq. (2–94).

of the circular streamlines has been transformed conformally into precisely the shape of S. The general mapping theorem says that such a transformation can always be carried out. By the superposition, a new irrotational flow has been created which satisfies the same boundary conditions, and it is obvious that uniqueness can be attained only by a specification of Γ.

Leaving aside the question of how the circulation was generated in the first place, it is possible to prove that Γ around any such body persists with time. This can be done by noting that the theorem of Kelvin, (1–12), is based on a result which can be generalized as follows:

$$\frac{D}{Dt} \int_A^B \mathbf{Q} \cdot d\mathbf{s} = -\int_A^B \frac{dp}{\rho} + \left[\Omega + \frac{Q^2}{2}\right]_A^B. \tag{2–94a}$$

The fluid here may even be compressible, and A and B are any two points in a continuous flow field, regardless of the degree of connectivity. See Fig. 2–7. If now we bring A and B together in such a way that the closed path is not simply connected, and if we assume that there is a unique relation between pressure and density, we are led to

$$\frac{D\Gamma}{Dt} = 0. \tag{2–94b}$$

Let us now again adopt the restriction to constant-density fluid and examine the question of how flows with multiply connected regions and nonzero circulations might be generated. We follow Kelvin in imagining at any instant that a series of barriers or diaphragms are inserted so as to make the original region simply connected by the specification that no path may cross any such barrier. Some examples for which the classical results agree with Hayes are shown in Fig. 2–8.

The insertion of such barriers, which are similar to cuts in the theory of the complex variable, renders Φ single-valued. We recall the physical interpretation [cf. (2–15)]

$$\mathbf{P} = -\rho\Phi \tag{2–95}$$

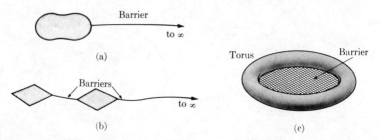

FIG. 2–8. Three examples of multiply connected flow fields, showing barriers that can be inserted to make Φ unique. In (a) and (b), the shaded shapes are two-dimensional.

of Φ as the impulsive pressure required to generate a flow from rest. If such impulsive pressures are applied only over the surfaces of the bodies and the boundary at infinity in regions like those of Fig. 2–8, a flow without circulation will be produced. But suppose, additionally, that discontinuities in **P** of the amount

$$\Delta \mathbf{P} = -\rho \, \Delta \Phi = -\rho \Gamma \qquad (2\text{--}96)$$

are applied across each of the barriers. Then a circulation can be produced around each path obstructed only by that particular barrier. The generation of a smoke ring by applying an impulse over a circular area is an obvious example. Note that $\Delta\Phi$ (or Γ) is constant all over any given barrier, but the location of the barriers themselves presents an element of arbitrariness.

2–8 Examples of Constant-Density Flows Where Circulation May Be Generated

An elementary illustration of the ideas of the foregoing section is provided by a two-dimensional vortex pair. We work here in terms of real variables rather than the complex variable, although it should be obvious to those familiar with two-dimensional flow theory that the results we obtain could be more conveniently derived by the latter approach. Consider a pair of vortices, which are equal and opposite and may be thought of as wrapped around very small circular cylindrical cores which constitute the boundaries S (Fig. 2–9). This motion can be generated by applying a downward force per unit area

$$\Delta \mathbf{P} = -\rho \, \Delta \Phi = \rho \Gamma \qquad (2\text{--}97)$$

across the barrier shown in the picture. The total Kelvin impulse per

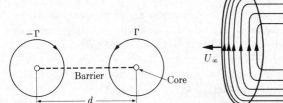

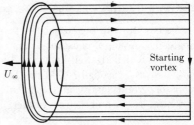

Fig. 2–9. Two equal and opposite line vortices separated a distance d.

Fig. 2–10. Vortex pattern simulating flow around a wing of finite span.

unit distance normal to the page is directed downward and may be written

$$\xi = -\mathbf{j}(\rho \Gamma d), \tag{2–98}$$

where $\mathbf{j}$ is a unit upward vector and d is the instantaneous separation of the vortex cores.

If, for instance, one of the two vortices is bound to a wing moving to the left with velocity U_∞ while the other remains at rest in the fluid in the manner of a starting vortex, the force exerted by the fluid on the supporting bodies is

$$\mathbf{F}_{\text{body}} = -\frac{d\xi}{dt} = -\frac{d}{dt}(-\mathbf{j}\rho\Gamma d)$$

$$= \mathbf{j}\rho\Gamma\frac{d}{dt}(d) = \rho U_\infty \Gamma \mathbf{j}. \tag{2–99}$$

This may be recognized as the two-dimensional lift called for by the theorem of Kutta and Joukowsky.

A more complicated system of vortices is used as an indirect means of representing the influence of viscosity on the flow around a lifting wing of finite span (Fig. 2–10). For any one of the infinite number of elongated vortex elements, the Kelvin impulse is directed downward and equals

$$\xi = -\mathbf{j}\rho\,\Delta\Phi \times [\text{area}]. \tag{2–100}$$

The area here changes at a rate dependent on the forward speed U_∞. The reaction to the force producing the increased impulses of the various vortices adds up to the instantaneous lift on the wing. Moreover, from the spanwise distribution of vortex strengths the spanwise distribution of lift is obtainable, and the energy in the vortex system is connected with the induced drag of the wing. It is evident that these vortices could not be generated in the first place except through the action of viscosity in producing a boundary layer on the wing, yet we can obtain much useful information about the loading on the system without actually attempting a full solution of the equations of Navier and Stokes.

2–9 Two-Dimensional, Constant-Density Flow: Fundamental Ideas

We now turn to the subject of two-dimensional, irrotational, steady or unsteady motion of constant-density fluid. We begin by listing a number of results which are well-known and may be found developed, for instance, in Chapters 5 through 7 of Milne-Thompson (1960). The combined conditions of irrotationality and continuity assure the existence of a velocity potential $\Phi(\mathbf{r}, t)$ and a stream function $\Psi(\mathbf{r}, t)$, such that

$$\mathbf{Q} = \nabla\Phi = \nabla \times (\mathbf{k}\Psi). \tag{2-101}$$

Here

$$\mathbf{r} = x\mathbf{i} + y\mathbf{j}. \tag{2-102}$$

If (2–101) is written out in component form, we obtain.

$$U = \frac{\partial\Phi}{\partial x} = \frac{\partial\Psi}{\partial y}\,; \qquad V = \frac{\partial\Phi}{\partial y} = -\frac{\partial\Psi}{\partial x}. \tag{2-103}$$

The latter equalities will be recognized as the Cauchy-Riemann relations. For constant-density fluid,

$$\nabla \cdot \mathbf{Q} = \nabla^2\Phi \tag{2-104}$$

is the volume divergence, whereas for a rotational flow,

$$\nabla \times \mathbf{Q} = -\mathbf{k}\nabla^2\Psi \tag{2-105}$$

is the vorticity vector. Therefore, in the case under consideration,

$$\nabla^2\Phi = 0 = \nabla^2\Psi. \tag{2-106}$$

Among other ways of constructing solutions to the two-dimensional Laplace equation, a function of either

$$Z \equiv x + iy = re^{i\theta} \tag{2-107}$$

or

$$\overline{Z} \equiv x - iy = re^{-i\theta} \tag{2 108}$$

alone will be suitable.* To be more specific, the Cauchy-Riemann relations are necessary and sufficient conditions for Φ and Ψ to be the real and imaginary parts, respectively, of th same analytic function of Z. This function we call the complex potential,

$$\mathcal{W}(Z) = \Phi + i\Psi. \tag{2-109}$$

---

* The bar over any symbol will be employed to designate a complex conjugate for the remainder of this chapter.

The following formulas for particle velocity and speed are easily derived:

$$U - iV = [U_r - iU_\theta]e^{-i\theta} = \frac{dw}{dZ}, \qquad (2\text{–}110)$$

$$Q^2 = \left|\frac{d\mathcal{W}}{dZ}\right|^2 = \frac{d\mathcal{W}}{dZ}\frac{d\overline{\mathcal{W}}}{dZ}. \qquad (2\text{–}111)$$

The lines $\Phi = $ const and $\Psi = $ const form orthogonal networks of equipotentials and streamlines in the x,y-plane, which is usually referred to as the Z-plane.

An interesting parallelism between the imaginary unit $i = \sqrt{-1}$ and the vector operator $\mathbf{k}\times$ is discussed in Milne-Thompson (1960), and some readers may find it helpful to study this more physical interpretation of a quantity which has unfortunately been given a rather formidable name.

The fact that the complex potential is a function of a single variable has many advantages. Differentiation is of the ordinary variety and can be conveniently cascaded or inverted. Also, it makes little difference whether we operate with the functional relationship $\mathcal{W}(Z)$ or $Z(\mathcal{W})$; many flows are more conveniently described by the latter.

We recall that many fundamental flow patterns are associated with simple singular forms of the complex potential. Thus $\ln (Z)$ implies a point source or point vortex, $1/Z$ is a doublet, and Z^α corresponds to various fluid motions with linear boundaries meeting at angles related to α. A failure of one or more of the underlying physical assumptions occurs at the singular point location. Nevertheless, the singular solutions are useful in constructing flows of practical interest in regions away from their centers. Forces and moments can be expressed in terms of contour integrals around the singularities and are therefore connected with residues at poles.

The complex potential is itself a kinematical concept. To find pressures and resultant forces in steady and unsteady flows, further information is required. Thus Bernoulli's equation, (1–63), is our tool for pressure calculation. The necessary quantities are taken from (2–111) and

$$\frac{\partial \Phi}{\partial t} = \mathbf{Re}\left\{\frac{\partial \mathcal{W}(Z)}{\partial t}\right\}, \qquad (2\text{–}112)$$

where the operator on the right means to take the real part of the quantity in braces.

For forces and moments on a single closed figure in steady flow, we have available the classical Blasius equations

$$F_x - iF_y = \frac{i\rho}{2}\oint_C \left(\frac{d\mathcal{W}}{dZ}\right)^2 dZ, \qquad (2\text{–}113)$$

$$M_0 = \mathbf{Re}\left\{-\frac{\rho}{2}\oint_C Z\left(\frac{d\mathcal{W}}{dZ}\right)^2 dZ\right\}, \qquad (2\text{–}114)$$

where M_0 is the counterclockwise moment exerted by the fluid on the profile about an axis through the origin, and C is a contour that surrounds the body but no other singularities of the flow field, if such exist. In the absence of external singularities, the contour may be enlarged indefinitely. Then if $\mathcal{W}(Z)$ can be expanded into an inverse power series in Z, which is nearly always the case, we identify all forces as coming from the $1/Z$ term and all moments as coming from the $1/Z^2$ term. We conclude that an effective source or vortex, plus a uniform stream, will lead to a resultant force. Moreover, a doublet may give rise to a moment, as can certain other combinations of source-like and vortex-like singular solutions.

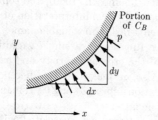

FIG. 2–11. Pressure force acting on a short segment of body surface in two-dimensional flow.

In a limited way, (2–113) and (2–114) can be extended to apply to unsteady flows. The development follows Section 6.41 of Milne-Thompson (1960), but a restriction is required which is not carefully stated there. Let us consider a two-dimensional body whose position is fixed and whose contour does not change with time but which is in an accelerated stream $U_\infty(t)$ or otherwise unsteady regime. See Fig. 2–11. We derive the force equation and simply write down its analog for the moment. Let C_B be a contour coinciding with the fixed body surface. We note that

$$dF_x - i\,dF_y = -p\,dy - ip\,dx = -ip\,d\overline{Z}. \qquad (2\text{–}115)$$

However,

$$p = -\rho\left[\frac{Q^2}{2} + \frac{\partial\Phi}{\partial t}\right] + \left[\begin{array}{c}\text{nonessential}\\ \text{increment}\end{array}\right]$$

$$= -\frac{\rho}{2}\frac{d\mathcal{W}}{dZ}\frac{d\overline{\mathcal{W}}}{dZ} - \rho\,\frac{\partial\Phi}{\partial t}. \qquad (2\text{–}116)$$

The quantity labeled "nonessential increment" is dropped from (2–116) since even a time function will contribute nothing to the total force or moment. We substitute into (2–115)

$$dF_x - i\,dF_y = i\,\frac{\rho}{2}\frac{d\mathcal{W}}{dZ}\,d\overline{\mathcal{W}} + i\rho\,\frac{\partial\Phi}{\partial t}\,d\overline{Z}. \qquad (2\text{–}117)$$

On the body surface,

$$\Psi = \Psi_B(t) \qquad (2\text{–}118)$$

independent of the space coordinates because of the assumed fixed position (i.e., the body is always an instantaneous streamline). Noting that the integration of force is carried out for a particular instant of time, we may write, following the contour C_B,

$$d\overline{W} = d\Phi = dW = \frac{dW}{dZ}\,dZ, \qquad (2\text{–}119)$$

$$\frac{\partial \Phi}{\partial t} = \frac{d\overline{W}}{\partial t} + i\frac{d\Psi_B}{dt}. \qquad (2\text{–}120)$$

The latter holds true because

$$\overline{W} = \Phi - i\Psi. \qquad (2\text{–}121)$$

Finally, we integrate around the contour C_B and observe that the integral of the quantity $d\Psi_B/dt$ must vanish.

$$F_x - iF_y = i\frac{\rho}{2}\oint_{C_B}\left(\frac{dW}{dZ}\right)^2 dZ + i\rho\oint_{C_B}\frac{d\overline{W}}{\partial t}\,d\overline{Z}$$

$$= \frac{i\rho}{2}\oint_C\left(\frac{dW}{dZ}\right)^2 dZ + i\rho\frac{\partial}{\partial t}\oint_C \overline{W}\,d\overline{Z}. \qquad (2\text{–}122)$$

In this latter form the integrals are carried out around the contour C because each integrand is recognized as an analytic function of the variable of integration only, and contour deformation is permitted in the usual fashion. Of course, no pole singularities may be crossed during this deformation, and branch points must be handled by putting suitable cuts into the field. The extended Blasius equation for moments in unsteady flow reads

$$M_0 = -\mathbf{Re}\left\{\frac{\rho}{2}\oint_C Z\left(\frac{dW}{dZ}\right)^2 dZ + \rho\frac{\partial}{\partial t}\oint_{C_B}[\overline{W} + i\psi_B(t)]Z\,d\overline{Z}\right\}. \qquad (2\text{–}123)$$

Here the second contour may not be deformed from C_B since Z is not an analytic function of $\overline{Z}$.

We close this section by setting down, without proof, the two-dimensional counterpart of (2–28). For an arbitrary field point x, y, this theorem expresses the velocity potential as follows:

$$\Phi(x, y; t) = \frac{1}{2\pi}\oint_{C_B}\left[\frac{\partial \Phi}{\partial n}\ln r - \Phi\frac{\partial}{\partial n}(\ln r)\right]ds. \qquad (2\text{–}124)$$

Here the fluid must be at rest at infinity; line integration around the body contour is carried out with respect to dummy variables x_1, y_1; and

$$r = \sqrt{(x - x_1)^2 + (y - y_1)^2}. \qquad (2\text{–}125)$$

The natural logarithm of r is the potential of a two-dimensional line source centered at point x_1, y_1. When differentiated with respect to outward normal n, it is changed into a line doublet with its axis parallel to n.

2–10 Two-Dimensional, Constant-Density Flow: Conformal Transformations and Their Uses

A consequence of the mapping theorem of Riemann is that the exterior of any given single closed figure, such as an airfoil, in the complex Z-plane can be mapped into the exterior of any other closed figure in the ζ-plane by an analytic relation of the form

$$Z = f(\zeta). \tag{2–126}$$

See Fig. 2–12. It is frequently convenient to choose a circle for the ζ-figure. The angle between any two intersecting lines is preserved by the transformation; for example, a set of orthogonal trajectories in one plane also turns out to be a set in the other. If the point at infinity is to remain unchanged, the transformation can always be expanded at large distances into something of the form

$$Z = \zeta + \frac{a_1}{\zeta} + \frac{a_2}{\zeta^2} + \cdots, \tag{2–127}$$

where the a_n are complex constants.

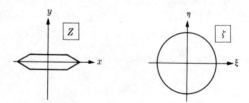

FIG. 2–12. An illustration of the Z- and ζ-planes connected through conformal transformation.

It is occasionally pointed out that a special case of conformal transformation is the complex potential itself, (2–109), which can be regarded as a mapping of the streamlines and equipotentials into a $(\Phi + i\Psi)$-plane, where they become equidistant, horizontal and vertical straight lines, respectively.

The practical significance of the mapping theorem is that it can be used to transform one irrotational, constant-density flow $\mathcal{W}_1(\zeta)$ with elementary boundaries into a second flow

$$\mathcal{W}(Z) = \mathcal{W}_1[f^{-1}(Z)], \tag{2–128}$$

which has a more complicated boundary shape under the control of the transformer. Contours of engineering interest, such as a prescribed wing section, are readily obtained by properly choosing the function in (2–126). Conditions far away from the two figures can be kept the same, thus allowing for a prescribed flight condition.

Velocities, and consequently pressures, can be transformed through the relation

$$U - iV = \frac{d\mathcal{W}}{dZ} = \frac{d\mathcal{W}_1}{d\zeta}\frac{d\zeta}{dZ} = \frac{U_\xi - iU_\eta}{dZ/d\zeta}. \tag{2–129}$$

It follows that

$$Q = \frac{|U_\xi - iU_\eta|}{|dZ/d\zeta|} = \frac{Q_\zeta}{|df/d\zeta|}, \tag{2–130}$$

ξ and η being the real and imaginary parts of ζ, respectively. The equations of Blasius can be employed to determine resultant force and moment in either of the two planes, and the transformation of variable itself is helpful when determining this information for the Z-figure. To provide starting points, many elementary complex potentials are known which characterize useful flows with circular boundaries.

Several transformations have proved either historically or currently valuable for constructing families of airfoil shapes and other two-dimensional figures with aeronautical applications. The reader is presumed to be familiar with the Joukowsky transformation, and much can be done with very minor refinements to the original investigations of Kutta and Joukowsky. No effort is made to expose in detail the various steps that have been carried out by different investigators, but we do list below a number of the more important transformations and something about their consequences.

1. The Joukowsky-Kutta Transformation

$$Z = \zeta + \frac{l^2}{\zeta}. \tag{2–131}$$

Here l is a positive real constant, and the so-called singular points of the transformation where the $dZ/d\zeta = 0$ are located at $\zeta = \pm l$, corresponding to $Z = \pm 2l$. When applied to suitably located circles in the ζ-plane, (2–131) is well known to produce ellipses, flat plates, circular arc profiles of zero thickness, symmetrical and cambered profiles with their maximum thickness far forward and with approximately circular-arc camber lines.

The shape obtained actually depends on the location of the circle relative to the aforementioned singular points. A cusped trailing edge is produced by passing the circle through the singular point on the downstream side of the circle.

2. The von Mises Transformations.
These transformations are special cases of the series, (2–127), in which it is truncated to a finite number of terms:

$$Z = \zeta + \frac{a_1}{\zeta} + \cdots + \frac{a_n}{\zeta^n}. \tag{2–132}$$

When an airfoil is being designed, the series is constructed by starting from the singular point locations at

$$\zeta = \zeta_{s_1}, \zeta_{s_2}, \ldots, \zeta_{s_{n+1}}. \tag{2–133}$$

The derivative of the transformation is written

$$\frac{dZ}{d\zeta} = \left[1 - \frac{\zeta_{s_1}}{\zeta}\right]\left[1 - \frac{\zeta_{s_2}}{\zeta}\right] \cdots \left[1 - \frac{\zeta_{s_{n+1}}}{\zeta}\right] \tag{2–134}$$

and subsequently integrated in closed form. By the rather laborious process of trial-and-error location of singular points, many practical airfoils were developed during the 1920's. It is possible to adjust the thickness and camber distributions in a very general way. Interesting examples of von Mises and other airfoils will be found discussed in a recent book by Riegels (1961).

3. The von Kármán-Trefftz Transformation.
This method derives from a scheme for getting rid of the cusp at the trailing edge, produced by the foregoing classes of transformations, and replacing it by a corner with a finite angle τ. To see how it accomplishes this, consider a transformation with a singular point at $\zeta = \zeta_0$, corresponding to a point $Z = Z_0$. In the vicinity of this particular singular point, it is easy to show that the transformation can be approximated by

$$Z - Z_0 = A[\zeta - \zeta_0]^n, \quad n > 1, \tag{2–135}$$

where A is some complex constant. Evidently, the quantity

$$\frac{dZ}{d\zeta} = \frac{d(Z - Z_0)}{d(\zeta - \zeta_0)} = nA[\zeta - \zeta_0]^{n-1} \tag{2–136}$$

vanishes at the point for $n > 1$, as it is expected to do. Let r_0 and θ_0 be the modulus and argument of the complex vector emanating from the point ζ_0. Equation (2–136) can be written

$$d(Z - Z_0) = nAr_0^{n-1}e^{i(n-1)\theta_0} d(\zeta - \zeta_0). \tag{2–137}$$

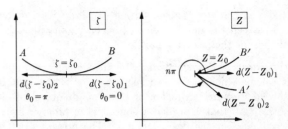

Fig. 2–13. A continuous element of arc AB, passing through a singular point in the ζ-plane, is transformed by (2–137) into a broken line $A'B'$.

This is to say, the element of arc $d(\zeta - \zeta_0)$ is rotated through an angle $(n - 1)\theta_0$ in passing from the ζ-plane to the Z-plane, if we overlook the effect of the constant A which rotates any line through ζ_0 by the same amount. Figure 2–13 demonstrates what this transformation does to a continuous curve passing through the point $\zeta = \zeta_0$ in the ζ-plane. By proper choice of n, the break in the curve which is produced on the Z-plane can be given any desired value between π and 2π.

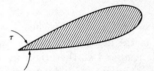

Fig. 2–14. Trailing-edge angle τ of the sort produced by the transformation equation (2–139).

We have no trouble in showing that the Joukowsky and von Mises transformations are cases of $n = 2$. For instance, the Joukowsky can be manipulated into

$$\frac{Z - 2l}{Z + 2l} = \left[\frac{\zeta - l}{\zeta + l}\right]^2. \tag{2–138}$$

Von Kármán and Trefftz (1918) suggested replacing (2–138) as follows:

$$\frac{Z - 2l}{Z + 2l} = \left[\frac{\zeta - l}{\zeta + l}\right]^{2 - (\tau/\pi)} \tag{2–139}$$

As in Fig. 2–14, the outer angle between the upper and lower surfaces of the airfoil at its trailing edge (T.E.) is now $(2\pi - \tau)$, so a finite interior angle has been introduced and can be selected at will.

In a similar fashion a factor

$$\left(1 - \frac{\zeta_{s_{T.E.}}}{\zeta}\right)^{1 - (\tau/\pi)}$$

can be included in the von Mises equation (2–134). Then if the circle in the ζ-plane is passed through the point $\zeta = \zeta_{s_{\mathrm{T.E.}}}$, an adjustable trailing edge is provided for the resulting profile.

4. The Theodorsen Transformation.* Theodorsen's method and the several extensions which have been suggested for it are capable of constructing the flow around an airfoil or other single object of completely arbitrary shape. All that is needed is some sort of table or equation providing the ordinates of the desired figure. The present brief discussion will emphasize the application to the airfoil.

The transformation is actually carried out in two steps. First the airfoil is located in the Z-plane as close as possible to where a similarly shaped Joukowsky airfoil would fall. It can be proved that this will involve locating the Joukowsky singular points $Z = \pm 2l$ halfway between the nose and the trailing edge and their respective centers of curvature. (Of course, if the trailing edge is pointed, the singularity $Z = -2l$ falls right on it.) By applying Joukowsky's transformation in reverse, the airfoil is transformed into a "pseudocircle" in the Z'-plane

$$Z = Z' + \frac{l^2}{Z'}. \tag{2–140}$$

The second procedure is the conversion of the pseudocircle to an exact circle centered at the origin in the ζ-plane by iterated determination of the coefficients in the transformation series

$$Z' = \zeta \exp\left(\sum_{n=1} \frac{C_n}{\zeta^n}\right). \tag{2–141}$$

Here the C_n are complex constants. The two steps of the process are illustrated in Fig. 2–15.

Let us consider the two steps in a little more detail. We write

$$Z = x + iy, \tag{2–142}$$

$$Z' = le^{\psi + i\theta}. \tag{2–143}$$

That is, the argument of Z' is

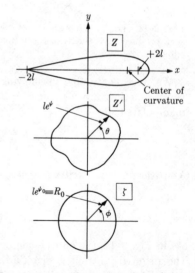

Fig. 2–15. An illustration of the three stages of Theodorsen's transformation.

---

* For further details on this procedure, see Theodorsen (1931), Theodorsen and Garrick (1933), and Abbott, von Doenhoff, and Stivers (1945).

denoted by θ and its modulus is le^{ψ}. It is not difficult to derive the direct and inverse relationships between the coordinates of the airfoil and pseudocircle:

$$\begin{aligned} x &= 2l \cosh \psi \cos \theta \\ y &= 2l \sinh \psi \sin \theta, \end{aligned} \tag{2-144}$$

$$\begin{aligned} 2 \sin^2 \theta &= p + \sqrt{p^2 + (y/l)^2} \\ 2 \sinh^2 \psi &= -p + \sqrt{p^2 + (y/l)^2}, \end{aligned} \tag{2-145}$$

where

$$p = 1 - \left(\frac{x}{2l}\right)^2 - \left(\frac{y}{2l}\right)^2. \tag{2-146}$$

Note that ψ will be quite a small number for profiles of normal thickness and camber.

The function $\psi(\theta)$ for the pseudocircle may be regarded as known at as many points as desired. We then write

$$\zeta = R_0 e^{i\phi} = le^{\psi_0} e^{i\phi}. \tag{2-147}$$

For points on the two contours only, the transformation, (2–141), can be manipulated as follows:

$$\frac{Z'}{\zeta} = \frac{le^{\psi + i\theta}}{le^{\psi_0 + i\phi}} = \exp\left[(\psi - \psi_0) + i(\theta - \phi)\right], \tag{2-148}$$

$$Z' = \zeta \exp\left[(\psi - \psi_0) + i(\theta - \phi)\right] = \zeta \exp\left(\sum_{n=1} \frac{C_n}{\zeta^n}\right) \tag{2-149}$$

Theodorsen (1931) adopted the symbol ϵ to denote the shift in argument going from the Z'- to the ζ-plane,

$$\epsilon = \phi - \theta \qquad \text{or} \qquad \phi = \theta + \epsilon. \tag{2-150}$$

Writing

$$C_n = A_n + iB_n, \tag{2-151}$$

we are able to derive

$$\psi - \psi_0 = \sum_{n=1} \left[\frac{A_n}{R_0^n} \cos n\phi + \frac{B_n}{R_0^n} \sin n\phi\right], \tag{2-152}$$

$$\epsilon = \sum_{n=1} \left[-\frac{B_n}{R_0^n} \cos n\phi + \frac{A_n}{R_0^n} \sin n\phi\right]. \tag{2-153}$$

These two equations imply that $(\psi - \psi_0)$ and ϵ are conjugate quantities. They are expressed as Fourier series in the variable ϕ, so that the standard formulas for individual Fourier coefficients could be employed if these quantities were known in terms of ϕ. Moreover, the conjugate property can be used to relate the functions ϵ and ψ directly, as follows:

$$\epsilon(\phi') = -\frac{1}{2\pi} \oint_0^{2\pi} \psi(\phi) \cot\left(\frac{\phi - \phi'}{2}\right) d\phi, \qquad (2\text{-}154)$$

$$\psi(\phi') = \psi_0 + \frac{1}{2\pi} \oint_0^{2\pi} \epsilon(\phi) \cot\left(\frac{\phi - \phi'}{2}\right) d\phi. \qquad (2\text{-}155)$$

In actuality, only $\psi(\theta)$ is available to begin with; θ can be regarded as a first approximation to ϕ, however, and (2-154) employed to get a first estimate of ϵ. Equation (2-150) then yields an improved approximation to ϕ and to $\psi(\phi)$, so that (2-154) may be used in an iterative fashion to obtain converged formulas for the desired quantities. For typical airfoils it is found that this process converges very rapidly, and the numerical integration of (2-154) need be iterated only once or twice. Of course, it is necessary to be careful about the pole singularity at $\phi = \phi'$.

As described in the references, Theodorsen's so-called ϵ-method has many uses in the theory of low-speed airfoils. For instance, one can generate families of profiles from assumed forms of the function $\epsilon(\phi)$. Approximate means have been developed, starting from an airfoil of known shape and pressure distribution, for adjusting this pressure distribution in a desired fashion. This scheme formed the basis for the laminar flow profiles

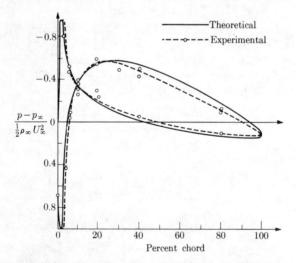

Fig. 2-16. Comparison between predicted and measured pressure distribution over the upper and lower surfaces of a Clark Y airfoil at an effective geometrical incidence of -1 deg 16 min. Theoretical incidence chosen to give approximately the measured value of total lift. [Adapted from Theodorsen (1931).]

which played such an important role in the early 1940's. Their shapes sustain a carefully adjusted, favorable pressure distribution to assure the longest possible laminar run prior to transition in the boundary layer. In wind tunnel tests, they achieve remarkable reductions in friction drag; unfortunately, these same reductions cannot usually be obtained in engineering practice, and some of the profiles have undesirable characteristics above the critical flight Mach number.

Figure 2–16 shows a particularly successful example of the comparison between pressure distribution measured on an airfoil and predicted by this so-called "ε-method."

5. The Schwarz-Christoffel Transformations. These transformations are described in detail in any advanced text on functions of a complex variable. They furnish a useful general technique for constructing flows with boundaries which are made up of straight-line segments.

2–11 The Kutta Condition and Lift

As is familiar to every student of aerodynamics, Joukowsky and Kutta discovered independently the need for circulation to render the two-dimensional, constant-density flow around a figure with a pointed trailing edge physically reasonable. This is a simple example of a scheme for fixing the otherwise indeterminate circulation around an irreducible path in a doubly connected region. It is one of a number of ways in which viscosity can be introduced at least indirectly into aerodynamic theory without actually solving the equations of Navier and Stokes. The circulation gives rise to a lift, which is connected with the continually increasing Kelvin impulse of the vortex pair, one of the vortices being the circulation bound to the airfoil, while the other is the "starting vortex" that was generated at the instant the motion began.

With respect to the lifting airfoil, we reproduce a few important results from Sections 7.40–7.53 of Milne-Thompson (1960). Let the profile and the circle which is being transformed into it be related as shown in Fig. 2–17.

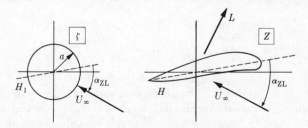

FIG. 2–17. Circle transformed into an airfoil at incidence in an incompressible two-dimensional stream. Direction of lift L on the airfoil is shown.

It is assumed that the transformation

$$Z = \zeta + \frac{a_1}{\zeta} + \frac{a_2}{\zeta^2} + \cdots, \tag{2-156}$$

which takes the circle into the airfoil, is known. It can be proved that the resulting force is normal to the oncoming stream and equal to

$$L = \rho U_\infty \Gamma. \tag{2-157}$$

Here Γ is the circulation bound to the airfoil, which incidentally may or may not satisfy the full Kutta condition of smooth flow off from the trailing edge. If this condition is entirely met, which is equivalent to neglecting the effect of displacement thickness of the boundary layer and the wake thickness at the trailing edge, then the circulation is given by

$$\Gamma = 4\pi U_\infty a \sin \alpha_{\text{Z.L.}}. \tag{2-158}$$

All the quantities here are defined in the figure. In particular, $\alpha_{\text{Z.L.}}$ is the angle of attack between the actual stream direction and the zero-lift (Z.L.) direction, determined as a line parallel to one between the center of the circle in the ζ-plane and the point H_1 which transforms into the airfoil trailing edge. Combining (2–157) and (2–158), we compute the lift

$$L = 4\pi\rho U_\infty^2 a \sin \alpha_{\text{Z.L.}}. \tag{2-159}$$

From this we find that the lift-curve slope, according to the standard aeronautical definition, is slightly in excess of 2π, reducing precisely to 2π when the airfoil becomes a flat plate of zero thickness, i.e., when the radius a of the circle becomes equal to a quarter of the chord.

The airfoil is found to possess an aerodynamic center (A.C.), a moment axis about which the pitching moment is independent of angle of attack. This point is located on the Z-plane as shown in Figs. 7.52 and 7.53 of Milne-Thompson (1960). The moment about the aerodynamic center is

$$M_{\text{A.C.}} = -2\pi\rho U_\infty^2 l^2 \sin 2\gamma, \tag{2-160}$$

where

$$a_1 = (le^{-i\gamma})^2. \tag{2-161}$$

Figure 2–18 gives some indication of the accuracy with which lift and moment can be predicted. The theoretical value of zero for drag in two dimensions is the most prominent failure of inviscid flow methods. It represents a nearly achievable ideal, however, as evidenced by the lift-to-drag ratio of almost 300 from a carefully arranged experiment, which is reported on page 8 of Jones and Cohen (1960).

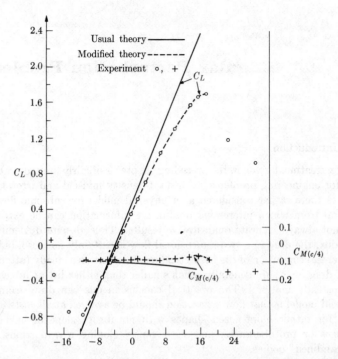

Fig. 2–18. Comparisons between predicted and measured lift coefficients and quarter-chord moment coefficients for an NACA 4412 airfoil. "Usual theory" refers to Theodorsen's procedure, whereas the modifications involve changing the function $\epsilon(\phi)$ so as to make circulation agree with the measured lift at a given angle of attack. [Adapted from Fig. 9 of Pinkerton (1936).]

More information will be found in Chapter 4 of Thwaites (1960) on refined ways of calculating constant-density flow around two-dimensional airfoils. In particular, these include reference to a modern theory by Spence and others which makes allowance for the boundary layer thickness and thus is able to carry the calculation of loading up to much higher angles of attack, even approaching the stall.

3

Singular Perturbation Problems

3–1 Introduction

The treatment given in the preceding chapter is of fairly limited practical use for engineering problems. Constant-density inviscid and irrotational flow is there rather considered as a physical model for subsonic flows in general from which interesting qualitative information can be extracted but not always accurate quantitative results. Thus, despite d'Alembert's paradox, the drag in a two-dimensional flow is certainly not zero, but the proper interpretation of the theoretical result is that, in a steady (attached) flow, drag forces are generally much smaller than either lift or forces due to unsteady motion. The practical conclusion one can draw from the inviscid model is that flow separation should be avoided at all costs. This calls for rather blunt-nosed shapes with no abrupt slope or curvature changes or protuberances, and with gently sloping rear portions, i.e., "streamlined" bodies.

Apart from the drag which is dominated by viscosity—the very thing that was neglected in the simplified model—the constant-density theory is able, in many practical cases, to produce remarkably good approximations to pressure distributions for speeds less than, say, half the speed of sound. Unfortunately, the calculation of inviscid flow for shapes of engineering interest is usually so difficult that one is forced to make some further approximation in order to obtain a result. For all the simple shapes considered in Chapter 2 (with the exception of certain airfoils and the ellipsoids with large fineness ratio) the nonviscous solutions happen to be almost completely useless, since in reality the flow will separate and the simplified model then loses its validity. The example emphasizes the fact that extreme caution must be exercised when using the physical model to obtain approximate engineering results rather than just to gain a general qualitative understanding of the physical situation and the mathematical structure of the problem. Considerable insight is generally required to judge when a simplified model will provide a useful first approximation to an actual physical situation.

The discussion may suffice to emphasize the basic difference between physical models and approximate solutions; for the former, one seeks an exact solution to a simplified and very often an artificial problem whereas, loosely speaking, in the latter, one seeks a simplified solution to a real problem. The distinction should always be kept in mind although it is not always clear-cut. For example, the continuum model of gas is also a very good approximation for all the problems that we will consider.

There are a variety of methods to obtain useful approximations. We shall discuss two different methods which are the ones mostly used in aerodynamic problems. One is the expansion in powers of a small parameter. Often the first term in this expansion may by itself be considered as a physical model. Some of the expansions that will be discussed and the physical models derived in this manner are illustrated in Fig. 3–1.

A great majority of problems in fluid mechanics have been successfully attacked by series-expansion methods. The series obtained are usually only semiconvergent, i.e., asymptotic, and also frequently not uniformly valid. These features, which are closely associated with the so-called singular perturbation nature of such problems, will be thoroughly discussed in the following section.

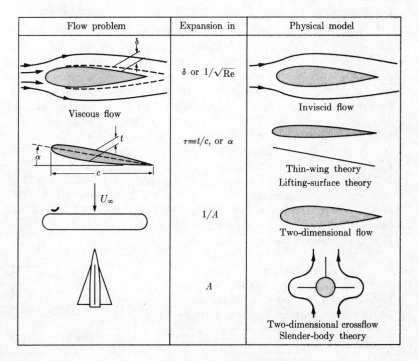

FIG. 3–1. Use of series expansions for obtaining approximate solutions to aerodynamic problems.

The second method of approximation that will be considered is the purely numerical one. In the future this will undoubtedly become of increasing importance as the full potentialities of modern computing devices are realized among aerodynamicists. There are many difficulties associated with numerical solutions. First of all, the equations for fluid flow are so complicated that no one has as yet succeeded in a step-by-step integration of the full gas dynamic equations even assuming a perfect, inviscid gas. Therefore, the examples given will concern the numerical solution of linearized problems. Second, it is very hard to estimate the error induced by the approximation scheme employed. In contrast, this could in principle always be done in the analytical series solution by estimating the first neglected term. For this purpose, and for checking out the computational program, the analytical solutions for the limiting cases are extremely useful. Thus, far from making them obsolete, the new possibilities for numerical solutions give the analytical solutions extended practical usefulness.

3–2 Expansion in a Small Parameter; Singular Perturbation Problems

As will be seen, the majority of the problems that will be considered subsequently are characterized mathematically by the property that, in the limit as the small parameter vanishes, one or more highest-order-derivative terms in the governing differential equation drop out so that the differential equation degenerates. Hence not all of the original boundary conditions of the problem can be satisfied. This type of problem is known as a singular perturbation problem. In recent years a very powerful systematic method to treat such problems has been developed by the Caltech school, primarily by Kaplun (1954, 1957) and Lagerstrom (1957).* The method is known by various names; the most frequently used one is "the inner and outer expansion method." Another one is "the method of matched asymptotic expansions" suggested by Bretherton (1962). This name has been adopted in a recent book by Van Dyke (1964) and we will follow here his usage of terminology. The reader is referred to this book for more details on the method.

In order to introduce the method and its basic ideas we will first, as is customary in the literature on the subject [see, e.g., Erdelyi (1961)], consider a simple problem involving only an ordinary differential equation. The problem chosen may be stated in physical terms as follows: "Given a mass m on a spring of spring constant k with a viscous damper of damping

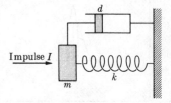

Fig. 3–2. Mass-spring-damper system considered.

* See also Lagerstrom and Cole (1955) and Friedrichs (1953, 1954).

constant d. (See Fig. 3-2.) At time $t = 0$ the mass is given an impulse I (for example, by shooting off a charge to the left). What is the subsequent motion of the mass when m is very small?"*

This problem will be solved in essentially three different ways. First, an exact solution may easily be obtained. The governing differential equation and the associated boundary conditions are

$$m \frac{d^2x}{dt^2} + d \frac{dx}{dt} + kx = 0, \tag{3-1}$$

$$u(0) = \left(\frac{dx}{dt}\right)_{t=0} = I/m, \tag{3-2}$$

$$x(0) = 0. \tag{3-3}$$

The solution is found to be

$$x = (I/d)(e^{-\lambda_2 t} - e^{-\lambda_1 t})/\sqrt{1 - 4mk/d^2}, \tag{3-4}$$

where

$$\lambda_{1,2} = -\frac{d}{2m}\left(1 \pm \sqrt{1 - 4mk/d^2}\right). \tag{3-5}$$

This will be the reference solution used in assessing the approximate solutions that follow.

We shall now obtain an approximate solution valid for small mass m by use of simple physical reasoning. From the boundary condition (3-2) it follows that the initial velocity will be very high for small m, hence the damping force will be the main decelerating force in the initial stages. The restoring force due to the spring, on the other hand, will be comparatively much smaller because initially x is small. Therefore, the initial motion of the mass is governed approximately by the following equation:

$$\frac{du}{dt} = -\left(\frac{d}{m}\right)u. \tag{3-6}$$

Now

$$x = \int_0^t u\, dt, \qquad dt = \frac{dt}{du}\, du = \frac{du}{du/dt},$$

that is,

$$x = -\left(\frac{m}{d}\right)\int_{u(0)}^u du = -\left(\frac{m}{d}\right)(u - u(0)) = \frac{I}{d} - \left(\frac{m}{d}\right)u. \tag{3-7}$$

---

* This problem was used by Prandtl to illustrate and explain his boundary layer theory (see Schlichting, 1960, p. 63).

The mass will have reached its maximum deviation when $u(t) = 0$. From (3–7) we thus obtain, approximately,

$$x_{max} = \frac{I}{d}. \tag{3–8}$$

After the mass has reached its maximum deviation, the spring will force it relatively slowly back to its original position. Since the mass is so small, the motion will then be dominated by the spring and the damper. Consequently, the following equation will approximately describe its subsequent motion:

$$d\frac{dx}{dt} + kx = 0. \tag{3–9}$$

This has the solution

$$x = Ae^{-kt/d}. \tag{3–10}$$

In order to determine the integration constant A approximately, we notice that the initial phase of the motion as described by (3–6)–(3–8) takes place almost instantaneously for vanishing m. Therefore, it would appear to a slow observer as if at time $t = 0+$ the mass were suddenly displaced to x_{max} and then released. Hence, the slow phase of the motion would approximately be described by (3–10) with A given by (3–8):

$$x \cong \frac{I}{d} e^{-(k/d)t}. \tag{3–11}$$

We thus have arrived at two different approximate solutions. Integrating (3–6) with (3–2) and (3–3), we obtain

$$x \cong \frac{I}{d}[1 - e^{-(d/m)t}], \tag{3–12}$$

which is valid for small times and which will be called the *inner solution* and denoted by a superscript i. The approximation (3–11) valid for large times will be called the *outer solution* and denoted by superscript o. Thus

$$x^i = \frac{I}{d}[1 - e^{-(d/m)t}], \tag{3–13}$$

$$x^o = \frac{I}{d} e^{-(k/d)t}. \tag{3–14}$$

We shall now see how this basically intuitive method may be systematized to yield additional terms in a power series of an appropriate small dimensionless parameter. First, it is necessary to introduce dimensionless variables. A suitable set is

$$t^* = \frac{k}{d}t, \tag{3–15}$$

$$x^* = \frac{d}{I}x, \tag{3–16}$$

which transforms the original problem to

$$\epsilon \frac{d^2 x^*}{dt^{*2}} + \frac{dx^*}{dt^*} + x^* = 0, \tag{3-17}$$

$$\left(\frac{dx^*}{dt^*}\right)_{t^*=0} = \frac{1}{\epsilon}, \tag{3-18}$$

$$x^*(0) = 0, \tag{3-19}$$

where

$$\epsilon = mk/d^2. \tag{3-20}$$

We will now attempt to derive a series solution of x^* in the small parameter ϵ guided by the physical approach tried above. First, consider an *outer expansion* of the form

$$x^* = x^o = \sum x_n^o \epsilon^n. \tag{3-21}$$

Upon substituting in (3–17) and equating terms of like powers of ϵ, we obtain

$$\frac{dx_0^o}{dt^*} + x_0^o = 0, \tag{3-22}$$

$$\frac{dx_n^o}{dt^*} + x_n^o = - \frac{d^2 x_{n-1}^o}{dt^{*2}}, \qquad n > 0. \tag{3-23}$$

The solution of (3–22) is

$$x_0^o = A_0^o e^{-t^*}. \tag{3-24}$$

Obviously, this corresponds to the solution (3–10) for large times, the outer solution. The determination of the constant A_0^o must wait for the moment.

The inner solution happens in a very short time. Therefore, in order to be able to study the solution with some resolution, we need to "magnify" the region of interest. This is achieved by *stretching* the independent variable. A suitable stretching is in the present example obtained by introducing for the inner solution $x^* = x^i$ as a new independent variable

$$\bar{t} = t^*/\epsilon, \tag{3-25}$$

which transforms the differential equation (3–17) and boundary conditions (3–18) and (3–19) into

$$\frac{d^2 x^i}{d\bar{t}^2} + \frac{dx^i}{d\bar{t}} + \epsilon x^i = 0, \tag{3-26}$$

$$\left(\frac{dx^i}{d\bar{t}}\right)_{\bar{t}=0} = 1, \tag{3-27}$$

$$x^i(0) = 0. \tag{3-28}$$

Expansion in ϵ, keeping $\bar{t}$ fixed, now gives the following *inner expansion:*

$$x^i = \sum_0 x_n^i(\bar{t})\epsilon^n, \tag{3-29}$$

$$\frac{d^2 x_0^i}{d\bar{t}^2} + \frac{dx_0^i}{d\bar{t}} = 0, \tag{3-30}$$

$$\frac{d^2 x_n^i}{d\bar{t}^2} + \frac{dx_n^i}{d\bar{t}} = -x_{n-1}^i, \quad n > 0. \tag{3-31}$$

It is important to apply just the right amount of stretching in order to get a useful inner solution. In the present case we were guided by the physical insight into the problem which tells us that in the lowest-order term there should be a balance between inertia and damping terms, such as is retained in (3–30). If one applies too much stretching, for example, by setting instead of (3–25)

$$\bar{t} = t^*/\epsilon^2, \tag{3-32}$$

the features allowing one to match the inner and outer solutions (see below) would be lost. Thus with (3–32) the equation for the lowest-order term would be

$$\frac{d^2 x_1^i}{d\bar{t}^2} = 0$$

with the solution

$$x^i \sim \epsilon\bar{t}.$$

In other words, the "magnification ratio" is so large that only the initial linear portion of the solution can be kept in view. This is illustrated in Fig. 3–3. The amount of stretching necessary for each problem is usually evident from the physics of the problem; however, a check on this will always be whether the expansion works.

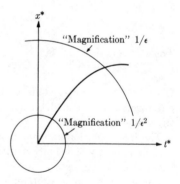

FIG. 3–3. Choice of stretching for inner solution.

The solution of (3–30) satisfying the boundary conditions (3–27) and (3–28) is

$$x_0^i = 1 - e^{-\bar{t}}, \tag{3-33}$$

which is equivalent to (3–12).

To complete the zeroth-order solution it remains to determine the constant A_0^o in (3–24). Let us assume that the validity of the inner and outer solutions overlaps in some region of t^* and that in this region we can find a $t^* = \delta(\epsilon)$ such that we have

$$\lim_{\epsilon \to 0} \delta = 0 \qquad (3\text{--}34)$$

and

$$\lim_{\epsilon \to 0} (\delta/\epsilon) = \infty. \qquad (3\text{--}35)$$

Such a choice would be, for example, $\delta = \sqrt{\epsilon}$. Requiring the two expansions to overlap in the limit gives (since the inner solution is expressed as a function of $\bar{t} = t^*/\epsilon$)

$$\lim_{\epsilon \to 0} [x_0^i(\delta/\epsilon)] = \lim_{\epsilon \to 0} [x_0^o(\delta)] \qquad (3\text{--}36)$$

or

$$x_0^i(\infty) = x_0^o(0). \qquad (3\text{--}37)$$

This is termed the *limit matching principle* which in words may be stated as follows:

The outer limit of the inner expansion = the inner limit of the outer expansion.

From (3–33) it follows that

$$x_0^i(\infty) \equiv x_0^{io} = 1$$

and from (3–24)

$$x_0^o(0) \equiv x_0^{oi} = A_0^o.$$

Applying the limit matching principle,

$$x_0^{io} = x_0^{oi}, \qquad (3\text{--}38)$$

gives

$$A_0^o = 1, \qquad (3\text{--}39)$$

which leads to an outer solution identical to the previous one obtained through intuitive reasoning.

Notice that the formal approach is nothing but a formalization of the intuitive one. However, the formal approach is capable of being extended to give higher-order terms to any order in ϵ, which the intuitive one is not. First, we may construct a *composite solution* that is uniformly valid to order ϵ over the whole region by setting

$$x_0^c \sim x_0^o + x_0^i - x_0^{oi}. \qquad (3\text{--}40)$$

It is seen that this solution, in view of the matching, approaches the inner and outer solutions in the inner and outer regions, respectively, and carries over smoothly in between them. In the problem considered, the zeroth-order composite solution becomes

$$x_0^c \sim e^{-t^*} - e^{-\bar{t}} = e^{-t^*} - e^{-t^*/\epsilon}. \qquad (3\text{--}41)$$

To proceed to the next higher approximation it follows from (3–23) and (3–24), (3–39) that the next term in the outer expansion is a solution of

$$\frac{dx_1^o}{dt^*} + x_1^o = -e^{-t^*}, \tag{3-42}$$

which has the general solution

$$x_1^o = -t^* e^{-t^*} + A_1^o e^{-t^*} \tag{3-43}$$

The first-order inner solution must, according to (3–31) and (3–33), satisfy

$$\frac{d^2 x_1^i}{d\bar{t}^2} + \frac{dx_1^i}{d\bar{t}} = e^{-\bar{t}} - 1. \tag{3-44}$$

The general solution of this equation is

$$x_1^i = A_1^i + B_1^i e^{-\bar{t}} - \bar{t} - \bar{t}e^{-\bar{t}} \tag{3-45}$$

The constants A_1^i and B_1^i are to be determined such that the inner boundary conditions (3–27) and (3–28) are fulfilled. Since the lowest-order term has already taken care of these, x_1^i and its first derivative must both be zero. This gives

$$A_1^i = -B_1^i = 2, \tag{3-46}$$

and hence

$$x_1^i = 2(1 - e^{-\bar{t}}) - \bar{t}(1 + e^{-\bar{t}}). \tag{3-47}$$

The two-term inner expansion is thus

$$x^i \sim x_0^i + \epsilon x_1^i = 1 - e^{-\bar{t}} + \epsilon[2(1 - e^{-\bar{t}}) - \bar{t}(1 + e^{-\bar{t}})] \tag{3-48}$$

or, expressed in the outer (physical) variable

$$x^i \sim 1 - t^* - (1 + t^*)e^{-t^*/\epsilon} + 2\epsilon(1 - e^{-t^*/\epsilon}). \tag{3-49}$$

The two-term outer solution is obtained from (3–24) and (3–41) to be

$$x^o \sim x_0^o + \epsilon x_1^o = e^{-t^*} - \epsilon t^* e^{-t^*} + \epsilon A_1^o e^{-t^*} \tag{3-50}$$

The behavior of this near the inner limit may be obtained by series expansion in t^*. The first terms in such an expression are

$$x^{oi} \sim 1 - t^* - \epsilon(t^* - A_1^o). \tag{3-51}$$

In the inner expansion, on the other hand, the exponential term will be negligible in the outer limit and thus

$$x^{io} \sim 1 - t^* + 2\epsilon. \tag{3-52}$$

It is evident from comparison of (3–50) and (3–52) that the two expressions match if

$$A_1^o = 2. \tag{3–53}$$

The procedure may be formalized as follows: Express the n-term inner expansion in outer variables and take the m-term outer expansion of this. In our case take $n = m = 2$. Then the two-term outer expansion of the two-term inner expansion as obtained from (3–49) is

$$x^{io} \sim 1 - t^* + 2\epsilon. \tag{3–54}$$

Next, express the two-term outer expansion in inner variables. This gives

$$x^o = e^{-\epsilon i} - \epsilon^2 \bar{i} e^{-\epsilon i} + \epsilon A_1^o e^{-\epsilon i}. \tag{3–55}$$

Take the two-term expansion of this. This yields

$$x^{oi} \sim 1 - \epsilon \bar{i} + \epsilon A_1^o. \tag{3–56}$$

Reexpress this in outer variables. Hence

$$x^{oi} \sim 1 - t^* + \epsilon A_1^o. \tag{3–57}$$

Equating (3–57) and (3–54) gives $A_1^o = 2$ as before. We have shown an application of the *asymptotic matching principle of* Kaplun and Lagerstrom (1957) which may be stated as follows (Van Dyke, 1964):

> *The m-term outer expansion of (the n-term inner expansion) = the n-term inner expansion of (the m-term outer expansion).*

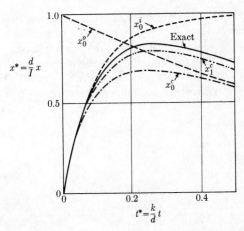

FIG. 3–4. Comparisons of various approximations for $\epsilon = mk/d^2 = 0.1$.

This principle should hold for any combination of n and m, not only when they are equal as in the present case.

In the problems that will be treated in the following, mostly the limit matching principle will be used. We may now construct a composite expansion valid to first order by setting

$$x^c \sim x^i + x^o - x^{io} = (1 + 2\epsilon)(e^{-t^*} - e^{-t^*/\epsilon})$$
$$- t^*(e^{-t^*/\epsilon} + \epsilon e^{-t^*}). \quad (3\text{--}58)$$

In Fig. 3–4 are shown the various approximate solutions for $\epsilon = 0.1$ together with the exact solution. As seen, the first-order composite solution gives a rather good approximation to the exact solution everywhere, and the next term will probably account for practically all of the remaining difference.

We summarize now the main elements of the method:

(1) Writing the differential equations in a nondimensional form.

(2) Straightforward power series expansion of the differential equation and the associated boundary conditions using the physical variables. This gives the *outer expansion*.

(3) Suitable *stretching* of the independent variable to magnify the inner region sufficiently to be able to discern the details of the inner solution. Power series expansion of the solution in the small parameter keeping the stretched coordinate fixed gives the *inner expansion*.

(4) *Matching* the inner and outer expansions asymptotically.

(5) Constructing the *composite expansion*.

The method of matched asymptotic expansions has been used successfully in a wide variety of fluid flow problems as well as in the theory cf elasticity, and for some problems in rigid-body dynamics.

In the subsequent application of the method we will seldom proceed beyond the lowest-order approximation, in which case the limit matching principle usually suffices. Also composite expansions will only rarely be considered.

4

Effects of Viscosity

4-1 Introduction

Viscous flows at high Reynolds numbers constitute the most obvious example of singular perturbation problems. The viscosity multiplies the highest-order derivative terms in the Navier-Stokes equations, and these will therefore in the limit of zero viscosity degenerate to a lower order. The boundary condition of zero tangential velocity on the body (or of continuous velocity in the stream) is therefore lost. This necessitates the introduction of a thin boundary layer next to the body constituting the inner region where the inviscid equations are not uniformly valid. Unfortunately, only a very restricted class of viscous-flow problems can be analyzed by the direct use of the method of matched asymptotic expansions. First, only for a very limited class of bodies will the boundary layer remain attached to the body surface. When separation occurs, the location of the region where viscosity is important is no longer known *a priori*. The second difficulty is that for very high Reynolds numbers the flow in the boundary layer becomes unstable and transition to turbulence occurs. As yet, no complete theory for predicting turbulent flow exists.

Before considering some of the model problems that may be analyzed, we will give a short description of the qualitative effects of viscosity.

4-2 Qualitative Effects of Viscosity

It is a common feature of most flows of engineering interest that the viscosity of the fluid is extremely small. The Reynolds number, $\mathrm{Re} = U_\infty l/\nu$, which gives an overall measure of the ratio of inertia forces to viscous forces, is in typical aeronautical applications of the order 10^6 or more. For large ships, Reynolds numbers of the order 10^9 are common. Viscosity can then only produce significant forces in regions of extremely high shear, i.e., in extremely thin shear layers where there is a substantial variation of velocity across the streamlines. The thickness of the laminar boundary layer on a flat plate of length l is approximately $\simeq 5l/\sqrt{\mathrm{Re}}$. For $\mathrm{Re} = 10^6$, $\delta/l \simeq 0.005$, which is so thin that it cannot even be illustrated in a figure without expanding the scale normal to the plate. A turbulent boundary layer has a considerably greater thickness, reflecting its higher drag and therefore larger momentum loss; an approximate formula given in Schlichting (1960), p. 38, is $\delta/l \simeq 0.37/\mathrm{Re}^{1/5}$. For

Re $= 10^6$ this gives $\delta/l = 0.023$, which is still rather small. Viscosity is only important in a very small portion of the turbulent boundary layer next to the surface, in the "viscous sublayer."

The transition from a laminar to a turbulent boundary layer is a very complicated process that depends on so many factors that precise figures for transition Reynolds numbers cannot be given. For a flat plate in a very quiet free stream (i.e., one having a rms turbulent velocity fluctuation intensity of 0.1% or less) transition occurs at approximately a distance from the leading edge corresponding to Re $\sim 3 \times 10^6$. With a turbulent intensity of only 0.3% in the oncoming free stream the transition Reynolds number decreases to about 1.5×10^6. These distances are far beyond that for which the boundary layer first goes unstable. Stability calculations show that on a flat plate this occurs at Re $\simeq 10^5$. The complicated series of events between the point where instability first sets in until transition occurs has only recently been clarified (see Klebanoff, Tidstrom, and Sargent, 1962). Transition is strongly influenced by the pressure gradient in the flow; a negative ("favorable") pressure gradient tends to delay it and a positive ("adverse") one tends to make it occur sooner. As a practical rule of thumb one can state that the laminar boundary layer can only be maintained up to the point of minimum pressure on the airfoil. On a so-called laminar-flow airfoil one therefore places this point as far back on the airfoil as possible in order to try to achieve as large a laminar region as possible. Laminar-flow airfoils work successfully for moderately high Reynolds numbers ($<10^7$) and low lift coefficients but require extremely smooth surfaces in order to avoid premature transition. At very high Reynolds numbers transition starts occurring in the region of favorable pressure gradient.

Both laminar and turbulent boundary layers will separate if they have to go through extensive regions of adverse pressure gradients. Separation will always occur for a subsonic flow at sharp corners, because there the pressure gradient would become infinite in the absence of a boundary layer. Typical examples of unseparated and separated flows are shown in Fig. 4–1. In the separated flow there will always be a turbulent wake

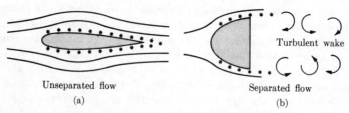

Unseparated flow
(a)

Turbulent wake

Separated flow
(b)

FIG. 4–1. Examples of high Reynolds number flows. Asterisks indicate regions in which viscosity is important.

behind the body. In principle one could have instead a region of fairly quiescent flow in the wake, separated from the outer flow by a thin laminar shear layer attached to the laminar boundary layer on a body. However, a free shear layer, lacking the restraining effect of a wall, will be highly unstable and will therefore turn turbulent almost immediately. Because of the momentum loss due to the turbulent mixing in the wake the drag of the body will be quite large. On a thin airfoil at a small angle of attack the boundary layer will separate at the sharp trailing edge but there will be a very small wake so that a good model for the flow is the attached flow with the Kutta condition for the inviscid outer flow determining the circulation.

4–3 Boundary Layer on a Flat Plate

We shall consider the viscous laminar high Reynolds number flow over a semi-infinite flat plate at zero angle of attack. This is the simplest case of a boundary layer that may serve as a model for the calculation of boundary layer effects on a thin airfoil. For incompressible flow the Navier-Stokes equations and the equation of continuity read

$$\frac{DQ}{Dt} = -\frac{\Delta p}{\rho} + \nu\nabla^2 Q$$

$$\nabla \cdot Q = 0.$$

(4–1)

The boundary conditions are that the velocity vanishes on the plate and becomes equal to U_∞ far away from the plate surface. Thus for two-dimensional flow

$$Q(x, 0) = 0 \quad \text{for } x > 0, \qquad (4\text{–}2)$$

$$Q = U_\infty i$$
$$p = p_\infty \quad \text{for } z \to \infty \cdot \qquad (4\text{–}3)$$

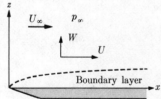

Fig. 4–2. Laminar boundary layer over a flat plate.

We have assumed that the leading edge of the plate is located at the origin (see Fig. 4–2). By assuming that the plate is semi-infinite one avoids the problem of considering upstream effects from the trailing edge. These are actually quite small and do not show up in the boundary layer approximations to be derived.

In the process of introducing nondimensional coordinates a minor difficulty is encountered because there is no natural length in the problem to which spatial coordinates could be referred. We will circumvent this by selecting an arbitrary reference length and thereby implicitly assume that the behavior of the solution for $x = 0(l)$ and $z = 0(l)$ will be studied.

The appropriate nondimensional variables to be used are thus

$$x^* = x/l$$
$$z^* = z/l$$
$$\mathbf{Q}^* = \mathbf{Q}/U_\infty$$
$$p^* = p/\rho U_\infty^2.$$
(4-4)

These transform the system (4–1)–(4–3) into

$$U^* U_{x^*}^* + W^* U_{z^*}^* = -p_{x^*}^* + \mathrm{Re}^{-1} \nabla^{*2} U^*,$$ (4-5)

$$U^* W_{x^*}^* + W^* W_{z^*}^* = -p_{z^*}^* + \mathrm{Re}^{-1} \nabla^{*2} W^*,$$ (4-6)

$$U_{x^*}^* + W_{z^*}^* = 0,$$ (4-7)

$$U^*(x^*, 0) = W^*(x^*, 0) = 0 \quad \text{for} \quad x^* > 0,$$ (4-8)

$$U^* = 1 \quad \text{for} \quad z^* \to \infty,$$ (4-9)

$$p^* = p_\infty^* \quad \text{for} \quad z^* \to \infty.$$ (4-10)

The asterisks will be dispensed with from now on. The equation of continuity may be eliminated at the outset by introducing the stream function Ψ so that

$$U = \Psi_z$$ (4-11)

$$W = -\Psi_x.$$ (4-12)

Substituting these into the equations above we obtain

$$\Psi_z \Psi_{zx} - \Psi_x \Psi_{zz} = -p_x + \mathrm{Re}^{-1} \nabla^2 \Psi_z,$$ (4-13)

$$-\Psi_z \Psi_{xx} + \Psi_x \Psi_{xz} = -p_z - \mathrm{Re}^{-1} \nabla^2 \Psi_x,$$ (4-14)

$$\Psi_z(x, 0) = \Psi_x(x, 0) = 0 \quad \text{for} \quad x > 0,$$ (4-15)

$$\Psi_z(x, \infty) = 1,$$ (4-16)

$$p(x, \infty) = p_\infty.$$ (4-17)

The next task is to identify an appropriate small parameter ϵ. This we do by considering an inner expansion of the form

$$\Psi^i(x, \bar{z}) = \epsilon \Psi_1^i + \epsilon^2 \Psi_2^i + \cdots,$$ (4-18)

$$p^i(x, \bar{z}) = p_0^i + \epsilon p_1^i + \cdots,$$ (4-19)

$$\bar{z} = z/\epsilon.$$ (4-20)

In the expansion (4–18) there cannot be any term of zeroth order because otherwise the U-component

$$U^i = (1/\epsilon) \Psi_{\bar{z}}^i$$ (4-21)

would become infinite as $\epsilon \to 0$. [There can be no term of the form $\Psi_0^i(x)$ because of the boundary condition (4–15).] Upon introducing (4–18) and (4–19) into (4–13) and retaining only the lowest-order terms we obtain

$$\Psi_{1\bar{z}}^i \Psi_{1\bar{z}x}^i - \Psi_{1x}^i \Psi_{1\bar{z}\bar{z}}^i = -p_{0x}^i + (\epsilon^2 \, \text{Re})^{-1} \Psi_{1\bar{z}\bar{z}\bar{z}}^i. \qquad (4\text{–}22)$$

The effect of viscosity is given by the last term. In order to retain it in the limit of $\epsilon \to 0$ we need to set

$$\epsilon = (\text{Re})^{-1/2}. \qquad (4\text{–}23)$$

Having established ϵ we introduce the series (4–18) and (4–19) into (4–14), and thus to lowest order

$$p_{0\bar{z}}^i = 0. \qquad (4\text{–}24)$$

This shows that p_0^i is a function of x only; in other words, to the lowest order, the pressure is constant across the boundary layer. We now turn to the outer flow and set

$$\Psi^o = \Psi_0^o + \epsilon \Psi_1^o + \cdots, \qquad (4\text{–}25)$$

$$p^o = p_0^o + \epsilon p_1^o + \cdots \qquad (4\text{–}26)$$

Upon substitution into the Navier-Stokes equations we obtain

$$\begin{aligned}
(\Psi_{0z}^o + \epsilon \Psi_{1z}^o + \cdots)(\Psi_{0xz}^o + \epsilon \Psi_{1xz}^o + \cdots) & \\
- (\Psi_{0x}^o + \epsilon \Psi_{1x}^o + \cdots)(\Psi_{0zz}^o + \epsilon \Psi_{1zz}^o + \cdots) & \\
= - p_{0x}^o - \epsilon p_{1x}^o + \epsilon^2 \nabla^2 \Psi_{0z}^o + \cdots, &
\end{aligned} \qquad (4\text{–}27)$$

$$\begin{aligned}
- (\Psi_{0z}^o + \epsilon \Psi_{1z}^o + \cdots)(\Psi_{0xx}^o + \epsilon \Psi_{1xx}^o + \cdots) & \\
+ (\Psi_{0x}^o + \epsilon \Psi_{1x}^o + \cdots)(\Psi_{0xz}^o + \epsilon \Psi_{1xz}^o + \cdots) & \\
= -p_{0z}^o - \epsilon p_{1z}^o - \epsilon^2 \nabla^2 \Psi_{0x}^o - \cdots &
\end{aligned} \qquad (4\text{–}28)$$

Thus, the terms of order ϵ^0 give

$$\Psi_{0z}^o \Psi_{0xz}^o - \Psi_{0x}^o \Psi_{0zz}^o = -p_{0x}^o, \qquad (4\text{–}29)$$

$$- \Psi_{0z}^o \Psi_{0xx}^o + \Psi_{0x}^o \Psi_{0xz}^o = -p_{0z}^o. \qquad (4\text{–}30)$$

Equations (4–29) and (4–30) are precisely Euler's equations for an inviscid flow. Since the flow must be irrotational far upstream, it therefore must be irrotational everywhere to zeroth order, so that

$$\nabla^2 \Psi_0^o = 0. \qquad (4\text{–}31)$$

By substitution into (4–27) and (4–28) it is seen that the last term vanishes and hence the flow is irrotational to first order. Upon continuing the process it is found that the flow is irrotational to any order, hence

$$\nabla^2 \Psi_n^o = 0. \qquad (4\text{–}32)$$

An integral of the inviscid equations for a potential flow is given by Bernoulli's equation

$$p^o = p_\infty - \tfrac{1}{2}[(\Psi_x^o)^2 + (\Psi_z^o)^2 - 1], \qquad (4\text{--}33)$$

which is given here in its nondimensional form.

To determine the outer solution uniquely we need to specify the normal velocity on the boundary (the plate). However, this is located inside the inner region and thus outside the region of validity of the outer solution. This boundary condition must therefore be obtained by matching the outer and inner solutions for W. We have

$$W^o = -\Psi_{0x}^o - \epsilon\Psi_{1x}^o - \epsilon^2\Psi_{2x}^o - \cdots \qquad (4\text{--}34)$$

and

$$W^i = -\epsilon\Psi_{1x}^i - \epsilon^2\Psi_{2x}^i - \cdots \qquad (4\text{--}35)$$

Using the limit matching principle we then find

$$\left.\begin{aligned}\Psi_{0x}^o(x, 0) &= 0 \\ \Psi_{1x}^o(x, 0) &= \Psi_{1x}^i(x, \infty)\end{aligned}\right\} \quad \text{for} \quad x > 0. \qquad \begin{aligned}(4\text{--}36)\\(4\text{--}37)\end{aligned}$$

It follows directly from (4–36) that the outer flow must be a parallel undisturbed one:

$$\Psi_0^o = z, \qquad (4\text{--}38)$$

$$p_0^o = p_\infty \qquad (4\text{--}39)$$

as is of course also evident from the physics of the problem. From (4–24) and matching of the pressures it follows that

$$p_0^i = p_\infty \qquad (4\text{--}40)$$

and (4–22) thus takes the following form:

$$\Psi_{1\bar{z}}^i\Psi_{1\bar{z}x}^i - \Psi_{1x}^i\Psi_{1\bar{z}\bar{z}}^i = \Psi_{1\bar{z}\bar{z}\bar{z}}^i. \qquad (4\text{--}41)$$

Notice that this equation contains only x-derivatives of first order. It is a parabolic differential equation, like the equation of heat conduction (with x replacing time). A characteristic feature of such an equation is that its solutions admit no upstream influence. Thus the assumption of an infinite plate introduces actually no practical restriction on the result, since it will be equally valid for a finite plate.

The inner boundary conditions for Ψ^i are those given by (4–15),

$$\Psi_{1\bar{z}}^i(x, 0) = \Psi_{1x}^i(x, 0) = 0 \quad \text{for} \quad x > 0. \qquad (4\text{--}42)$$

The remaining boundary condition required for the third-order equation (4–41) is lost in the outer region and must be obtained through matching of U. We have

$$U^i = \Psi^i_{1\bar{z}}(x, \bar{z}) + \epsilon\Psi^i_{2\bar{z}}(x, \bar{z}) + \cdots, \qquad (4\text{--}43)$$

$$U^o = \Psi^o_{0z}(x, z) + \epsilon\Psi^o_{1z}(x, z) + \cdots \qquad (4\text{--}44)$$

Upon matching by setting $U^i(x, \infty) = U^o(x, 0)$ and using (4–38), we obtain in the limit as $\epsilon \to 0$

$$\Psi^i_{1\bar{z}}(x, \infty) = 1. \qquad (4\text{--}45)$$

The three boundary conditions (4–42) and (4–45) completely specify the solution of (4–41). It is possible to solve this nonlinear equation in the present case by seeking a self-similar solution of the form (see Schlichting, 1960, p. 143)*

$$\Psi^i_1 = \sqrt{2x}\, f(\eta), \qquad (4\text{--}46)$$

where

$$\eta = \bar{z}/\sqrt{2x}.$$

This transformation gives an ordinary differential equation for f, namely:

$$f''' + ff'' = 0, \qquad (4\text{--}47)$$

$$f(0) = f'(0) = 0, \qquad (4\text{--}48)$$

$$f'(\infty) = 1. \qquad (4\text{--}49)$$

The solution of (4–47) must be obtained numerically. A table of f may be found, for example, in Schlichting (1960), p. 121. The solution (4–46) is Blasius' solution of Prandtl's boundary layer equations. Having Ψ^i_1 we may easily calculate Ψ^o_1. From the matching condition (4–37) and (4–46) it follows that

$$\Psi^o_{1x}(x, 0) = \frac{1}{\sqrt{2x}} \lim_{\eta \to \infty} (f - \eta f') = -0.865/\sqrt{x}, \qquad (4\text{--}50)$$

where the constant has been evaluated from the numerical solution. We need a solution of Laplace's equation satisfying (4–50) on the positive x-axis and giving vanishing velocities at infinity. The easiest way to find this is to employ complex variables. Let

$$Y = x + iz. \qquad (4\text{--}51)$$

Then a solution of Laplace's equation is given by

$$\Phi^o_1 + i\Psi^o_1 = F(Y), \qquad (4\text{--}52)$$

---

* The definitions (4–46) and (4–47) differ from those in Schlichting (1960) by the factors of 2 which are introduced in order to make (4–47)–(4–49) free from numerical factors.

where F is any analytic function. The velocity components are given by

$$U_1^o - iW_1^o = \frac{dF}{dY} = \Psi_{1z}^o + i\Psi_{1x}^o. \qquad (4\text{--}53)$$

It is easily verified that the appropriate solution is

$$F(Y) = -1.73iY^{1/2} \qquad (4\text{--}54)$$

or

$$\Psi_1^o = -1.73 \, \textbf{\textit{Re}} \, \{Y^{1/2}\}. \qquad (4\text{--}55)$$

From this, the first-order velocity perturbations in the outer flow may be calculated and hence the pressure 'from the Bernoulli equation, (4–33). An interesting result from (4–55) is that on the plate ($z = 0$) the U-perturbation is zero to first order, and the pressure perturbation on the plate is only of (at most) order ϵ^2.

The boundary condition (4–50) for the first-order outer flow may be obtained in a different and physically more instructive way. Consider the integral

$$\delta^* = \int_0^\infty (1 - U/U_\infty) \, dz. \qquad (4\text{--}56)$$

This is called the "displacement thickness," because the effect of the slowing down of the fluid in the boundary layer on the flow outside it is equivalent to that of a body of thickness δ^*. To zeroth order, a uniformly valid approximation to U is given in our problem by $\Psi_{1\hat{z}}^i$, and we thus obtain after changing integration variables to η

$$\delta^* = \epsilon\sqrt{2x} \int_0^\infty (1 - f') \, d\eta = \epsilon\sqrt{2x} \lim_{\eta \to \infty} (\eta - f) = 1.73\epsilon\sqrt{x}. \quad (4\text{--}57)$$

The equivalence of the limiting value with that occurring in (4–50) is easily seen since $f'(\infty) = 1$. A flow that is tangent to a thin body of this thickness should to a first approximation have

$$W(x, 0) = U_\infty \frac{d\delta^*}{dx} = 0.865\epsilon/\sqrt{x},$$

which is identical to (4–50).

Since typical values of interest for the small parameter ϵ are 0.001–0.01, one would expect good agreement between the first-order theory and experiments. As seen from Fig. 4–3 this is indeed the case; the experimental points are almost indistinguishable from the theoretical curve.

The Blasius solution just described was the first successful attempt to solve Prandtl's more general boundary layer equations. These result as the first-order inner term in the present expansion scheme when the zeroth-order outer flow is not restricted to the trivial parallel flow as in the case

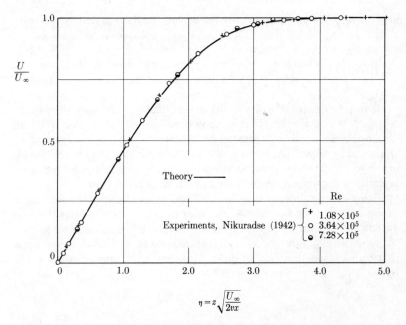

$$\eta = z \sqrt{\frac{U_\infty}{2vx}}$$

FIG. 4–3. Velocity distribution in a flat-plate boundary layer. [Adapted from *Boundary Layer Theory* by H. Schlichting, Copyright © (1960) McGraw-Hill Book Company. Used by permission of McGraw-Hill Book Company.]

considered. Assume for simplicity that the plate is still flat and located along the x-axis, but that some outside disturbance is causing a pressure distribution different from zero on the plate. Again, the zeroth-order outer flow gives a uniformly valid approximation to the pressure, also in the boundary layer, and the differential equation for the first-order inner solution becomes

$$\Psi^i_{1\bar{z}} \Psi^i_{1\bar{z}x} - \Psi^i_{1x} \Psi^i_{1\bar{z}\bar{z}} = -p^o_x(x,0) + \Psi^i_{1\bar{z}\bar{z}\bar{z}}, \qquad (4\text{--}58)$$

which is equivalent to Prandtl's boundary layer equations. One can easily see that this will also give a first-order approximation for a curved body, provided x is taken along the surface and $\bar{z}$ normal to it. The matching condition is

$$U^i(x, \infty) = U^o(x, 0), \qquad (4\text{--}59)$$

that is,

$$\Psi^i_{1\bar{z}}(x, \infty) = \Psi^o_{0z}(x, 0) = U^o(x, 0). \qquad (4\text{--}60)$$

But now $U^o(x, 0)$ is related through Bernoulli's equation to the pressure distribution

$$U^o = \sqrt{1 - 2(p^o - p_\infty)} \qquad (4\text{--}61)$$

and no longer a constant. Only for a very restricted class of pressure distributions is it possible to find self-similar solutions and hence reduce the partial differential equation (4–58) to an ordinary one. When this is not possible one has to resort to some approximate method of solution.

A discussion of some proposed methods may be found in Schlichting (1960).

Having evaluated the first-order inner solution one may proceed to calculate the first-order effect on pressure. As was seen, this could be obtained in an equivalent manner by considering the original body to be thickened by an amount equal to the displacement thickness of the boundary layer. Since the latter is extremely small, the first-order effect on pressure is usually negligible from a practical point of view.

Higher-order solutions may be obtained by continuing the expansion procedure using the asymptotic matching principle. It turns out that the expansion becomes nonunique after two terms, presumably because of a higher-order long-range effect from disturbances at the leading edge. For a thorough discussion of higher-order boundary layer approximations, the reader is referred to Van Dyke (1964).

5

Thin-wing Theory

5–1 Introduction

In this chapter we shall derive the equations of motion governing the sub- or supersonic flow around thin wings. Although this problem may be handled by simpler regular perturbation methods, we shall here instead use the method of matched asymptotic expansions for basically two reasons. First, the present method gives a clearer picture of the usual linearized formulation as that for an *outer* flow for which the wing in the limit collapses onto a plane (taken to be $z = 0$). Secondly, the similarities and dissimilarities between the airfoil problem and that for a slender body of revolution, which will be treated in the following chapter, will be more readily apparent; in fact, as will later be seen, the slender-body theory formulation can be obtained from the two-dimensional airfoil case by means of a simple modification of the inner solution.

The procedure will be carried out in detail for two-dimensional flow only, but the extension to three dimensions is quite straightforward. The first term in the series expansion leads to the well-known linearized wing theory. In this chapter, the solutions for two-dimensional flow are given as examples. Three-dimensional wings will be considered in later chapters. The case of transonic flow requires special treatment and will be deferred to Chapter 12.

5–2 Expansion Procedure for the Equations of Motion

Anticipating the three-dimensional wing case, for which the standard choice is to orient the wing in the x, y-plane, we shall this time consider two-dimensional flow in the x, z-plane around a thin airfoil located mainly along the x-axis with the free stream of velocity U_∞ in the direction of the positive x-axis as before. Thus, referring to Fig. 5–1, we let the location of the upper and lower airfoil surfaces be given by

$$z_u = \epsilon \bar{f}_u(x) = \tau \bar{g}(x) + \theta \bar{h}(x) - \alpha x$$
$$z_l = \epsilon \bar{f}_l(x) = -\tau \bar{g}(x) + \theta \bar{h}(x) - \alpha x, \tag{5-1}$$

where ϵ is a small dimensionless quantity measuring the maximum cross-wise extension of the airfoil, τ its thickness ratio, α the angle of attack, and θ is a measure of the amount of camber. The functions $\bar{g}(x)$ and $\bar{h}(x)$

81

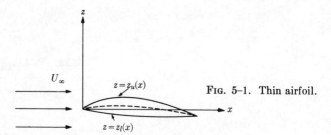

FIG. 5-1. Thin airfoil.

define the distribution of thickness and camber, respectively, along the chord. It will be assumed that $\bar{g}$ and $\bar{h}$ are both smooth and that $\bar{g}'$ and $\bar{h}'$ are of order of unity everywhere along the chord. A blunt leading edge is thus excluded. In the limit of $\epsilon \to 0$ the airfoil collapses to a segment along the x-axis assumed to be located between $x = 0$ and $x = c$. We will seek the leading terms in a series expansion in ϵ of Φ to be used as an approximation for thin airfoils with small camber and angle of attack.

For two-dimensional steady flow the differential equation (1–74) for Φ simplifies to

$$(a^2 - \Phi_x^2)\Phi_{xx} + (a^2 - \Phi_z^2)\Phi_{zz} - 2\Phi_x\Phi_z\Phi_{xz} = 0, \qquad (5\text{–}2)$$

where the velocity of sound is given by (1–67), which, for the present case, simplifies to

$$a^2 = a_\infty^2 - \frac{\gamma - 1}{2}(\Phi_x^2 + \Phi_z^2 - U_\infty^2). \qquad (5\text{–}3)$$

From Φ the pressure can be obtained using (1–64):

$$C_p = \frac{2}{\gamma M^2}\left\{\left[1 - \frac{\gamma - 1}{2a_\infty^2}(\Phi_x^2 + \Phi_z^2 - U_\infty^2)\right]^{\gamma/(\gamma-1)} - 1\right\}. \qquad (5\text{–}4)$$

The boundary conditions are that the flow is undisturbed at infinity and tangential to the airfoil surface. Hence

$$\Phi_z/\Phi_x = \epsilon\frac{d\bar{f}_u}{dx} \quad \text{on} \quad z = \epsilon\bar{f}_u$$

$$\Phi_z/\Phi_x = \epsilon\frac{d\bar{f}_l}{dx} \quad \text{on} \quad z = \epsilon\bar{f}_l. \qquad (5\text{–}5)$$

Additional boundary conditions required to make the solution unique for a subsonic flow are that the pressure is continuous at the trailing edge (Kutta-Joukowsky condition) and also everywhere outside the airfoil. In a supersonic flow, pressure discontinuities must satisfy the Rankine-Hugoniot shock conditions. However, to the approximation considered here, these are always automatically satisfied. Strictly speaking, a supersonic flow with curved shocks is nonisentropic, and hence nonpotential, but the effects of entropy variation will not be felt to within the approximation considered here, provided the Mach number is moderate.

We seek first an outer expansion of the form

$$\Phi^o = U_\infty[\Phi_0^o(x, z) + \epsilon\Phi_1^o(x, z) + \cdots]. \qquad (5\text{--}6)$$

The factor U_∞ is included for convenience; in this manner the first partial derivatives of the Φ_n-terms will be nondimensional. Since the airfoil in the limit of $\epsilon \to 0$ collapses to a line parallel to the free stream, the zeroth-order term must represent parallel undisturbed flow. Thus

$$\Phi_0^o = x. \qquad (5\text{--}7)$$

By introducing the series (5–6) into (5–2) and (5–3), and using (5–7), we obtain after equating terms of order ϵ

$$(1 - M^2)\Phi_{1xx}^o + \Phi_{1zz}^o = 0. \qquad (5\text{--}8)$$

The only boundary condition available for this so far is that flow perturbations must vanish at large distances,

$$\Phi_{1x}^o, \Phi_{1z}^o \to 0 \quad \text{for} \quad \sqrt{x^2 + z^2} \to \infty. \qquad (5\text{--}9)$$

The remaining boundary conditions belong to the inner region and are to be obtained by matching. The inner solution is sought in the form

$$\Phi^i = U_\infty[\Phi_0^i(x, \bar{z}) + \epsilon\Phi_1^i(x, \bar{z}) + \epsilon^2\Phi_2^i(x, \bar{z}) + \cdots], \qquad (5\text{--}10)$$

where

$$\bar{z} = z/\epsilon. \qquad (5\text{--}11)$$

This stretching enables us to study the flow in the immediate neighborhood of the airfoil in the limit of $\epsilon \to 0$ since the airflow shape then remains independent of ϵ (see Fig. 5–2) and the width of the inner region becomes of order unity. The zeroth-order inner term is that of a parallel flow, that is, $\Phi_0^i = x$, because the inner flow as well as the outer flow must be parallel in the limit $\epsilon \to 0$. (This could of course also be obtained from the matching procedure.) From the expression for the W-component,

$$W = \Phi_z = \frac{1}{\epsilon}\Phi_{\bar{z}}^i = U_\infty[\Phi_{1\bar{z}}^i(x, \bar{z})$$
$$+ \epsilon\Phi_{2\bar{z}}^i(x, \bar{z}) + \cdots], \qquad (5\text{--}12)$$

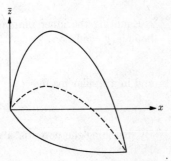

Fig. 5–2. Airfoil in stretched coordinate system.

we see directly that Φ_1^i must be independent of $\bar{z}$, say $\Phi_1^i = \bar{g}_1(x)$ (which in general is different above and below the airfoil), otherwise W would not vanish in the limit of zero ϵ. Hence (5–10) may directly be simplified to

$$\Phi^i = U_\infty[x + \epsilon\bar{g}_1(x) + \epsilon^2\Phi_2^i(x, \bar{z}) + \cdots]. \tag{5–13}$$

By substituting (5–13) into the differential equation (5–2) and the associated boundary condition (5–5) we find that

$$\Phi_{2\bar{z}\bar{z}}^i = 0, \tag{5–14}$$

$$\Phi_{2\bar{z}}^i = \frac{d\bar{f}_u}{dx} \quad \text{for} \quad \bar{z} = \bar{f}_u(x)$$

$$\Phi_{2\bar{z}}^i = \frac{d\bar{f}_l}{dx} \quad \text{for} \quad \bar{z} = \bar{f}_l(x). \tag{5–15}$$

Thus, the solution must be linear in $\bar{z}$,

$$\Phi_2^i = \bar{z}\frac{d\bar{f}_u}{dx} + \bar{g}_{2u}(x), \tag{5–16}$$

for $\bar{z} \geq \bar{f}_u$, with a similar expression for $\bar{z} \leq \bar{f}_l$. This result means that to lowest order the streamlines are parallel to the airfoil surface throughout the inner region.

The inner solution cannot, of course, give vanishing disturbances at infinity since this boundary condition belongs to the outer region. We therefore need to match the inner and outer solutions. This can be done in two ways; either one can use the limit matching principle for W or the asymptotic matching principle for Φ. From (5–16) it follows that W^i is independent of $\bar{z}$ to lowest order, hence in the outer limit $\bar{z} = \infty$

$$W^{io} = \frac{1}{\epsilon}\Phi_{\bar{z}}^i(x, \infty) = \epsilon\frac{d\bar{f}_u}{dx}. \tag{5–17}$$

Now

$$W^o = \epsilon\Phi_{1z}^o. \tag{5–18}$$

Equating the inner limit ($z = 0+$) to (5–17), we obtain the following boundary condition:

$$\Phi_{1z}^o(x, 0+) = \frac{d\bar{f}_u}{dx}, \tag{5–19}$$

and in a similar manner

$$\Phi_{1z}^o(x, 0-) = \frac{d\bar{f}_l}{dx}. \tag{5–20}$$

By matching the potential itself we find that

$$\bar{g}_{1u}(x) = \Phi_1^o(x, 0+). \tag{5–21}$$

To determine $\bar{g}_2$ it is necessary to go to a higher order in the outer solution.

To illustrate the use of the asymptotic matching principle we first express the two-term outer flow in inner variables,

$$\Phi^o = U_\infty[x + \epsilon\Phi_1^o(x, \epsilon\bar{z}) + \cdots], \qquad (5\text{--}22)$$

and then take the three-term inner expansion of this, namely

$$\Phi^o = U_\infty[x + \epsilon\Phi_1^o(x, 0+) + \epsilon^2\bar{z}\Phi_{1z}^o(x, 0+) + \cdots], \qquad (5\text{--}23)$$

which, upon reexpression in outer variables yields

$$\Phi^o = U_\infty[x + \epsilon\Phi_1^o(x, 0+) + \epsilon z\Phi_{1z}^o(x, 0+) + \cdots]. \qquad (5\text{--}24)$$

The three-term inner expansion, expressed in outer variables, reads

$$\Phi^i = U_\infty\left[x + \epsilon\bar{g}_{1u}(x) + \epsilon z\frac{d\bar{f}_u}{dx} + \epsilon^2\bar{g}_{2u}(x) + \cdots\right]. \qquad (5\text{--}25)$$

Thus the equating of the two-term outer expansion of the three-term inner expansion

$$\Phi^i = U_\infty\left[x + \epsilon\bar{g}_{1u}(x) + \epsilon z\frac{d\bar{f}_u}{dx} + \cdots\right], \qquad (5\text{--}26)$$

with the three-term inner expansion of the two-term outer expansion as given by (5–23) leads to

$$\bar{g}_{1u} = \Phi_1^o(x, 0+)$$

$$\Phi_{1z}^o(x, 0+) = \frac{d\bar{f}_u}{dx}$$

as before.

From the velocity components we may calculate the pressure coefficient by use of (5–4). Expanding in ϵ and using (5–21) we find that the pressure on the airfoil surface is given by

$$C_p = -2\epsilon\Phi_{1x}^o(x, 0\pm), \qquad (5\text{--}27)$$

where the plus sign is to be used for the upper surface and the minus sign for the lower one.

An examination of the expression (5–25) makes clear that such a simple inner solution cannot hold in regions where the flow changes rapidly in the x-direction as near the wing edges, or near discontinuities in airfoil surface slope. For a complete analysis of the entire flow field, these must be considered as separate inner flow regions to be matched locally to the outer flow. The singularities in the outer flow that are usually encountered at, for example, wing leading edges, do not occur in the real flow and should be interpreted rather as showing in what manner the perturbations due to the edge die off at large distances. For a discussion of edge effects on the basis of matched asymptotic expansions, see Van Dyke (1964).

In the following we will use the notation

$$\epsilon \Phi_1^o = \varphi, \tag{5-28}$$

where φ is commonly known as the velocity perturbation potential (the factor U_∞ is sometimes included in the definition). The above procedure is easily extended to three dimensions, and for φ we then obtain the following set of linearized equations of motion and boundary conditions:

$$(1 - M^2)\varphi_{xx} + \varphi_{yy} + \varphi_{zz} = 0, \tag{5-29}$$

$$\varphi_z = \frac{\partial z_u}{\partial x} \quad \text{on } z = 0+ \quad \text{for} \quad x, y \text{ on } S$$
$$\varphi_z = \frac{\partial z_l}{\partial x} \quad \text{on } z = 0- \quad \text{for} \quad x, y \text{ on } S, \tag{5-30}$$

$$C_p = -2\varphi_x, \tag{5-31}$$

where S is the part of the x, y-plane onto which the wing collapses as $\epsilon \to 0$.

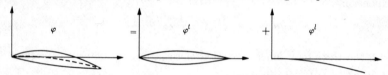

FIG. 5–3. Separation of thickness and lift problems.

Since the equations of motion and the boundary conditions are linear, solutions may be superimposed linearly. It is therefore convenient to write the solution as a sum of two terms, one giving the flow due to thickness and the other the flow due to camber and angle of attack (see Fig. 5–3). Thus we set

$$\varphi = \varphi^t + \varphi^l, \tag{5-32}$$

where

$$\varphi_z^t(x, y, 0\pm) = \pm\tau \frac{\partial \bar{g}}{\partial x}, \tag{5-33}$$

$$\varphi_z^l(x, y, 0\pm) = \theta \frac{\partial \bar{h}}{\partial x} - \alpha \tag{5-34}$$

on S. It follows from (5–33) and (5–34) that φ^t is symmetric in z whereas φ^l is antisymmetric. As indicated in Fig. 5–3, φ^t represents the flow around a symmetric airfoil at zero angle of attack whereas φ^l represents that around an inclined surface of zero thickness, a "lifting surface." It will be apparent from the simple examples to be considered below that the lifting problem is far more difficult to solve than the thickness problem, at least for subsonic flow.

Another consequence of the linearization of the problem is that complicated solutions can be built up by superposition of elementary singular solutions. The ones most useful for constructing solutions to wing problems are those for a source, doublet and elementary horseshoe vortex. For incompressible flow the first two have already been discussed in Chapter 2. The elementary horseshoe vortex consists of two infinitely long vortex filaments of unit strength but opposite signs located infinitesimally close together

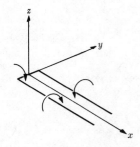

FIG. 5–4.   The elementary horseshoe vortex.

along the positive x-axis and joined together by an infinitesimal piece along the y-axis, also of unit strength (see Fig. 5–4).

The solution for this can be obtained by integrating the solution for a doublet in the x-direction. Thus

$$\varphi^\Gamma = \frac{1}{4\pi} \int_x^\infty \frac{z\, dx_1}{(x_1^2 + y_1^2 + z^2)^{3/2}} = \frac{1}{4\pi}\, \frac{z}{y^2 + z^2}\left(1 + \frac{x}{\sqrt{x^2 + y^2 + z^2}}\right).$$

$$(5\text{–}35)$$

To extend these solutions to compressible flow the simplest procedure is to notice that by introducing the stretched coordinates

$$\bar{y} = \beta y, \qquad \bar{z} = \beta z, \tag{5–36}$$

where $\beta = \sqrt{1 - M^2}$, the differential equation (5–29) transforms to the Laplace equation expressed in the coordinates x, $\bar{y}$, $\bar{z}$. Hence any solution to the incompressible flow problem will become a solution to the compressible flow if y and z are replaced by $\bar{y}$ and $\bar{z}$. This procedure gives as a solution for the simple source

$$\varphi^S = -\frac{1}{4\pi}\, \frac{1}{\sqrt{x^2 + \beta^2 r^2}}, \tag{5–37}$$

where

$$r^2 = y^2 + z^2.$$

This solution could be analytically continued to supersonic flow ($M > 1$). However, the solution will then be real only within the two Mach cones ($\sqrt{M^2 - 1}\,)r < |x|$ (see Fig. 5–5), and for physical reasons the solution within the upstream Mach cone must be discarded. Hence, since we must discard half the solution it seems reasonable that the coefficient in front of it should be increased by a factor of two in order for the total volume

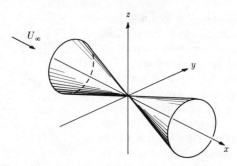

Fig. 5–5. Upstream and downstream Mach cones.

output to be the same. Thus the solution for a supersonic source would read

$$\varphi^S = -\frac{1}{2\pi} \frac{1}{\sqrt{x^2 - (M^2 - 1)r^2}}.$$ (5–38)

A check on the constant will be provided in Chapter 6.

The use of the elementary solutions to construct more complicated ones is a method that will be frequently employed later in connection with three-dimensional wing theories. This method is particularly useful for developing approximate numerical theories. However, in the two-dimensional cases that will be considered next as illustrations of the linearized wing theory a more direct analytical method is utilized.

5–3 Thin Airfoils in Incompressible Flow

Considering first the symmetric problem for an airfoil at $M = 0$ of chord c we seek a solution of

$$\varphi_{xx} + \varphi_{zz} = 0,$$ (5–39)

subject to the boundary conditions that

$$\varphi_z(x, 0\pm) = \pm\tau \frac{d\bar{g}}{dx} \quad \text{for} \quad 0 < x < c$$ (5–40)

and that the disturbance velocities are continuous outside the airfoil and vanish for $\sqrt{x^2 + z^2} \to \infty$. Since we are dealing with the Laplace equation in two dimensions, the most efficient approach is to employ complex variables. Let

$$Y = x + iz$$
$$\mathcal{W}(Y) = \varphi(x, z) + i\psi(x, z)$$ (5–41)
$$q = \frac{d\mathcal{W}}{dY} = u(x, z) - iw(x, z),$$

where q is the complex perturbation velocity vector made dimensionless through division by U_∞. In this way we assure that, provided Φ is analytic in Y, φ and ψ, as well as q, are solutions of the Laplace equation. Thus we may concentrate on finding a $q(Y)$ that satisfies the proper boundary conditions. In the nonlifting case we seek a $q(Y)$ that vanishes for $|Y| \to \infty$ and takes the value

$$q(x, 0\pm) = u(x, 0\pm) - iw(x, 0\pm)$$
$$= u_0(x) \mp iw_0(x), \quad \text{say,} \quad \text{for} \quad 0 \le x \le c, \tag{5–42}$$

where

$$w_0(x) = \tau \frac{d\bar{g}}{dx}. \tag{5–43}$$

The imaginary part of $q(Y)$ is thus discontinuous along the strip $z = 0$, $0 \le x \le c$, with the jump given by the tangency condition (5–40). In order to find the pressure on the airfoil we need to know u_0, because from (5–31)

$$C_p(x, 0) = -2\varphi_x(x, 0) = -2u_0. \tag{5–44}$$

To this purpose we make use of Cauchy's integral formula which states that given an analytic function $f(Y_1)$ in the complex plane $Y_1 = x_1 + iz_1$, its value in the point $Y_1 = Y$ is given by the integral

$$f(Y) = \frac{1}{2\pi i} \oint_C \frac{f(Y_1)\, dY_1}{Y_1 - Y}, \tag{5–45}$$

where C is any closed curve enclosing the point $Y_1 = Y$, provided $f(Y_1)$ is analytic everywhere inside C. We shall apply (5–45) with $f = q$ and an integration path C selected as shown in Fig. 5–6.

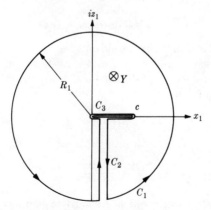

Fig. 5–6. Integration path in the complex plane.

The path was chosen so as not to enclose completely the slit along the real axis representing the airfoil because q is discontinuous, and hence nonanalytic, across the slit. Thus

$$q(Y) = \frac{1}{2\pi i} \oint_{C_1+C_2+C_3} \frac{q(Y_1)\, dY_1}{Y_1 - Y}. \tag{5-46}$$

In the limit of $R_1 \to \infty$ the integral over C_1 must vanish, since from the boundary conditions $q(Y_1) \to 0$ for $Y_1 \to \infty$. The integrals over the two paths C_2 cancel; hence

$$q(Y) = \frac{1}{2\pi i} \int_{C_3} \frac{q(Y_1)\, dY_1}{Y_1 - Y} = \frac{1}{2\pi i} \int_0^c \frac{\Delta q(x_1)\, dx_1}{x_1 - Y}, \tag{5-47}$$

where Δq is the difference in the value of q between the upper and lower sides of the slit. From (5–42) it follows that

$$\Delta q(x_1) = q(x_1, 0+) - q(x_1, 0-) = -2iw_0(x_1). \tag{5-48}$$

Hence, upon inserting this into (5–47), we find that

$$q(Y) = -\frac{1}{\pi} \int_0^c \frac{w_0(x_1)\, dx_1}{x_1 - Y}, \tag{5-49}$$

which, together with (5–43), gives the desired solution in terms of the airfoil geometry. Separation of real and imaginary parts gives

$$u(x, z) = \frac{1}{\pi} \int_0^c \frac{(x - x_1)w_0(x_1)\, dx_1}{(x - x_1)^2 + z^2}, \tag{5-50}$$

$$w(x, z) = \frac{1}{\pi} \int_0^c \frac{zw_0(x_1)\, dx_1}{(x - x_1)^2 + z^2}. \tag{5-51}$$

In the limit of $z \to 0+$ the second integral will receive contributions only from the region around $x_1 = x$ and is easily shown to yield w_0 as it should. To obtain a meaningful limit for the first integral, we divide the region of integration into three parts as follows:

$$u(x, z) = \frac{1}{\pi} \int_0^{x-\delta} \frac{(x - x_1)w_0(x_1)\, dx_1}{(x - x_1)^2 + z^2} + \frac{1}{\pi} \int_{x-\delta}^{x+\delta} \frac{(x - x_1)w_0(x_1)\, dx_1}{(x - x_1)^2 + z^2}$$

$$+ \frac{1}{\pi} \int_{x+\delta}^c \frac{(x - x_1)w_0(x_1)\, dx_1}{(x - x_1)^2 + z^2}, \tag{5-52}$$

where δ is a small quantity but is assumed to be much greater than z. We may, therefore, directly set $z = 0$ in the first and third integrals. In the second one, we may, for small δ, replace $w_0(x_1)$ by $w_0(x)$ as a first approximation, whereupon the integrand becomes antisymmetric in $x - x_1$ and the integral hence vanishes. The integral (5–50) is therefore in the limit of $z = 0$ to be interpreted as a Cauchy principal value integral (as indicated by the symbol C):

$$u_0(x) = \frac{1}{\pi} \oint_0^c \frac{w_0(x_1)\,dx_1}{x - x_1}, \tag{5–53}$$

which is therefore defined as

$$\oint_0^c \sim dx_1 = \lim_{\delta \to 0} \left\{ \int_0^{x-\delta} \sim dx_1 + \int_{x+\delta}^c \sim dx_1 \right\}. \tag{5–54}$$

Turning now to the lifting case, we recall that u is antisymmetric in z, and w symmetric. Consequently, on the airfoil,

$$w(x, 0) = w_0(x) = \theta \frac{d\overline{h}}{dx} - \alpha \tag{5–55}$$

is the same top and bottom, and in (5–45) then

$$\Delta q = u(x, 0+) - u(x, 0-) = 2u_0(x) = \gamma(x), \tag{5–56}$$

where $\gamma(x)$ is the nondimensional local strength of the vortices distributed along the chord. Hence, (5–47) will yield the following integral formula

$$q(Y) = -\frac{1}{2\pi i} \int_0^c \frac{\gamma(x_1)\,dx_1}{Y - x_1}, \tag{5–57}$$

whose imaginary part gives, in the limit of $z = 0$,

$$w_0(x) = -\frac{1}{2\pi} \oint_0^c \frac{\gamma(x_1)\,dx_1}{x - x_1}. \tag{5–58}$$

This is the integral equation of thin airfoil theory first considered by Glauert (1924). Instead of attacking (5–58) we will use analytical techniques similar to those used above to obtain directly a solution of the complex velocity $q(Y)$. This solution will then, of course, also provide a solution of the singular integral equation (5–58). For a more general treatment of this kind we refer to the book by Muskhelishvili (1953).

Again, we shall start from Cauchy's integral formula (5–45) but this time we instead choose

$$f(Y) = q(Y)h(Y), \tag{5–59}$$

where $h(Y)$ is an analytic function assumed regular outside the slit and sufficiently well behaved at infinity so that $f(Y) \to 0$ for $|Y| \to \infty$.

Substitution of (5–59) into (5–45) and selection of the path C in the same way as before gives

$$q(Y) = \frac{1}{2h(Y)\pi i} \int_0^c \frac{\Delta f(x_1)\, dx_1}{x_1 - Y}, \tag{5–60}$$

where $\Delta f(x)$ is the jump in the function $f = q(Y)h(Y)$ between the upper and lower sides of the slit. In order to obtain a solution expressed as an integral over the known quantity w_0 rather than the unknown u_0, we need a function h with a jump across the slit canceling the jump in u_0. Thus we require that

$$h(x, 0-) = -h(x, 0+) \quad \text{for} \quad 0 \leq x \leq c, \tag{5–61}$$

which upon insertion into (5–60) gives

$$q(Y) = -\frac{1}{\pi h(Y)} \int_0^c \frac{w_0(x_1) h(x_1, 0+)\, dx_1}{x_1 - Y}. \tag{5–62}$$

It remains to find an appropriate function $h(Y)$. A little speculation will convince us that a function possessing the property (5–61) must have branch points at $Y = 0$ and $Y = c$ and be of the general form

$$h(Y) = Y^{m+1/2}(c - Y)^{n+1/2}, \tag{5–63}$$

where m and n are integers. That (5–63) satisfies (5–61) can be seen by setting $Y = |Y|e^{i\theta_1}$ and $c - Y = |c - Y|e^{i\theta_2}$. At the upper side of the slit $\theta_1 = \theta_2 = 0$, and hence

$$h(x, 0+) = x^{m+1/2}(c - x)^{n+1/2}. \tag{5–64}$$

In going to the lower side we must not cross the slit. At the passage around the branch point at $Y = 0$, θ_1 increases by 2π but θ_2 remains zero. Thus

$$h(x, 0-) = e^{2\pi i(m+1/2)} x^{m+1/2}(c - x)^{n+1/2} = -x^{m+1/2}(c - x)^{n+1/2}. \tag{5–65}$$

Introduction of (5–63) into (5–62) yields

$$q(Y) = \frac{1}{\pi Y^{m+1/2}(c - Y)^{n+1/2}} \int_0^c \frac{w_0(x_1) x_1^{m+1/2}(c - x_1)^{n+1/2}\, dx_1}{Y - x_1}. \tag{5–66}$$

For the integral to converge, m and n cannot be smaller than -1. Furthermore, since the integral for large $|Y|$ vanishes like Y^{-1} we must choose

$$m + n \geq -1 \tag{5–67}$$

in order for q to vanish at infinity. It follows from (5–66) that in the neighborhood of the leading edge

$$q \sim Y^{-m-1/2}, \tag{5–68}$$

whereas near the trailing edge

$$q \sim (c - Y)^{-n-1/2}. \tag{5–69}$$

From the latter it follows that the Kutta-Joukowsky condition of finite velocity at the trailing edge is fulfilled only if $n \leq -1$. Hence from what was said earlier the only possible choice is

$$n = -1. \tag{5–70}$$

From m we then find from (5–67) that it cannot be less than zero. It seems reasonable from a physical point of view that the lowest possible order of singularity of the leading edge should be chosen, namely

$$m = 0. \tag{5–71}$$

However, from a strictly mathematical point of view there is nothing in the present formulation that requires this choice; thus any order singularity could be admissible. In settling this point the method of matched asymptotic expansions again comes to the rescue. The present formulation holds strictly for the outer flow only, which was matched to the inner flow near the airfoil. However, as was pointed out in Section 5–2, the simple inner solution (5–25) obviously cannot hold near the leading edge since there the x-derivatives in the equation of motion will become of the same order as $\bar{z}$-derivatives. To obtain the complete solution we therefore need to consider an additional inner region around the leading edge which is magnified in such a manner as to keep the leading edge radius finite in the limit of vanishing thickness. Such a procedure shows (Van Dyke, 1964) that the velocity perturbations due to the lifting flow vanish as $Y^{-1/2}$ far away from the leading edge. Hence (5–71) is verified and consequently

$$q(Y) = \frac{1}{\pi} \left(\frac{c - Y}{Y} \right)^{1/2} \int_0^c \frac{w_0(x_1)}{Y - x_1} \sqrt{\frac{x_1}{c - x_1}} \, dx_1. \tag{5–72}$$

The real part of this gives for $z = 0+$

$$u(x, 0+) = \frac{1}{\pi} \left(\frac{c - x}{x} \right)^{1/2} \oint_0^c \frac{w_0(x_1)}{x - x_1} \sqrt{\frac{x_1}{c - x_1}} \, dx_1, \tag{5–73}$$

which is a solution of the integral equation (5–58).

As a simple illustration of the theory the case of an uncambered airfoil will be considered. Then for the lifting flow

$$w_0 = -\alpha \tag{5-74}$$

and for (5–73) we therefore need to evaluate the integral

$$I = \frac{1}{\pi} \oint_0^c \frac{dx_1}{x - x_1} \sqrt{\frac{x_1}{c - x_1}}. \tag{5-75}$$

This rather complicated integral may be handled most conveniently by use of the analytical techniques employed above. Using analytical continuation, (5–75) is first generalized by considering instead the complex integral

$$\mathfrak{g} = \frac{1}{\pi} \int_0^c \frac{dx_1}{Y - x_1} \sqrt{\frac{x_1}{c - x_1}}, \tag{5-76}$$

whose real part reduces to (5–75) for $z = 0+$. Now we employ Cauchy's integral formula (5–45) with

$$f(Y) = \left(\frac{Y}{c - Y}\right)^{1/2} \tag{5-77}$$

and the same path of integration as considered previously (see Fig. 5–6). Thus

$$\left(\frac{Y}{c - Y}\right)^{1/2} = \frac{1}{2\pi i} \oint_{C_1 + C_2 + C_3} \frac{dY_1}{Y_1 - Y} \left(\frac{Y_1}{c - Y_1}\right)^{1/2}. \tag{5-78}$$

Along the large circle C_1 we find by expanding the integrand in Y_1^{-1}

$$\frac{1}{2\pi i} \int_{C_1} \frac{dY_1}{Y_1 - Y} \left(\frac{Y_1}{c - Y_1}\right)^{1/2} = \frac{1}{2\pi i} \int_{C_1} \frac{dY_1}{Y_1} [i + O(Y_1^{-1})] = i.$$

The integral over C_2 cancels as before, whereas the contribution along C_3 becomes

$$\frac{1}{2\pi i} \int_{C_3} \frac{dY_1}{Y_1 - Y} \left(\frac{Y_1}{c - Y_1}\right)^{1/2} = -\frac{1}{\pi i} \int_0^c \frac{dx_1}{Y - x_1} \left(\frac{x_1}{c - x_1}\right)^{1/2} = i\mathfrak{g}. \tag{5-79}$$

Hence we have found that

$$\left(\frac{Y}{c - Y}\right)^{1/2} = i + i\mathfrak{g},$$

that is,

$$\mathfrak{g} = -1 - i\left(\frac{Y}{c - Y}\right)^{1/2}. \tag{5-80}$$

Taking the real part of this for $z = 0+$ we obtain

$$I = -1 \tag{5-81}$$

and, consequently, by introducing (5–74) into (5–73),

$$u(x, 0+) = \alpha \sqrt{\frac{c - x}{x}} \equiv u_0(x). \tag{5-82}$$

Hence the lifting pressure distribution

$$\Delta C_p = C_p(x, 0-) - C_p(x, 0+) = 4u_0(x) = 4\alpha \sqrt{\frac{c - x}{x}} \tag{5-83}$$

has a square-root singularity at the leading edge and goes to zero at the trailing edge as the square root of the distance to the edge. The same behavior near the edges may be expected also for three-dimensional wings.

The total lift is easily obtained by integrating the lifting pressure over the chord. An alternative procedure is to use Kutta's formula

$$L = \rho U_\infty \Gamma. \tag{2-157}$$

The total circulation Γ around the airfoil can be obtained by use of (5–72). Thus

$$\Gamma = -U_\infty \oint q \, dY = -\frac{U_\infty}{\pi} \oint \left(\frac{c - Y}{Y} \right)^{1/2} dY \int_0^c \frac{w_0(x_1)}{Y - x_1} \sqrt{\frac{x_1}{c - x_1}} \, dx_1.$$

The path of integration around the airfoil is arbitrary. Taking it to be a large circle approaching infinity we find that

$$\Gamma = -\frac{U_\infty}{\pi} \oint dY \left[-\frac{i}{Y} + O(Y^{-2}) \right] \int_0^c w_0(x_1) \sqrt{\frac{x_1}{c - x_1}} \, dx_1$$

$$= -2U_\infty \int_0^c w_0(x_1) \sqrt{\frac{x_1}{c - x_1}} \, dx_1. \tag{5-84}$$

For the flat plate this leads to the well-known result

$$C_L = 2\pi\alpha. \tag{5-85}$$

In view of the linearity of camber and angle-of-attack effects, the lift-curve slope should be equal to 2π for any thin profile. Most experiments show a somewhat smaller value (by up to about 10%). This discrepancy is usually attributed to the effect of finite boundary layer thickness near the trailing edge, which causes the rear stagnation point to move a small distance upstream on the upper airfoil surface from the trailing edge with an accompanying loss of circulation and lift. This effect is very sensitive to trailing-edge angle. For airfoils with a cusped trailing edge (= zero trailing-edge angle), carefully controlled experiments give very nearly the full theoretical value of lift-curve slope.

According to thin-airfoil theory, the lifting pressure distribution is given by (5–83) for all uncambered airfoils. Figure 5–7 shows a comparison between this theoretical result and experiments for an NACA 0015 airfoil performed by Graham, Nitzberg, and Olson (1945). The lowest Mach number considered by them was $M = 0.3$, and the results have therefore been corrected to $M = 0$ using the Prandtl-Glauert rule (see Chapter 7). The agreement is good considering the fairly large thickness (15%), except near the trailing edge. The discrepancy there is mainly due to viscosity as discussed above. It is interesting to note that the theory is accurate very close to the leading edge despite its singular behavior at $x = 0$ discussed earlier. In reality, ΔC_p must, of course, be zero right at the leading edge.

With the aid of the Prandtl-Glauert rule the theory is easily extended to the whole subsonic region (see Section 7–1). The first-order theory has

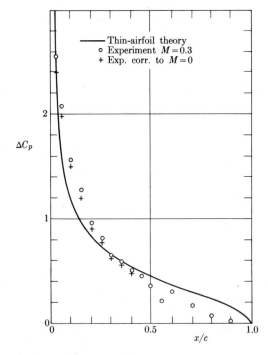

FIG. 5–7. Comparison of theoretical and experimental lifting pressure distributions on a NACA 0015 airfoil at 6° angle of attack. [Based on experiments by Graham, Nitzberg, and Olson (1945).]

been extended by Van Dyke (1956) to second order. He found that it is then necessary to handle the edge singularities appearing in the first-order solution carefully, using separate inner solutions around the edges; otherwise an incorrect second-order solution would be obtained in the whole flow field.

5–4 Thin Airfoils in Supersonic Flow

For $M > 1$ the differential equation governing φ may be written

$$-B^2\varphi_{xx} + \varphi_{zz} = 0, \qquad (5\text{–}86)$$

where $B = \sqrt{M^2 - 1}$. This equation is hyperbolic, which greatly simplifies the problem. A completely general solution of (5–86) is easily shown to be

$$\varphi = F(x - Bz) + G(x + Bz). \qquad (5\text{–}87)$$

Notice the great similarity of (5–87) to the complex representation (5–41) of φ in the incompressible case; in fact (5–87) may be obtained in a formal way from (5–41) simply by replacing z by $\pm iBz$. The lines

$$\begin{aligned} x - Bz &= \text{const} \\ x + Bz &= \text{const} \end{aligned} \qquad (5\text{–}88)$$

are the characteristics of the equation, in the present context known as Mach lines. Disturbances in the flow propagate along the Mach lines. (This can actually be seen in schlieren pictures of supersonic flow.) In the first-order solution the actual Mach lines are approximated by those of the undisturbed stream.

Since the disturbances must originate at the airfoil, it is evident that in the solution (5–87) G must be zero for $z > 0$, whereas $F = 0$ for $z < 0$. The solution satisfying (5–30) is thus

$$\begin{aligned} \varphi &= -\frac{1}{B}z_u(x - Bz) \quad \text{for} \quad z > 0 \\ \varphi &= \frac{1}{B}z_l(x + Bz) \quad \text{for} \quad z < 0. \end{aligned} \qquad (5\text{–}89)$$

From (5–31) it therefore follows that

$$C_{pu} = \frac{2}{B}\frac{\partial z_u}{\partial x}. \qquad (5\text{–}90)$$

This formula was first given by Ackeret (1925). Comparisons of this simple result with experiments are shown in Fig. 5–8. It is seen that the first-order theory tends to underestimate the pressure and in general is

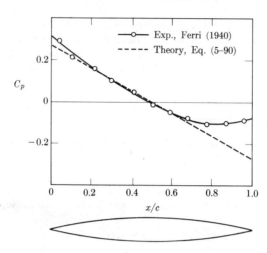

FIG. 5–8. Comparison of theoretical and experimental supersonic pressure distribution at $M = 1.85$ on a 10% thick biconvex airfoil at 0° angle of attack.

less accurate than that for incompressible flow. The deviation near the trailing edge is due to shock wave-boundary layer interaction which tends to make the higher pressure behind the oblique trailing-edge shocks leak upstream through the boundary layer.

The much greater mathematical simplicity of supersonic-flow problems than subsonic ones so strikingly demonstrated in the last two sections is primarily due to the absence of upstream influence for $M > 1$. Hence the flows on the upper and lower sides of the airfoil are independent and there is no need to separate the flow into its thickness and lifting parts. In later chapters cases of three-dimensional wings will be considered for which there is interaction between the two wing sides over limited regions.

Slender-Body Theory

6–1 Introduction

We shall now study the flow around configurations that are "slender" in the sense that all their crosswise dimensions like span and thickness are small compared to the length. Such a configuration could, for example, be a body of revolution, a low-aspect ratio wing, or a low-aspect ratio wing-body combination. The formal derivation of the theory may be thought of as a generalization to three dimensions of the thin airfoil theory; however, the change in the structure of the inner solution associated with the additional dimension introduces certain new features into the problem with important consequences for the physical picture.

The simplest case of a nonlifting body of revolution will be considered first in Sections 6–2 through 6–4 and bodies of general shape in Sections 6–5 through 6–7.

6–2 Expansion Procedure for Axisymmetric Flow

We shall consider the flow around a slender nonlifting body of revolution defined by

$$r = R(x) = \epsilon \overline{R}(x) \tag{6-1}$$

for small·values of the thickness ratio ϵ. For steady axisymmetric flow the differential equation (1–74) for Φ reads

$$(a^2 - \Phi_x^2)\Phi_{xx} + (a^2 - \Phi_r^2)\Phi_{rr} + \frac{a^2}{r}\Phi_r - 2\Phi_x\Phi_r\Phi_{xr} = 0, \tag{6-2}$$

where

$$a^2 = a_\infty^2 - \frac{\gamma - 1}{2}(\Phi_x^2 + \Phi_r^2 - U_\infty^2). \tag{6-3}$$

The requirement that the flow be tangent to the body surface gives the following boundary condition:

$$\frac{\Phi_r}{\Phi_x} = \epsilon \frac{d\overline{R}}{dx} \quad \text{at} \quad r = \epsilon \overline{R}. \tag{6-4}$$

We shall consider an outer expansion of the form

$$\Phi^o = U_\infty[x + \epsilon \Phi_1^o(x, r) + \epsilon^2 \Phi_2^o(x, r) + \cdots] \tag{6-5}$$

and an inner expansion

$$\Phi^i = U_\infty[x + \epsilon\Phi_1^i(x, \bar{r}) + \epsilon^2\Phi_2^i(x, \bar{r}) + \cdots], \tag{6-6}$$

where

$$\bar{r} = r/\epsilon. \tag{6-7}$$

As in the thin airfoil case Φ_1^i must be a function of x, only, because otherwise the radial velocity component

$$U_r = \Phi_r^i = U_\infty[\Phi_{1\bar{r}}^i + \epsilon\Phi_{2\bar{r}}^i + \cdots] \tag{6-8}$$

will not vanish in the limiting case of zero body thickness. Substituting (6–6) into (6–2) and (6–3) and retaining only terms of order ϵ^0, we obtain

$$\Phi_{2\bar{r}\bar{r}}^i + \frac{1}{\bar{r}}\Phi_{2\bar{r}}^i = 0. \tag{6-9}$$

From (6–4) we find the following boundary condition:

$$\Phi_{2\bar{r}}^i(x, \bar{R}) = \frac{d\bar{R}}{dx} = \bar{R}'(x). \tag{6-10}$$

The solution of (6–9) satisfying (6–10) is easily shown to be

$$\Phi_2^i = \bar{R}\bar{R}' \ln \bar{r} + \bar{g}_2(x), \tag{6-11}$$

in which the function $\bar{g}_2$ must be found by matching to the outer flow. From (6–11) it follows that the radial velocity component is

$$\frac{U_r}{U_\infty} = \frac{\epsilon\bar{R}\bar{R}'}{\bar{r}} + \cdots = \frac{\epsilon^2\bar{R}\bar{R}'}{r} + \cdots \tag{6-12}$$

Matching this to the velocity component from the outer flow

$$U_r/U_\infty = \epsilon\Phi_{1r}^o + \epsilon^2\Phi_{2r}^o + \cdots, \tag{6-13}$$

we find that Φ_{1r}^o must be zero as $r \to 0$. The only solution for Φ_1^o that will, in addition, satisfy the condition of vanishing perturbations at infinity, is a constant which is taken to be zero. The perturbation velocities in the outer flow are thus of order ϵ^2 as compared to ϵ in both the two-dimensional and finite-wing cases. That the flow perturbations are an order of magnitude smaller for a body of revolution is reasonable from a physical point of view since the flow has one more dimension in which to get around the body.

Since $\Phi_1^i = 0$ it follows by matching that also $\Phi_1^o = 0$. This will have the consequence that all higher-order terms of odd powers in ϵ will be zero, and the series expansion thus proceeds in powers of ϵ^2. With $\Phi_1^o = 0$, substitution of the series for the outer flow into (6–2) and (6–3) gives for the lowest-order term

$$(1 - M^2)\Phi^o_{2xx} + \frac{1}{r}\Phi^o_{2r} + \Phi^o_{2rr} = 0. \tag{6–14}$$

The matching of the radial velocity component requires according to (6–12) that in the limit of $r \to 0$

$$U^o_r = \Phi^o_r = \frac{\epsilon^2 U_\infty \overline{R}\overline{R}'}{r} + O(\epsilon^3). \tag{6–15}$$

Hence

$$\Phi^o_{2r} = \frac{\overline{R}\overline{R}'}{r} \quad \text{as} \quad r \to 0$$

or

$$\lim_{r \to 0} (r\Phi^o_{2r}) = \overline{R}\overline{R}'. \tag{6–16}$$

The boundary condition at infinity is that Φ^o_{2x} and Φ^o_{2r} vanish there. Matching of Φ^o_2 itself with Φ^i_2 as given by (6–11) yields

$$\bar{g}_2(x) = \lim_{r \to 0} [\Phi^o_2 - \overline{R}\overline{R}' \ln r] + \overline{R}\overline{R}' \ln \epsilon. \tag{6–17}$$

The last term comes from the replacement of $\bar{r}$ by r/ϵ in (6–11). It follows that the inner solution is actually of order $\epsilon^2 \ln \epsilon$ rather than ϵ^2 as was assumed in the derivation. However, from a practical point of view, we may regard $\ln \epsilon$ as being of order unity since $\ln \epsilon$ is less singular in the limit of $\epsilon \to 0$ than any negative fractional power of ϵ, however small.

In calculating the pressure in the inner flow it is necessary to retain some terms beyond those required in the thin-airfoil case. By expanding (1–64) for small flow disturbances we find that

$$C_p = -\frac{\Phi_x^2 + \Phi_r^2 - U_\infty^2}{U_\infty^2} + \frac{M^2}{4}\left(\frac{\Phi_x^2 + \Phi_r^2 - U_\infty^2}{U_\infty^2}\right)^2 + \cdots, \tag{6–18}$$

which, upon introduction of the inner expansion, gives

$$C_p = -\epsilon^2[2\Phi^i_{2x} + (\Phi^i_{2\bar{r}})^2] + \cdots \tag{6–19}$$

The terms neglected in (6–19) are of order $\epsilon^4 \ln \epsilon$, or higher.

As was done in the case of a thin wing, we introduce a perturbation velocity potential φ, in the present case defined as

$$\varphi = \epsilon^2 \Phi^o_2. \tag{6–20}$$

The equations derived above then become for the outer flow

$$(1 - M^2)\varphi_{xx} + \frac{1}{r}\varphi_r + \varphi_{rr} = 0, \tag{6–21}$$

$$\lim_{r \to 0} (2\pi r\varphi_r) = S'(x), \tag{6–22}$$

where $S(x) \equiv \pi R^2(x)$ is the cross-sectional area of the body. The pressure near the body surface is given by

$$C_p = -(2\varphi_x + \varphi_r^2) \tag{6-23}$$

to be evaluated at the actual position r (for $r = 0$ it becomes singular). From the result of the inner expansion it follows that in the region close to the body

$$\varphi \simeq \frac{1}{2\pi} S'(x) \ln r + g(x), \tag{6-24}$$

where $g(x)$ is related to $\bar{g}_2(x)$ in an obvious manner.

In (6-24), the first term represents the effect of local flow divergence in the crossflow plane due to the rate of change of body cross-sectional area. According to the slender-body theory this effect is thus seen to be approximately that of a source in a two-dimensional constant-density flow in the y,z-plane. Hence, the total radial mass outflow in the inner region is independent of the radius r, as is indeed implied in the boundary condition (6-22). The second term, $g(x)$, contains the Mach number dependence and accounts for the cumulative effects of distant sources in a manner that will be further discussed in the next section.

6-3 Solutions for Subsonic and Supersonic Flows

The outer flow is easily built up from a continuous distribution of sources along the x-axis. The solution for a source in a subsonic flow is given by (5-37). Thus, for a distribution of strength $f(x)$ per unit axial distance,

$$\varphi = -\frac{1}{4\pi} \int_{-\infty}^{\infty} \frac{f(x_1)\,dx_1}{\sqrt{(x - x_1)^2 + \beta^2 r^2}}. \tag{6-25}$$

The source strength must be determined such that the boundary condition (6-22) is satisfied. It follows directly from (6-22) that the volumetric outflow per unit length should be equal to the streamwise rate of change of cross-sectional area (multiplied by U_∞). Hence we have

$$f = S'(x). \tag{6-26}$$

The result thus becomes

$$\varphi = -\frac{1}{4\pi} \int_0^l \frac{S'(x_1)\,dx_1}{\sqrt{(x - x_1)^2 + \beta^2 r^2}}. \tag{6-27}$$

We need to expand the solution for small r in order to determine an inner solution of the form (6-24). This can be done in a number of ways, for example by Fourier transform techniques (Adams and Sears, 1953) or integration by parts. Here we shall select a method used by Oswatitsch and Keune (1955) for its physical perspicuity. It is seen that the kernel in the integral (6–27) for small r is approximately

$$\frac{1}{\sqrt{(x - x_1)^2 + \beta^2 r^2}} \simeq \frac{1}{|x - x_1|} \tag{6–28}$$

except for $|x - x_1| = 0(r)$. Therefore, we recast (6–27) as follows:

$$\varphi = -\frac{1}{4\pi} \int_0^l \frac{S'(x)\, dx_1}{\sqrt{(x - x_1)^2 + \beta^2 r^2}} - \frac{1}{4\pi} \int_0^l \frac{S'(x_1) - S'(x)}{\sqrt{(x - x_1)^2 + \beta^2 r^2}}\, dx_1. \tag{6–29}$$

In the first term $S'(x)$ may be taken outside the integral, which may then be evaluated:

$$\int_0^l \frac{dx_1}{\sqrt{(x - x_1)^2 + \beta^2 r^2}} = -\ln\left[x - l + \sqrt{(x - l)^2 + \beta^2 r^2}\,\right]$$
$$+ \ln\left[x + \sqrt{x^2 + \beta^2 r^2}\,\right]. \tag{6–30}$$

Now, for small r, and $x - l < 0$,

$$x - l + \sqrt{(x - l)^2 + \beta^2 r^2}$$
$$\simeq -(l - x) + (l - x)\left[1 + \frac{\beta^2 r^2}{2(l - x)^2} + \cdots\right]$$
$$= \frac{\beta^2 r^2}{2(l - x)} + O(r^4).$$

Also, for $x > 0$,

$$x + \sqrt{x^2 + \beta^2 r^2} \simeq x + x\left(1 + \frac{\beta^2 r^2}{2x^2} + \cdots\right) = 2x + O(r^2).$$

Thus

$$\int_0^l \frac{dx_1}{\sqrt{(x - x_1)^2 + \beta^2 r^2}} = -\ln\frac{\beta^2 r^2}{2(l - x)} + \ln 2x + O(r^2). \tag{6–31}$$

In the second term of (6–29) we may use the approximation (6–28) because the numerator tends to zero for $x_1 \to x$ (the error actually turns out to be of order $r^2 \ln r$). Thus, collecting all the terms in (6–29), we find that for small r

$$\varphi \simeq \frac{S'(x)}{2\pi} \ln r + \frac{S'(x)}{4\pi} \ln\frac{\beta^2}{4x(l - x)} - \frac{1}{4\pi} \int_0^l \frac{S'(x_1) - S'(x)}{|x - x_1|}\, dx_1, \tag{6–32}$$

which is equivalent to (6–24) with

$$g(x) = \frac{S'(x)}{4\pi} \ln\frac{\beta^2}{4x(l - x)} - \frac{1}{4\pi} \int_0^l \frac{S'(x_1) - S'(x)}{|x - x_1|}\, dx_1. \tag{6–33}$$

The last term gives the effect due to variation of source strength at body stations fairly far away from station x. This form of the integral is particularly convenient when the cross-sectional area distribution is given as a polynomial, since then the integrand will become a polynomial in x_1. We may obtain an alternate form by performing an integration by parts in (6–33). This gives

$$g(x) = \frac{S'(x)}{2\pi} \ln \frac{\beta}{2} - \frac{1}{4\pi} \int_0^x S''(x_1) \ln (x - x_1) \, dx_1$$

$$+ \frac{1}{4\pi} \int_x^l S''(x_1) \ln (x_1 - x) \, dx_1, \qquad (6\text{–}34)$$

where we have assumed that $S'(0) = S'(l) = 0$, that is, the body has a pointed nose and ends in a point or in a cylindrical portion.

The solution for supersonic flow can be found in the same manner. Using (5–38) we obtain

$$\varphi = -\frac{1}{2\pi} \int_0^{x-Br} \frac{S'(x_1) \, dx_1}{\sqrt{(x - x_1)^2 - B^2 r^2}}, \qquad (6\text{–}35)$$

where

$$B = \sqrt{M^2 - 1}.$$

The upper integration limit follows because each source can only be felt inside its downstream Mach cone; hence the rearmost source that can influence the flow in the point x, r is located at $x_1 = x - Br$. Rewriting of (6–35) in a similar manner as (6–30) gives

$$\varphi = -\frac{S'(x)}{2\pi} \int_0^{x-Br} \frac{dx_1}{\sqrt{(x - x_1)^2 - B^2 r^2}}$$

$$- \frac{1}{2\pi} \int_0^{x-Br} \frac{S'(x_1) - S'(x)}{\sqrt{(x - x_1)^2 - B^2 r^2}} \, dx_1. \qquad (6\text{–}36)$$

For the first term we obtain

$$\int_0^{x-Br} \frac{dx_1}{\sqrt{(x - x_1)^2 - B^2 r^2}}$$

$$= -\ln Br + \ln [x + \sqrt{x^2 - B^2 r^2}] \simeq -\ln \frac{Br}{2x}. \qquad (6\text{–}37)$$

In the second integral we may replace the square root by $|x - x_1|$ as before. In addition, the upper integration limit may be replaced by x for small r, the error being consistent with that incurred in (6–37). Thus

$$\varphi \simeq \frac{S'(x)}{2\pi} \ln r + \frac{S'(x)}{2\pi} \ln \frac{B}{2x} - \frac{1}{2\pi} \int_0^x \frac{S'(x_1) - S'(x)}{x - x_1} \, dx_1. \quad (6\text{–}38)$$

That the correct factor $1/2\pi$ (cf. 6–24) was obtained for the first term confirms the constant for the supersonic source solution (5–38) selected by intuitive reasoning. For the supersonic case we thus have

$$g(x) = \frac{S'(x)}{2\pi} \ln \frac{B}{2x} - \frac{1}{2\pi} \int_0^x \frac{S'(x_1) - S'(x)}{x - x_1} \, dx_1. \quad (6\text{–}39)$$

As with subsonic flow, an alternate form can be obtained by integrating the last term by parts. This yields

$$g(x) = \frac{S'(x)}{2\pi} \ln \frac{B}{2} - \frac{1}{2\pi} \int_0^x S''(x_1) \ln (x - x_1) \, dx_1, \quad (6\text{–}40)$$

where we have assumed that $S'(0) = 0$. This form will be used later for the calculation of drag.

It is interesting to note how $g(x)$ changes from subsonic to supersonic flow, as seen by comparing (6–34) with (6–40). First, β is replaced by B. Secondly, the integral

$$\frac{1}{4\pi} \int_x^l S''(x_1) \ln (x_1 - x) \, dx_1,$$

which represents the upstream influence in subsonic flow, changes to

$$-\frac{1}{4\pi} \int_0^x S''(x_1) \ln (x - x_1) \, dx_1,$$

that is, becomes equal to half the total downstream influence. To understand this behavior, consider the disturbance caused by a source in one cross section x as it is felt on the body at other cross sections. The disturbance will spread along two wave fronts, one wave moving downstream with a velocity of (approximately) $a_\infty + U_\infty$ and the other either upstream or downstream with a velocity of $a_\infty - U_\infty$, depending on whether the flow is subsonic or supersonic. The effect of fast-moving waves is given by the first integral in (6–34), whereas that of slow-moving waves is given by the second integral. In the supersonic case, the fast and slow waves each contribute half of the integral in (6–40). Because of the small crosswise dimensions of the body, the curvature of the waves may be neglected in the present approximation. Hence their fronts may be treated as plane, the total effect being given by a function of x only.

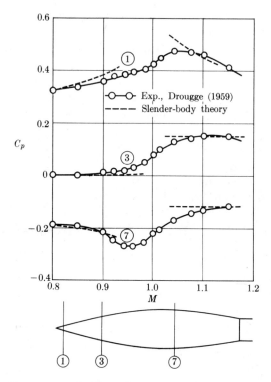

Fig. 6–1. Pressures on the forward portion on a body of revolution. [Adapted from Drougge (1959). Courtesy of Aeronautical Research Institute of Sweden.]

A comparison of calculated and measured pressure distributions given by Drougge (1959) is shown in Fig. 6–1. The excellent agreement despite the fairly large thickness ratio ($\tau = \frac{1}{6}$) demonstrates the higher accuracy of slender-body theory than thin-airfoil theory. In the former theory the error term is of order ϵ^4 (or, rather, $\epsilon^4 \ln \epsilon$), whereas in the latter it is of order ϵ^2. In assessing the accuracy of slender-body theory for practical cases, however, one must remember that the body considered in Fig. 6–1 is very smooth, with small second derivative of the cross-sectional area distribution, and should therefore be ideally suited for the theory.

The weak Mach number dependence of $\ln |1 - M^2|$ as compared to $|1 - M^2|^{-1/2}$ in the thin-airfoil case, with the associated weaker singularity at $M = 1$, is significant. It indicates that the linearized slender-body theory generally holds closer to $M = 1$ than does the thin-airfoil theory for the same thickness ratio, i.e., the true transonic region should be much smaller. For the body in Fig. 6–1, the linearized theory gives accurate pressure distributions for $M \leq 0.90$ and $M \geq 1.10$. In the transonic region the slow-moving waves will have time to interact and accumulate on the body, thus creating nonlinear effects that cannot be treated with the present linearized theory. The transonic case will be further discussed in Chapter 12.

Implicit in the derivation of the theory was the assumption that S' is continuous everywhere, as is also evident from the results which show that φ becomes logarithmically singular at discontinuities of S' and the pressure thus singular as the inverse of the distance. However, the slender-body theory may be considered as the correct "outer" solution away from the discontinuity with a separate "inner" solution required in its immediate neighborhood. Such a theory has in effect been developed by Lighthill (1948).

6–4 General Slender Body

For a general slender body we assume that the body surface may be defined by an expression of the following form

$$B(x, y, z) = \overline{B}(x, y/\epsilon, z/\epsilon) = 0, \tag{6–41}$$

where ϵ is the slenderness parameter (for example, the aspect ratio in the case of a slender wing and the thickness ratio in the case of a slender body of revolution). With the definition (6–41), a class of affine bodies with a given cross-sectional shape is studied for varying slenderness ratio ϵ, and the purpose is to develop the solution for the flow in an asymptotic series in ϵ with the lowest-order term constituting the slender-body approximation. In the stretched coordinate system

$$\overline{y} = y/\epsilon, \qquad \overline{z} = z/\epsilon \tag{6–42}$$

the cross-sectional shape for a given x becomes independent of ϵ. From the results of Section 6–4 it is plausible that the inner solution would be of the form

$$\Phi^i = U_\infty[x + \epsilon^2 \Phi_2^i(x, \overline{y}, \overline{z}) + \cdots], \tag{6–43}$$

that is, there will be no first-order term. The correctness of (6–43) will become evident later from the self-consistency of the final result. For a steady motion the condition of tangential flow at the body surface requires that the outward normal to the surface be perpendicular to the flow velocity vector:

$$\text{grad } B \cdot \text{grad } \Phi = 0. \tag{6–44}$$

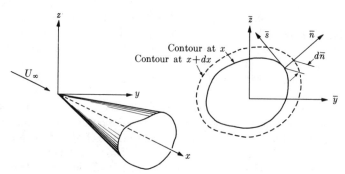

FIG. 6–2. General slender configuration.

Or, upon introducing (6–42) and (6–43) and dropping higher-order terms,

$$\overline{B}_x + \overline{B}_{\overline{y}}\Phi^i_{2\overline{y}} + B^i_{\overline{z}}\Phi^i_{2\overline{z}} = 0. \tag{6–45}$$

This relation may be put into a physically more meaningful form in the following manner. Introduce, temporarily, for each point on the contour considered, a coordinate system $\overline{n}$, $\overline{s}$ such that $\overline{n}$ is in the direction normal to, and $\overline{s}$ tangential to, the contour at the point, as shown in Fig. 6–2. Obviously, (6–45) then takes the form

$$\overline{B}_x + \overline{B}_{\overline{n}}\Phi^i_{2\overline{n}} = 0. \tag{6–46}$$

Let $d\overline{n}$ denote the change, in the direction of $\overline{n}$, of the location of the contour when going from the cross section at x to the one at $x + dx$. Moving along the body surface with $d\overline{s} = 0$ we then have

$$d\overline{B} = \overline{B}_x\,dx + \overline{B}_{\overline{n}}\,d\overline{n} = 0. \tag{6–47}$$

Upon combining (6–46) and (6–47) we obtain

$$\Phi^i_{2\overline{n}} = \frac{d\overline{n}}{dx}, \tag{6–48}$$

a condition that simply states that the streamline slope must equal the surface slope in the plane normal to the surface.

By introducing (6–43) into the differential equation (1–74) for Φ we find that Φ^i_2 must satisfy the Laplace equation in the $\overline{y}$, $\overline{z}$-plane:

$$\Phi^i_{2\overline{y}\overline{y}} + \Phi^i_{2\overline{z}\overline{z}} = 0. \tag{6–49}$$

A formal solution may be obtained by applying (2–124). (This solution was deduced by using Green's theorem in two dimensions.) Thus

$$\Phi^i_2 = \frac{1}{2\pi}\oint\left(\Phi^i_{2\overline{n}} - \Phi^i_2\frac{\partial}{\partial\overline{n}}\right)\ln\overline{r}_1\,d\overline{s}_1 + \overline{g}_2(x), \tag{6–50}$$

where index 1 denotes dummy integration variables as usual, and

$$\bar{r}_1 = \sqrt{(\bar{y} - \bar{y}_1)^2 + (\bar{z} - \bar{z}_1)^2}.$$

As in the body-of-revolution case the function $\bar{g}_2(x)$ must be obtained by matching. Note that (6–50) is in general not useful for evaluating Φ_2^i, since only the first term in the integrand is known from the boundary condition on the body. Nevertheless, it can be used to determine Φ_2^i for large $\bar{r}$, since then $\partial/\partial\bar{n} \ln \bar{r}_1$ may be neglected compared to $\ln \bar{r}_1$ and, furthermore, $\bar{r}_1$ may be approximated by $\bar{r}$. Hence the outer limit of the inner solution becomes

$$\Phi_2^{io} \sim \frac{1}{2\pi} \ln \bar{r} \oint \Phi_{2\bar{n}}^i \, d\bar{s}_1 + \bar{g}_2(x). \qquad (6\text{–}51)$$

Applying the boundary condition (6–48) and noting (see Fig. 6–2) that

$$\oint \frac{d\bar{n}}{dx} \, d\bar{s}_1 = \frac{d\bar{S}}{dx}, \qquad (6\text{–}52)$$

where

$$\bar{S}(x) = \frac{1}{\epsilon^2} S(x)$$

is the reduced cross-sectional area, we find that

$$\Phi_2^{io} \sim \frac{\bar{S}'(x)}{2\pi} \ln \bar{r} + \bar{g}_2(x), \qquad (6\text{–}53)$$

which is the same as the solution (6–11) for an axisymmetric body having the same cross-sectional area distribution as the actual slender body. We shall, following Oswatitsch and Keune (1955), term this body *the equivalent body of revolution*. By matching it will then follow that $\bar{g}_2(x)$ must be identical to that for the equivalent body of revolution. We have by this proved the following *equivalence rule*, which was first explicitly stated by Oswatitsch and Keune (1955) for transonic flow, but which was also implicit in an earlier paper by Ward (1949) on supersonic flow:

(a) *Far away from a general slender body the flow becomes axisymmetric and equal to the flow around the equivalent body of revolution.*

(b) *Near the slender body, the flow differs from that around the equivalent body of revolution by a two-dimensional constant-density crossflow part that makes the tangency condition at the body surface satisfied.*

Proofs similar to the one given here have been given by Harder and Klunker (1957) and by Guderley (1957). The equivalence rule allows great simplifications in the problem of calculating the perturbation velocity potential

$$\varphi = \epsilon^2 \Phi_2. \qquad (6\text{–}54)$$

First, the outer axisymmetric flow is immediately given by the results of the previous section. Secondly, the inner problem is reduced to one of two-dimensional constant-density flow for which the methods of Chapter 2 may be applied. The following composite solution valid for the whole flow field has been suggested by Oswatitsch and Keune [cf. statement (b) above]: Let φ_e denote the solution for the equivalent body of revolution and φ_2 the inner two-dimensional crossflow solution that in the outer limit becomes $\varphi_2 \sim (1/2\pi)S'(x) \ln r$. Then the composite solution

$$\varphi^c = \varphi_e + \varphi_2 - \frac{1}{2\pi} S'(x) \ln r \qquad (6\text{--}55)$$

holds in the whole flow field (to within the slender-body approximation).

As in the case of a body of revolution, quadratic terms in the crossflow velocity components must be retained in the expression for the pressure, so that (cf. Eq. 6–23)

$$C_p = -2\varphi_x - \varphi_y^2 - \varphi_z^2. \qquad (6\text{--}56)$$

In view of the fact that the derivation given above did not require any specification of the range of the free-stream speed, as an examination of the expansion procedure for the inner flow will reveal, it should also be valid for transonic flow. As will be discussed in Chapter 12, the difference will appear in the outer flow which then, although still axisymmetric, must be obtained from a nonlinear equation rather than from the linearized (6–21) as in the sub- or supersonic case. The form of the differential equation for the outer flow does not affect the statements (a) and (b) above, however, and it turns out that the validity of the equivalence rule is less restricted for transonic than for sub- or supersonic flow so that it can then also be used for configurations of moderate aspect ratio provided the flow perturbations are small.

6–5 Examples of Lifting Slender-Body Flow

A particularly fortunate consequence of the equivalence rule is that the outer flow is needed only for the calculation of $g(x)$, which is a symmetric term that only influences the drag but not transverse forces and moments. For the calculation of lifting flows one therefore seeks the inner constant-density two-dimensional crossflow which is independent of the Mach number and which may be obtained, for example, by using complex variables. Some of the results of the classical two-dimensional theory may be directly applied. Thus, the flow around a circular cylinder applies to the lifting slender body of revolution, and the solution for a flat plate normal to the stream can be used for the flow around a thin, slender wing. A simple, yet practically useful configuration that incorporates these as special cases is that of a mid-winged body of revolution (see Fig. 6–3).

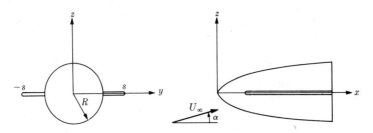

Fig. 6-3. Wing-body configuration.

To determine the perturbation velocity potential ($\varphi = \epsilon^2 \Phi_2^i$ in this case) it is, in this problem, convenient to align the x-axis with the body axis and let the free-stream vector be inclined by the angle α to the x-axis. That we then consider the flow in cross sections normal to the body axis instead of normal to the free stream will only introduce differences of order α^2, which will be negligible in the present approximation; they will only be of importance for the calculation of higher-order terms. The problem becomes that of finding a two-dimensional constant-density flow having a nondimensional vertical velocity of

$$w_\infty = \frac{W_\infty}{U_\infty} = \sin \alpha \simeq \alpha \qquad (6\text{-}57)$$

at infinity and zero normal velocity component at the body contour. Let the velocity potential corresponding to this flow be φ':

$$\varphi' = \varphi + \alpha z. \qquad (6\text{-}58)$$

We may obtain φ' from a complex potential:

$$\mathcal{W}'(X) = \varphi'(y, z) + i\psi'(y, z), \qquad (6\text{-}59)$$

where

$$X = y + iz. \qquad (6\text{-}60)$$

The complex potential $\mathcal{W}'$ will be constructed in steps using conformal transformation. First, the Joukowsky transformation

$$X_1 = X + R^2/X \qquad (6\text{-}61)$$

maps the outside of the contour onto the outside of a slit along the y_1-axis (see Fig. 6-4) of width $2s_1$, where

$$s_1 = s + R^2/s. \qquad (6\text{-}62)$$

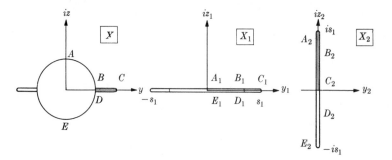

FIG. 6–4. Mapping of wing-body cross section onto a slit.

Corresponding points are marked in the figure. A second transformation

$$X_2 = (X_1^2 - s_1^2)^{1/2} \qquad (6\text{–}63)$$

transforms the horizontal slit to a vertical slit of width $2s_1$. Since in both transformations the plane remains undistorted at infinity, the flow in the X_2-plane is simply

$$\mathcal{W}'(X_2) = -i\alpha X_2, \qquad (6\text{–}64)$$

that is, an undisturbed vertical flow of (nondimensional) velocity α. By substituting (6–61)–(6–63) into (6–64) we obtain

$$\mathcal{W}'(X) = -i\alpha \left[\left(X + \frac{R^2}{X} \right)^2 - \left(s + \frac{R^2}{s} \right)^2 \right]^{1/2}. \qquad (6\text{–}65)$$

Thus for the complex velocity perturbation potential,

$$\mathcal{W}(X) = \mathcal{W}'(X) + i\alpha X = -i\alpha \left\{ \left[\left(X + \frac{R^2}{X} \right)^2 - \left(s + \frac{R^2}{s} \right)^2 \right]^{1/2} - X \right\}. \qquad (6\text{–}66)$$

This solution was given by Spreiter (1950). It is a straightforward process to derive from it the cases of body alone ($R = s$) and wing along ($R = 0$).

It is interesting to note that the crossflow considered above has no physical significance in a truly two-dimensional case, since then the flow will separate and the flow becomes rotational and nonpotential. In the slender-body case the axial flow keeps the crossflow from separating so that the potential-flow solution gives a realistic result. However, for large angles of attack the flow will separate, particularly when the aspect ratio is very small. The type of flow that then will be encountered is illustrated in Fig. 6–5. The flow separates at the leading edges and forms two stationary, more-or-less concentrated vortices above the wing. Separation gives rise to an additional "drag" in the crossflow plane, which is equivalent to

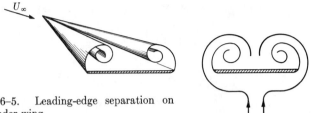

FIG. 6–5.   Leading-edge separation on a slender wing.

increased lift and drag on the vehicle. Simplified models of this type of flow have been considered by, among others, Bollay (1937), Legendre (1952), and Mangler and Smith (1956).

The calculation of total lift and moments on slender bodies will be considered in Section 6–7.

6–6 The Pressure Drag of a Slender Body in Supersonic Flow

The pressure drag acting on a body in supersonic flow can be thought of as composed of two parts, the wave drag and the vortex drag (see further Chapter 9). If the body has a blunt base, there is, in addition, a base drag. The wave drag results from the momentum carried away by the pressure waves set up by the body as it travels at a speed greater than the speed of sound. In a subsonic flow there is, of course, no wave drag, since no standing pressure waves are possible. The vortex drag arises from the momentum carried away by the vortices trailing from a lifting body and is governed by the same relations in both supersonic and subsonic flow.

The pressure drag of a slender body in a supersonic flow is most easily calculated by considering the flow of momentum through a control surface surrounding the body. We shall here follow essentially the approach taken by Ward (1949), which gives the total pressure drag but does not specify how the drag is split up into wave drag and vortex drag.

In Section 1–6 it was shown that, by considering the flow of momentum through a control surface S surrounding the body, the force on the body is given by

$$\mathbf{F}_{\text{body}} = -\oiint_{S} [p\mathbf{n} + \rho\mathbf{Q}(\mathbf{Q} \cdot \mathbf{n})]\, dS, \tag{1–51}$$

where $\mathbf{n}$ is the outward unit normal to the surface S, and $\mathbf{Q}$ is the velocity vector. It is convenient to introduce into (1–51) the perturbation velocity $U_\infty\mathbf{q} = \mathbf{Q} - i U_\infty$, which gives

$$\mathbf{F}_{\text{body}} = -\oiint_{S} [p\mathbf{n} + \rho U_\infty^2 (\mathbf{q} + \mathbf{i})(\mathbf{q} \cdot \mathbf{n} + \mathbf{i} \cdot \mathbf{n})]\, dS. \tag{6–67}$$

This may be simplified somewhat by use of the equation of continuity (cf. 1–45)

$$\oiint_S \rho(\mathbf{q} + \mathbf{i}) \cdot \mathbf{n} \, dS = 0. \qquad (6\text{–}68)$$

Hence, since S is a closed surface,

$$\mathbf{F}_{\text{body}} = -\oiint_S [(p - p_\infty)\mathbf{n} + \rho U_\infty^2 \mathbf{q}(\mathbf{q} \cdot \mathbf{n} + \mathbf{i} \cdot \mathbf{n})] \, dS, \qquad (6\text{–}69)$$

which is the form we are going to use.

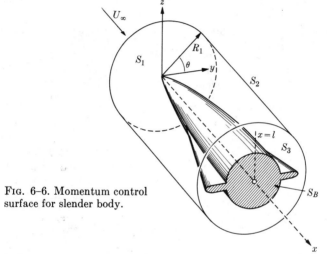

FIG. 6–6. Momentum control surface for slender body.

It is convenient in the present case to choose S in the manner shown in Fig. 6–6. Thus the surface S consists of three parts: S_1, S_2, and S_3, of which S_1 and S_3 are circular disks and S_2 a cylinder parallel to the main flow whose radius R_1 will be chosen so that S_2 is at the outer limit of the inner region. It is assumed that S_1 lies ahead of the body so that the flow is undisturbed there, and S_3 is located at the base section of the body. The body may have a blunt base, but the linearized theory is, of course, not valid for the calculation of the pressure on the base. We assume that the base pressure p_B is known, so that the base drag contribution

$$D_B = (p_\infty - p_B)S_B \qquad (6\text{–}70)$$

to the total drag is given. In supersonic flow the effect of the blunt base will not be felt upstream of the base section. Hence the linearized theory may be used to calculate the flow ahead of the base section and thus the pressure drag on the remainder of the body.

Since S is located in the inner region, the flow is essentially incompressible so that, on S, ρ may be considered constant and equal to its free-stream value. The term $\mathbf{q} \cdot \mathbf{n} + \mathbf{i} \cdot \mathbf{n}$ in (6–69) is simply the nondimensional velocity component normal to S. Thus, by taking the x-component of (6–69) we obtain

$$D - D_B = -\rho_\infty U_\infty^2 \iint_{S_2} \varphi_x \varphi_r \, dS_2 - \iint_{S_3 - S_B} (p - p_\infty) \, dS_3$$
$$-\rho_\infty U_\infty^2 \iint_{S_3 - S_B} \varphi_x (1 + \varphi_x) \, dS_3. \tag{6–71}$$

Now, according to (6–56),

$$p - p_\infty = \tfrac{1}{2}\rho_\infty U_\infty^2 C_p = -\rho_\infty U_\infty^2 \varphi_x - \tfrac{1}{2}\rho_\infty U_\infty^2 (\varphi_y^2 + \varphi_z^2) \tag{6–72}$$

in the inner region. Hence (6–71) simplifies further to

$$D - D_B = -\rho_\infty U_\infty^2 \iint_{S_2} \varphi_x \varphi_r \, dS_2 + \tfrac{1}{2}\rho_\infty U_\infty^2 \iint_{S_3 - S_B} (\varphi_y^2 + \varphi_z^2 - 2\varphi_x^2) \, dS_3. \tag{6–73}$$

In the last term $2\varphi_x^2$ may be neglected compared to $\varphi_y^2 + \varphi_z^2$, and hence

$$\frac{D - D_B}{\rho_\infty U_\infty^2} = -\iint_{S_2} \varphi_x \varphi_r \, dS_2 + \tfrac{1}{2} \iint_{S_3 - S_B} (\varphi_y^2 + \varphi_z^2) \, dS_3. \tag{6–74}$$

Since S_2 is located at the outer limit of the inner solution, the equivalence rule tells us that on S_2

$$\varphi \simeq \frac{1}{2\pi} S'(x) \ln r \Big|_{r=R_1} + g(x) \tag{6–75}$$

or

$$\varphi_r = \frac{S'}{2\pi R_1}; \qquad \varphi_x = \frac{S''}{2\pi} \ln R_1 + g'(x). \tag{6–76}$$

Also

$$dS_2 = R_1 \, d\theta \, dx;$$

thus

$$\iint_{S_2} \varphi_x \varphi_r \, dS_2 = \int_0^l S' \left[\frac{S''}{2\pi} \ln R_1 + g'(x) \right] dx, \tag{6–77}$$

where it has been assumed that the base section is located at $x = l$. The first term can be directly integrated and the second term integrated by parts. This gives

$$\iint_{S_2} \varphi_x \varphi_r \, dS_2 = \frac{1}{4\pi} [S'(l)]^2 \ln R_1 + S'(l)g(l) - \int_0^l S'' g \, dx. \tag{6–78}$$

The integral over S_3 may be simplified by use of the form (2–7) of Green's theorem which in two dimensions becomes

$$\oint_C \varphi \frac{\partial \varphi}{\partial n} \, ds = \iint_S (\nabla 2\varphi \cdot \nabla 2\varphi + \varphi \nabla^2 2\varphi) \, dS, \qquad (6–79)$$

where C is the contour bounding S, n the outward normal, and ∇_2^2 the two-dimensional crossflow Laplace operator. In the present case the contour consists of two portions, the base section contour C_B and a circle of radius R_1. At the circle we may again make the substitution (6–75) and (6–76), whereupon that portion of the line integral may be evaluated. The integral over S_3 consequently becomes

$$\iint_{S_3 - S_B} (\varphi_y^2 + \varphi_z^2) \, dS_3 = S'(l)\left[\frac{1}{2\pi} S'(l) \ln R_1 + g(l)\right] - \oint_{C_B} \varphi \frac{\partial \varphi}{\partial n} \, ds.$$

$$(6–80)$$

The minus sign in the last term appears because n is now assumed to be the normal outward from C_B and is thus the inward normal to S_3. Addition of the two integrals (6–78) and (6–80) gives for the drag

$$\frac{D - D_B}{\rho_\infty U_\infty^2} = \int_0^l S''(x)g(x) \, dx - \tfrac{1}{2}S'(l)g(l) - \tfrac{1}{2}\oint_{C_B} \varphi \frac{\partial \varphi}{\partial n} \, ds. \quad (6–81)$$

Substituting the expression (6–40) for $g(x)$ into (6–81) we finally obtain

$$\frac{D - D_B}{\rho_\infty U_\infty^2} = -\frac{1}{2\pi} \int_0^l S''(x) \, dx \int_0^x S''(x_1) \ln (x - x_1) \, dx_1$$

$$+ \frac{1}{4\pi} S'(l) \int_0^l S''(x_1) \ln (l - x_1) \, dx_1 - \tfrac{1}{2}\oint_{C_B} \varphi \frac{\partial \varphi}{\partial n} \, ds. \quad (6–82)$$

Note that φ and its derivative normal to the cross section are needed only at the base section to evaluate the last integral. Setting

$$\varphi = \varphi_2 + g(x), \qquad (6–83)$$

where φ_2 is defined as in (6–55), and using the expression (6–40) for $g(x)$, we may rewrite (6–82) as follows:

$$\frac{D - D_B}{\rho_\infty U_\infty^2} = -\frac{1}{2\pi} \int_0^l S''(x) \, dx \int_0^x S''(x_1) \ln (x - x_1) \, dx_1$$

$$+ \frac{1}{2\pi} S'(l) \int_0^l S''(x_1) \ln (l - x_1) \, dx_1 - \frac{1}{4\pi} [S'(l)]^2 \ln \frac{B}{2}$$

$$- \tfrac{1}{2}\oint_{C_B} \varphi_2 \frac{\partial \varphi_2}{\partial n} \, ds. \qquad (6–84)$$

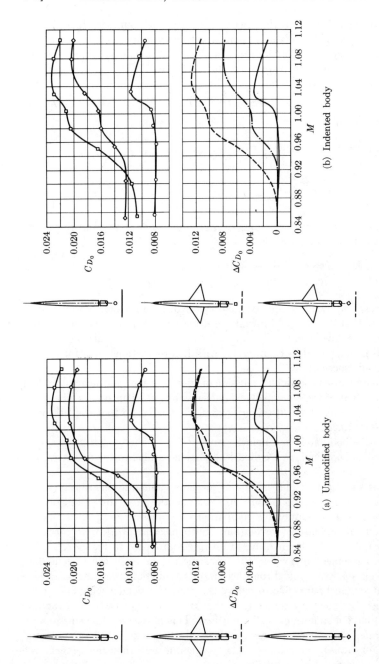

Fig. 6–7. Drag rise at zero lift for a wing-body combination, for body alone, and for equivalent body. [Adapted from Whitcomb (1956). Courtesy of the National Aeronautics and Space Administration.]

This formula shows how the drag varies with Mach number. If $S'(l) = 0$, that is, if the body ends in a point or with a cylindrical portion, the drag becomes independent of the Mach number. Of particular interest is the drag of the equivalent body of revolution, for which

$$\varphi_2 = \frac{1}{2\pi} S'(x) \ln r. \qquad (6\text{--}85)$$

Hence

$$\frac{D_{\text{eq. body}} - D_B}{\rho_\infty U_\infty^2} = -\frac{1}{2\pi} \int_0^l S''(x)\, dx \int_0^x S''(x_1) \ln (x - x_1)\, dx_1$$

$$+ \frac{1}{2\pi} S'(l) \int_0^l S''(x_1) \ln (l - x_1) - \frac{1}{4\pi} [S'(l)]^2 \ln \frac{BR(l)}{2}. \qquad (6\text{--}86)$$

The difference between the drag of a general slender body and that of its equivalent body of revolution is thus

$$\frac{D - D_{\text{eq. body}}}{\rho_\infty U_\infty^2} = -\tfrac{1}{2} \oint_{C_B} \varphi_2 \frac{\partial \varphi_2}{\partial n}\, ds + \frac{1}{4\pi} [S'(l)]^2 \ln R(l). \qquad (6\text{--}87)$$

This is the equivalence rule for the pressure drag. It is a fairly easy matter to show that it must hold for all speed regimes whenever the equivalence rule for the flow is valid. Thus it will also hold for transonic speeds, and, as will be explained in Chapter 12, with less restrictions than for sub- or supersonic speeds.

In many cases the right-hand side of (6–87) is zero and hence the drag equal to that of the equivalent body of revolution. This occurs whenever:

(a) The body ends with an axisymmetric portion so that the two parts in (6–87) cancel.

(b) The body ends in a point.

(c) The body ends in a cylindrical portion parallel to the free stream so that $\partial\varphi/\partial n$ and S' are zero.

Most practical slender missile or airplane configurations satisfy (a) or (c). For such a body one can thus experimentally test the validity of the equivalence rule simply by comparing the pressure drag with that of the equivalent body of revolution. Such measurements were made by Whitcomb (1956). Some of his results are reproduced in Fig. 6–7.

The agreement is good in the transonic region when the viscous drag has been separated out. From these results Whitcomb drew the conclusion that it should be possible to reduce the drag of a slender wing-body combination by indenting the body so that the equivalent body of revolution would have a smooth area distribution. This is the well-known *transonic area rule*, which has been used successfully to design low-drag configurations for transonic airplanes. The savings in drag that can be achieved are demonstrated in Fig. 6–7(b).

6–7 Transverse Forces and Moments on a Slender Body

The transverse forces and moments (lift, side force, pitching moment, etc.) on a slender body can be obtained by considering the flow of momentum through a control surface surrounding the body, as in the preceding section. However, we shall instead use a different method that makes use of results previously deduced for unsteady constant-density flow.

Let x_1, y_1, z_1 be a coordinate system fixed with the fluid so that

$$x_1 = x - U_\infty t, \qquad y_1 = y, \qquad z_1 = z. \tag{6–88}$$

An observer in this coordinate system will see the body moving past with a velocity U_∞ in the negative x_1-direction. Consider now the fluid motion in a slab of width dx_1 perpendicular to the free stream as the body moves past. The crossflow in the neighborhood of the body will be governed by the equations for two-dimensional constant-density flow in the crossflow plane, but the flow will now be unsteady since consecutive cross sections of the body pass through the slab as the body travels by. The incompressible crossflow in the slab will thus be that around a two-dimensional body that changes shape and translates with time (and also rotates if rolling of the slender body is considered). This situation is illustrated in Fig. 6–8.

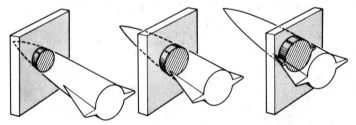

Fig. 6–8. Unsteady flow in a plane slab due to penetration of slender body.

Since the flow has no circulation in the crossflow plane we can directly apply the methods developed in Section 2–4. [The Blasius' equation (2–122) for unsteady flow is not applicable since it was derived under the assumption that the cross section does not change with time.] Thus, if ξ is the crossflow momentum vector per unit body length, the force acting on the body cross section of width dx is, according to (2–61),

$$d\mathbf{F} = \mathbf{j}Y + \mathbf{k}L = -dx_1 \frac{d\xi}{dt}, \tag{6–89}$$

where L is the lift and Y is the side force. The momentum vector ξ is given by (2–54). Thus, in two-dimensional flow,

$$\xi = -\rho_\infty U_\infty \oint \varphi_1 \mathbf{n} \, ds, \tag{6–90}$$

where φ_1 is φ expressed in (x_1, y_1, z_1, t) and the integral is to be taken around the (instantaneous) body cross section. The factor U_∞ comes from the definition of φ. It follows from Fig. 6–9 that

$$\mathbf{n}\, ds = (\mathbf{j} \cos \theta + \mathbf{k} \sin \theta)\, ds$$

$$= \mathbf{j}\, dz_1 - \mathbf{k}\, dy_1. \qquad (6\text{–}91)$$

Hence

$$\boldsymbol{\xi} = -\rho_\infty U_\infty \mathbf{j} \int \Delta\varphi_1(z_1)\, dz_1$$

$$- \rho_\infty U_\infty \mathbf{k} \int \Delta\varphi_1(y_1)\, dy_1, \qquad (6\text{–}92)$$

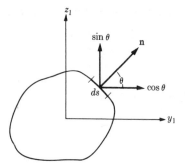

Fig. 6–9. Instantaneous body cross section.

where $\Delta\varphi_1(y_1)$ is the difference in φ_1 between the upper and lower surfaces of the cross section and $\Delta\varphi_1(z_1)$ is the difference in φ_1 between the right and left surfaces of the cross section. Introducing (6–92) into (6–89) we thus obtain

$$\frac{dL}{dx_1} = \rho_\infty U_\infty \frac{d}{dt} \int \Delta\varphi_1(y_1)\, dy_1 \qquad (6\text{–}93)$$

and

$$\frac{dY}{dx_1} = \rho_\infty U_\infty \frac{d}{dt} \int \Delta\varphi_1(z_1)\, dz_1. \qquad (6\text{–}94)$$

Now, going back to the original coordinate system traveling with the body and remembering that for steady flow $d/dt = U_\infty\, d/dx$ we may write (6–93) and (6–94) as follows:

$$\frac{dL}{dx} = \rho_\infty U_\infty^2 \frac{d}{dx} \int \Delta\varphi(y)\, dy, \qquad (6\text{–}95)$$

$$\frac{dY}{dx} = \rho_\infty U_\infty^2 \frac{d}{dx} \int \Delta\varphi(z)\, dz. \qquad (6\text{–}96)$$

By integrating over the body length, total lift, pitching moment, etc. can then be obtained. Particularly simple are the expressions for the total forces, which become

$$L = \rho_\infty U_\infty^2 \int_{C_B} \Delta\varphi(y)\, dy, \qquad (6\text{–}97)$$

$$Y = \rho_\infty U_\infty^2 \int_{C_B} \Delta\varphi(z)\, dz, \qquad (6\text{–}98)$$

where C_B indicates that the integral is to be evaluated at the base cross section. In order to calculate total forces, one thus only needs the cross flow at the base. Frequently, the flow is given in complex variables, in which case it is convenient to work with the complex force combination

$$F = Y + iL = -i\rho_\infty U_\infty^2 \oint_{C_B} \varphi \, dX, \tag{6–99}$$

where

$$X = y + iz.$$

[This formula could, of course, also have been obtained by introducing the complex vector directly in (6–92).] The idea is then to introduce the complex potential $\mathcal{W}(X) = \varphi + i\psi$, which would reduce the problem to that of evaluating a closed-contour complex integral. However, a direct replacement of φ by $\mathcal{W}$ will generally lead to an incorrect result unless the stream function ψ happens to be zero, or constant, along the cross section contour. We therefore introduce, as in (6–59), the potential $\mathcal{W}'$ for the related flow having zero normal velocity at the contour and velocity components at infinity proportional to the side-slip angle and angle of attack, respectively, at the base section. Thus

$$\varphi = \mathbf{Re}\ \{\mathcal{W}'(X)\} - \alpha_B z - \beta_B y, \tag{6–100}$$

where α_B and β_B are the angle of attack and side-slip angle at the base (which would be different from the overall angle of attack and side slip if the body were cambered). Now ψ' is zero along the contour, and we may therefore set

$$F = -i\rho_\infty U_\infty^2 \oint \mathcal{W}' \, dX - i\rho_\infty U_\infty^2 S_B(\alpha_B - i\beta_B), \tag{6–101}$$

where S_B is the base area. The last term follows from simple geometrical considerations that give, for example, that

$$\oint z \, dX = -\int \Delta z \, dy = -S_B. \tag{6–102}$$

In the first integral of (6–101) we may choose any path of integration that encloses the base contour, the most convenient one being a large circle at infinity. Assuming that $\mathcal{W}'$ may be expressed by the following Laurent series for large $|X|$

$$\mathcal{W}' = a_0 X + \sum_1^\infty \frac{a_{-n}}{X^n}, \tag{6–103}$$

and a_{-1} being the residue at infinity, we obtain from (6–101)

$$F = 2\pi\rho_\infty U_\infty^2 a_{-1} - i\rho_\infty U_\infty^2 S_B(\alpha_B - i\beta_B). \tag{6–104}$$

As an example, we shall apply this formula to the lifting wing-body combination considered in Section 6–5. By expanding (6–65) in inverse powers of X,

$$\mathcal{w}' = -i\alpha \left\{ X - \frac{1}{2X}\left[\left(s + \frac{R^2}{s} \right)^2 - 2R^2 \right] + \cdots \right\}, \quad (6\text{--}105)$$

and inserting the residue as given by the second term of (6–105), the following result is obtained

$$L = \pi\rho_\infty U_\infty^2 \alpha (s^2 - R^2 + R^4/s^2)_B, \quad (6\text{--}106)$$

where index B refers to the base section. (The side force is, of course, zero in this case.) This result contains as special cases those of the wing alone and body alone. In the latter case, setting $s_B = R_B$ in (6–106) we obtain

$$L = \pi\rho_\infty U_\infty^2 R_B^2 \alpha, \quad (6\text{--}107)$$

that is, the lift coefficient based on the base area is simply

$$C_L = 2\alpha, \quad (6\text{--}108)$$

a result first derived by Munk (1924). An interesting conclusion from this is that on a body pointed at the rear no lift is exerted, only a pitching

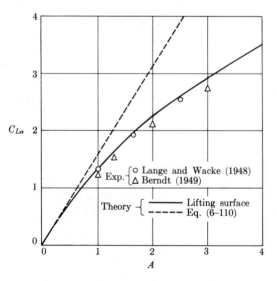

Fig. 6–10. Comparison of theoretical and experimental lift-curve slopes for low-aspect-ratio triangular wings. (Numerical lifting-surface theory, courtesy of Dr. Sheila Widnall. See also Chapter 7.)

moment. This is destabilizing, tending to increase the angle of attack (cf. ellipsoid example in Chapter 2 and Fig. 2–5). In reality, viscous effects will cause a small positive lift.

For the case of a wing alone ($R_B = 0$), (6–106) gives

$$L = \pi\rho_\infty U_\infty^2 s_B^2 \alpha. \tag{6–109}$$

Hence, any slender wing with a straight, unswept trailing edge will have a lift coefficient (R. T. Jones, 1946)

$$C_L = \frac{\pi}{2} A\alpha. \tag{6–110}$$

Comparisons for delta wings with experiments and a numerical lifting-surface theory, presented in Fig. 6–10, show that this simple formula overestimates the lift by 10% and more for $A > 1.0$.

It should be pointed out that (6–106) and (6–109) hold only for wings having monotonically increasing span from the pointed apex to the base section, otherwise sections forward of the base section will produce a wake that will influence the flow at the base, so that it no longer becomes independent of the flow in other cross sections. For the case of an uncambered wing with swept-forward trailing edges, one can easily show that the lift on sections behind that of the maximum span is zero in the slender-wing approximation, and hence (6–109) will hold if s_B is replaced by $s_{\max}$, the maximum semispan. In the case of swept-back trailing edges, as for an arrowhead or swallowtail wing, Mangler (1955) has shown that the determination of the flow requires the solution of an integral equation.

A practically useful formula to estimate the effect of a fuselage on total lift is obtained by dividing (6–106) by (6–109). Thus

$$\frac{L_{\text{wing+body}}}{L_{\text{wing alone}}} = 1 - \left(\frac{R_B}{s_B}\right)^2 + \left(\frac{R_B}{s_B}\right)^4, \tag{6–111}$$

which shows that the body interference tends to decrease the lift.

7

Three-Dimensional Wings in Steady, Subsonic Flow

7–1 Compressibility Corrections for Wings

This chapter deals with the application to finite, almost-plane wings of the linearized, small-perturbation techniques introduced in Chapter 5. By way of introduction, we first review the similarity relations which govern variations in the parameter M, the flight Mach number.

In the light of the asymptotic expansion procedure, the principal unknown, from which all other needed information can be calculated, is the first-order term* Φ_1^o in the outer expansion for the velocity potential. The term Φ_1^o is connected to the more familiar perturbation velocity potential $\varphi(x, y, z)$ by (5–28). The latter is governed by the differential equations and boundary conditions (5–29)–(5–30), which we reproduce here (see also Fig. 5–1):

$$(1 - M^2)\varphi_{xx} + \varphi_{yy} + \varphi_{zz} = 0, \tag{5–29}$$

$$\left.\begin{aligned}\varphi_z &= \frac{\partial z_u}{\partial x} \quad \text{at} \quad z = 0+ \\ \varphi_z &= \frac{\partial z_l}{\partial x} \quad \text{at} \quad z = 0-\end{aligned}\right\} \quad \text{for} \quad (x, y) \quad \text{on} \quad S. \tag{5–30}$$

The pressure coefficient at any point in the field, including the upper and lower wing surfaces $z = 0\pm$, is found from

$$C_p = -2\varphi_x. \tag{5–31}$$

Extending a procedure devised by Prandtl and Glauert for two-dimensional airfoils (see Fig. 7–1), Göthert (1940) introduced a transformation of independent and dependent variables which is equivalent to

$$\left.\begin{aligned}x_0 &= x/\beta \\ y_0 &= y \\ z_0 &= z \\ \varphi_0(x_0, y_0, z_0) &= \varphi(x, y, z)\end{aligned}\right\} \tag{7–1}$$

* The zeroth-order term is, of course, the free stream $\Phi_0^o = U_\infty x$.

124

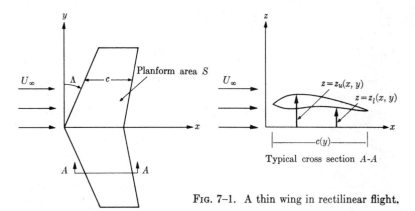

FIG. 7–1. A thin wing in rectilinear flight.

Where $\beta \equiv \sqrt{1 - M^2}$, as in Chapter 5. Equation (7–1) converts (5–29) into the constant-density perturbation equation

$$(\varphi_0)_{x_0 x_0} + (\varphi_0)_{y_0 y_0} + (\varphi_0)_{z_0 z_0} = 0. \qquad (7-2)$$

Some care must be observed when interpreting the transformed boundary condition at the wing surface. Thus, for example, the first of (5–30) states that just above the wing's projection on the x, y-plane the vertical velocity component produced by the sheet of singularities representing the wing's disturbance must have certain values, say $F_u(x, y)$. After transformation, we obtain

$$(\varphi_0)_{z_0} = F_u(\beta x_0, y_0) \equiv F_{u_0}(x_0, y_0) \qquad (7-3)$$

at $z_0 = 0+$, for (x_0, y_0) on S_0, where S_0 is an area of the x_0, y_0-plane whose lateral dimensions are the same as the original planform projection S, but which is stretched chordwise by a factor $1/\beta$. (See Fig. 7–2.)

Equation (7–3) and the equivalent form for the lower surface state, however, that the "equivalent" wing in zero-M, constant-density flow has (at corresponding stations) the *same* thickness ratio τ, fractional camber θ, and angle of attack α as the original wing in the compressible stream.

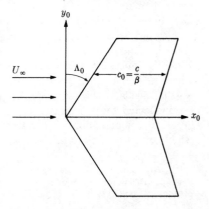

FIG. 7–2. Equivalent wing planform in zero-Mach-number flow. If sweep is present, $\tan \Lambda_0 = (1/\beta) \tan \Lambda$. The aspect ratio is $A_0 = \beta A$.

The similarity law might be abbreviated

$$\varphi(x, y, z; M, A, \tau, \theta, \alpha) = \varphi\left(\frac{x}{\beta}, y, z; M = 0, \beta A, \tau, \theta, \alpha\right), \qquad (7\text{-}4)$$

where the semicolon is used to separate the independent variables from the parameters.

By way of physical explanation,* Göthert's extended Prandtl-Glauert law states that to every subsonic, compressible flow over a thin wing there exists an equivalent flow of constant density liquid (at the same flight speed and free-stream ambient conditions) over a second wing, obtained from the first by a chordwise stretching $1/\beta$ without change of surface slope distribution. It is obvious from (5–31) that pressure coefficients at corresponding points in the two flows are related by

$$C_p = \frac{1}{\beta}(C_p)_{M=0}. \qquad (7\text{-}5)$$

Since they are all calculated from similar dimensionless chordwise and spanwise integrations of the C_p-distribution, quantities like the sectional lift and moment coefficients $C_l(y)$, $C_m(y)$, the total lift and moment coefficients C_L, C_M, and the lift-curve slope $\partial C_L/\partial \alpha$ are found from their constant-density counterparts by the same factor $1/\beta$ as in (7–5). It is of interest in connection with spanwise load distribution, however, that the total lift forces and running lifts per unit y-distance are equal on the two wings, because of the increased chordwise dimensions at $M = 0$.

Unfortunately, when one is treating a given three-dimensional configuration, the foregoing transformation requires that a different planform be analyzed (or a different low-speed model be tested) for each flight Mach number at which loading data are needed. This is not true for two-dimensional airfoils, since then the chordwise distortion at fixed α, etc., is no more than a change of scale on an otherwise identical profile; we have already seen (Section 1–4) that such a change has no effect on the physical flow quantities *at fixed M*.

Measurements like those of Feldman (1948) correlate with the Göthert-Prandtl-Glauert law rather well up to the vicinity of critical Mach number, where sonic flow first appears at the wing surface. They also verify what we shall see later theoretically, that the coefficient of induced drag should be unaffected by M-changes below M_{crit}. There exist, of course, more accurate compressibility corrections based on nonlinear considerations which are successful up to somewhat higher subsonic M.

---

* An illuminating discussion appears in Section A.3 of Jones and Cohen (1960). See also the extended development in Chapter 10 of Liepmann and Roshko (1957).

Inasmuch as (5–29) applies also to small-perturbation supersonic flow, $M > 1$, one might suspect that the foregoing considerations could be extended directly into that range. This is an oversimplification, however, since the boundary conditions at infinity undergo an essential change—disturbances are not permitted to proceed upstream but may propagate only downstream and laterally in the manner of an outward-going sound wave. (The behavior is connected with a mathematical alteration in the nature of the partial differential equation, from elliptical to hyperbolic or "wavelike.") What one does discover is the existence of a convenient reference Mach number, $M = \sqrt{2}$, which plays a role similar to $M = 0$ in the subsonic case. When $M = \sqrt{2}$, the quantity $B \equiv \sqrt{M^2 - 1}$ becomes unity and all flow Mach lines are inclined at 45° to the flight direction. Repetition of the previous reasoning leads to a supersonic similarity law

$$\varphi(x, y, z; M, A, \tau, \theta, \alpha) = \varphi\left(\frac{x}{B}, y, z; M = \sqrt{2}, BA, \tau, \theta, \alpha\right). \qquad (7\text{–}6)$$

Pressure coefficients at corresponding points, lift coefficients, etc., are related by

$$C_p = \frac{1}{B}(C_p)_{M=\sqrt{2}}. \qquad (7\text{–}7)$$

Once more the equivalent planform at $M = \sqrt{2}$ is obtained from the original by chordwise distortion, but now this involves a stretching if the original $M < \sqrt{2}$ and a shrinking if $M > \sqrt{2}$. The process has been likened to taking hold of all Mach lines and rotating them to 45°, while chordwise dimensions vary in affine proportion.

Clearly, Eqs. (7–5), (7–7), and the associated transformation techniques fail in the transonic range where $M \cong 1$. It has been speculated, because the equivalent aspect ratio approaches zero as $M \to 1$ and slender-body-theory results for lift are independent of Mach number (Chapter 6), that linearized results for three-dimensional wings might be extended into this range. This is, unfortunately, an oversimplification. Starting from the proper, nonlinear formulation of transonic small-disturbance theory, Chapter 12 derives the actual circumstances under which linearization is permissible and gives various similarity rules. It is found, for instance, that loading may be estimated on a linearized basis whenever the parameter $A\tau^{1/3}$ is small compared to unity.

7–2 Constant-Density Flow; the Thickness Problem

Having shown how steady, constant-density flow results are useful at all subcritical M, we now elaborate them for the finite wing pictured in Fig. 7–1. As discussed in Section 5–2 and elsewhere above, it is convenient to identify and separate portions of the field which are symmetrical and antisymmetrical in z, later adding the disturbance velocities and pressures in accordance with the superposition principle. The separation process involves rewriting the boundary conditions (5–30) as

$$\varphi_z = \tau \frac{\partial \bar{g}}{\partial x} + \theta \frac{\partial \bar{h}}{\partial x} - \alpha \quad \text{at} \quad z = 0+ \quad \left.\right\} \quad \text{for } (x, y) \text{ on } S, \tag{7-8a}$$

$$\varphi_z = -\tau \frac{\partial \bar{g}}{\partial x} + \theta \frac{\partial \bar{h}}{\partial x} - \alpha \quad \text{at} \quad z = 0- \tag{7-8b}$$

where $\bar{h}(x, y)$ is proportional to the ordinate of the mean camber surface, while $2\bar{g}(x, y)$ is proportional to the thickness distribution. The differential equation is, of course, the three-dimensional Laplace equation

$$\nabla^2 \varphi = 0. \tag{7-9}$$

As a starting point for the construction of the desired solutions, we adapt (2–28) to express the perturbation velocity potential at an arbitrary field point (x, y, z),

$$\varphi(x, y, z) = \oiint_S \left[\varphi \frac{\partial}{\partial n} - \frac{\partial \varphi}{\partial n} \right] \left(\frac{1}{4\pi r} \right) dS. \tag{7-10}$$

Here n is the normal directed into the field, and the integrals must be carried out over the upper and lower surfaces of S. Dummy variables (x_1, y_1, z_1) will be employed for the integration process, so the scalar distance is properly written

$$r = \sqrt{(x - x_1)^2 + (y - y_1)^2 + (z - z_1)^2}. \tag{7-11}$$

In wing problems $z_1 = 0$ generally.

Considering the thickness alone, we have

$$\varphi(x, y, z) = \varphi(x, y, -z) \tag{7-12}$$

for all z and the boundary condition

$$\varphi_z \equiv + \left(\frac{\partial \varphi}{\partial n} \right)_u = \tau \frac{\partial \bar{g}}{\partial x} \quad \text{at} \quad z = 0+ \quad \left.\right\} \quad \text{for } (x, y) \text{ on } S. \tag{7-13a}$$

$$\varphi_z \equiv - \left(\frac{\partial \varphi}{\partial n} \right)_l = -\tau \frac{\partial \bar{g}}{\partial x} \quad \text{at} \quad z = 0- \tag{7-13b}$$

Moreover, no discontinuities of φ or its derivatives are expected anywhere else on or off the x, y-plane. In (7–10), $dS = dx_1\, dy_1$, the values of $\varphi(x_1, y_1, 0+)$ and $\varphi(x_1, y_1, 0-)$ appearing in the integrals over the upper and lower surfaces are equal, while the values of $\partial/\partial n(1/4\pi r)$ are equal and opposite. Hence the contributions from the first term in brackets cancel, leaving us with

$$\varphi(x, y, z) = \oint\!\!\!\oint_S \left[-\left(\frac{\partial \varphi}{\partial n}\right)_u - \left(\frac{\partial \varphi}{\partial n}\right)_l \right] \frac{dS}{4\pi r}$$

$$= -\frac{\tau}{2\pi} \iint_S \frac{\partial \bar{g}(x_1, y_1)}{\partial x_1} \frac{dx_1\, dy_1}{\sqrt{(x - x_1)^2 + (y - y_1)^2 + z^2}}, \qquad (7\text{–}14)$$

where (7–13) and (7–11) have been employed. Physically, (7–14) states that the flow due to thickness can be represented by a source sheet over the planform projection, with the source strength per unit area being proportional to twice the thickness slope $\partial \bar{g}/\partial x$. [Compare the two-dimensional counterpart, (5–50).]

Examination of (7–14) leads to the conclusion that the thickness problem is a relatively easy one. In the most common situation when the shape of the wing is known and the flow field constitutes the desired information, one is faced with a fairly straightforward double integration. For certain elementary functions $\bar{g}(x, y)$ this can be done in closed form; otherwise it is a matter of numerical quadrature, with careful attention to the pole singularity at $x_1 = x$, $y_1 = y$, when one is analyzing points on the wing $z = 0$. The pressure can be found from (5–31) and (7–14) as

$$C_p = -2\varphi_x$$

$$= -\frac{\tau}{\pi} \iint_S \left[\frac{\partial \bar{g}(x_1, y_1)}{\partial x_1} \right] \left[\frac{x - x_1}{r^3} \right] dx_1\, dy_1. \qquad (7\text{–}15)$$

There is no net loading, since C_p has equal values above and below the wing. Also the thickness drag works out to be zero, in accordance with d'Alembert's paradox (Section 2–5). Finally, it should be mentioned that, for any wing with closed leading and trailing edges,

$$\int_{\text{chord}} \frac{\partial \bar{g}}{\partial x_1}\, dx_1 = \bar{g}_{\text{TE}} - \bar{g}_{\text{LE}} = 0. \qquad (7\text{–}16)$$

This means that the total strength of the source sheet in (7–14) is zero. As a consequence, the disturbance at long distances from the wing approaches that due to a doublet with its axis oriented in the flight direction, rather than that due to a point source.

7–3 Constant-Density Flow; the Lifting Problem

For the purely antisymmetrical case we have

$$\varphi(x, y, z) = -\varphi(x, y, -z) \qquad (7\text{–}17)$$

with corresponding behavior in the velocity and pressure fields. The principal boundary condition now reads

$$\varphi_z = \theta \frac{\partial \overline{h}}{\partial x} - \alpha \quad \text{at} \quad z = 0\pm \quad \text{for} \quad (x, y) \quad \text{on} \quad S. \quad (7\text{--}18)$$

We must make allowance for a discontinuity in φ not only on the planform projection S but over the entire wake surface, extending from the trailing edge and between the wingtips all the way to $x = +\infty$ on the x, y-plane. For this reason, the simplest approach proves to be the use of (7–10) not as a means of expressing φ itself but the dimensionless x-component of the perturbation velocity

$$u = \varphi_x, \quad (7\text{--}19)$$

where u is essentially the pressure coefficient, in view of (5–31). Equation (1–81) shows us that we are also working with a quantity proportional to the small-disturbance acceleration potential, and this is the starting point adopted by some authors for the development of subsonic lifting wing theory.

Obviously, u is a solution of Laplace's equation, since the operation of differentiation with respect to x can be interchanged with ∇^2 in (7–9). We may therefore write

$$u(x, y, z) = \oiint\limits_{S'} \left[u \frac{\partial}{\partial n} - \frac{\partial u}{\partial n} \right] \left(\frac{1}{4\pi r} \right) dS. \quad (7\text{--}20)$$

We specify for the moment that S' encompasses both wing and wake, since the derivation of (2–28) called for integration over all surfaces that are sources of disturbance and made no allowance for circulation around any closed curve in the flow external to the boundary. It is an easy matter to show, however, that the choice of u as dependent variable causes the first term in the (7–20) brackets to vanish except on S and the second term to vanish altogether. Because there can be no pressure jump except through a solid surface, u is continuous through $z = 0$ on the wake. But

$$\frac{\partial}{\partial n} \left(\frac{1}{4\pi r} \right) = \pm \left. \frac{\partial}{\partial z_1} \left(\frac{1}{4\pi r} \right) \right|_{z_1=0} \quad (7\text{--}21)$$

are equal and opposite on top and bottom everywhere over S, so the first-term contributions remain uncanceled only on S. There u jumps by an amount

$$\dot{\gamma}(x_1, y_1) = u_u - u_l. \quad (7\text{--}22)$$

(By antisymmetry, $u_l = -u_u$.)

As regards the second bracketed term in (7–20), the condition of irrotationality reveals that

$$\frac{\partial u}{\partial n} = \frac{\partial u}{\partial z_1} = \frac{\partial w}{\partial x_1} \quad (7\text{--}23a)$$

on top and one finds that

$$\frac{\partial u}{\partial n} = -\frac{\partial u}{\partial z_1} = -\frac{\partial w}{\partial x_1} \qquad (7\text{--}23b)$$

on the bottom of S'. Both w and its derivatives are continuous through all of S', and therefore the upper and lower integrations cancel throughout. One is left with

$$u(x, y, z) = \iint_S [u_u - u_l] \frac{\partial}{\partial n} \left(\frac{1}{4\pi r}\right) dS$$

$$= \frac{1}{4\pi} \iint_S \gamma(x_1, y_1) \frac{\partial}{\partial z_1} \left(\frac{1}{r}\right)\bigg|_{z_1=0} dx_1\, dy_1$$

$$= -\frac{1}{4\pi} \iint_S \gamma(x_1, y_1) \frac{\partial}{\partial z} \left(\frac{1}{r}\right) dx_1\, dy_1. \qquad (7\text{--}24)$$

We have inserted (7–21) and (7–22) here, along with the fact that

$$\frac{\partial}{\partial z_1} = +\frac{\partial}{\partial(z_1 - z)}$$

$$= -\frac{\partial}{\partial z} \qquad (7\text{--}25)$$

when applied to a quantity which is a function of these two variables only in the combination $(z - z_1)$.

The modification of (7–24) into a form suitable for solving lifting-wing problems can be carried out in several ways. Perhaps the most direct is to observe that nearly always $w(x, y, 0)$ is known over S, and (7–24) should therefore be manipulated into an expression for this quantity. This we do by noting that $u = \varphi_x$ and $w = \varphi_z$, so that

$$w(x, y, z) = \frac{\partial}{\partial z} \int_{-\infty}^{x} u(x_0, y, z)\, dx_0, \qquad (7\text{--}26)$$

where account has been taken that $\varphi(-\infty, y, z) = 0$. Inserting (7–26) into (7–24) and interchanging orders of differentiation and integration, we get

$$w(x, y, z) = -\frac{1}{4\pi} \iint_S \gamma(x_1, y_1)$$

$$\times \left\{ \frac{\partial^2}{\partial z^2} \int_{-\infty}^{x} \frac{dx_0}{\sqrt{(x_0 - x_1)^2 + (y - y_1)^2 + z^2}} \right\} dx_1\, dy_1. \qquad (7\text{--}27)$$

After considerable algebra, we find that (7–27) can be reduced, as $z \to 0$, to one or the other of the following forms:

$$w(x, y, 0) = \theta \frac{\partial \overline{h}}{\partial x} - \alpha$$

$$= \frac{1}{4\pi} \oiint_S \frac{\gamma(x_1, y_1)}{(y - y_1)^2}$$

$$\times \left[1 + \frac{(x - x_1)}{\sqrt{(x - x_1)^2 + (y - y_1)^2}} \right] dx_1 \, dy_1$$

$$= -\frac{1}{4\pi} \oiint_S \frac{\partial \gamma}{\partial y_1} \frac{1}{(y - y_1)} , \tag{7–28a}$$

$$\times \left[1 + \frac{\sqrt{(x - x_1)^2 + (y - y_1)^2}}{(x - x_1)} \right] dx_1 \, dy_1. \tag{7–28b}$$

When deriving (7–28b), an integration by parts* on y_1 is carried out at finite z. The singularity encountered as $z \to 0$ is then similar to the one in upwash calculation at a two-dimensional vortex sheet and can be handled by the well-known Cauchy principal value. The integral in (7–28a) containing $\gamma(x_1, y_1)$ itself is, however, the more direct and directly useful form. In the process of arriving at it, we find ourselves confronted with the following steps:

$$\frac{\partial^2}{\partial z^2} \int_{-\infty}^{x} \frac{dx_0}{\sqrt{(x_0 - x_1)^2 + (y - y_1)^2 + z^2}}$$

$$= -\frac{\partial}{\partial z} \left\{ \frac{z}{(y - y_1)^2 + z^2} \left[1 + \frac{(x - x_1)}{\sqrt{(x - x_1)^2 + (y - y_1)^2 + z^2}} \right] \right\}$$

$$= -\frac{\partial}{\partial z} \left\{ \frac{z}{(y - y_1)^2 + z^2} \right\} \left[1 + \frac{(x - x_1)}{\sqrt{(x - x_1)^2 + (y - y_1)^2 + z^2}} \right]$$

$$+ z^2 \{\text{additional terms}\}. \tag{7–29}$$

For $z = 0$ all terms here will vanish formally except the one arising from the z-derivative of the numerator which will give a nonintegrable singularity of the form $1/(y - y_1)^2$. It is precisely with such limits, however, that Mangler's study of improper integrals [Mangler (1951)] is concerned. Indeed, if we examine Eqs. (33) and (34) of his paper, replacing his ξ with our y_1, we observe that our y_1-integral should be evaluated in accordance

---

* The integrated portion vanishes, if we assert that the area of integration extends an infinitesimal distance beyond the boundaries of S, because γ vanishes there.

with Mangler's principal-value technique and thereupon assumes a perfectly reasonable, finite value. The result implies, of course, that the self-induced normal velocity on a vortex sheet should not be infinite if it is calculated properly.

This recipe for computing the y_1-integral may be summarized

$$\oint_a^b \frac{F(y_1)}{(y - y_1)^2} \, dy_1$$
$$= \lim_{\epsilon \to 0} \left\{ \int_a^{y-\epsilon} \frac{F(y_1)}{(y - y_1)^2} \, dy_1 + \int_{y+\epsilon}^b \frac{F(y_1)}{(y - y_1)^2} \, dy_1 - \frac{2F(y)}{\epsilon} \right\}.$$
$$(7\text{--}30)$$

Equation (7–30) always yields a finite result. In fact, if an indefinite integral can be found for the integrand, the answer is obtained simply by inserting the limits $y_1 = a$ and $y_1 = b$, provided any logarithm of $(y - y_1)$ that appears is interpreted as $\ln |y - y_1|$. The validity of Mangler's principal value depends on the condition that the integrand, prior to letting $z \to 0$, be a solution of the two-dimensional Laplace equation. It is clear that this is true, in the present case, of the function that causes the singularity in the y_1-integration of (7–29), since

$$\left(\frac{\partial^2}{\partial y^2} + \frac{\partial^2}{\partial z^2} \right) \left[\frac{z}{(y - y_1)^2 + z^2} \right] = 0. \qquad (7\text{--}31)$$

It is of incidental interest that, if the velocity potential of the lifting flow is constructed by eliminating the operation $\partial/\partial z$ in (7–26), we obtain

$$\varphi(x, y, z) = \int_{-\infty}^x u(x_0, y, z) \, dx_0 = \frac{1}{4\pi} \iint_S \frac{z\gamma(x_1, y_1)}{(y - y_1)^2 + z^2}$$
$$\times \left[1 + \frac{(x - x_1)}{\sqrt{(x - x_1)^2 + (y - y_1)^2 + z^2}} \right] dx_1 \, dy_1.$$
$$(7\text{--}32)$$

Equation (7–32) provides confirmation for (5–35).

The question of exact or approximate solution of (7–28) is deferred to later sections. We note here that, when the angle of attack α and camber ordinates $\bar{h}(x, y)$ of the wing are given and the load distribution is required, (7–28) is a singular double integral equation for the unknown γ. Thus the problem is much more difficult mathematically than the corresponding thickness problem embodied in (7–14)–(7–15). On the other hand, when the loading is given and the shape of the wing to support it desired, the potential and upwash distributions are available by fairly straightforward integrations from (7–31) and (7–28), respectively. Finally, the thickness shape $\bar{g}(x, y)$ to generate a desired *symmetrical* pressure distribution must be determined by solving the rather complicated integral equation which results from z-differentiation of (7–14).

We finish this section with some further discussion of the lift, drag, and nature of the wake. From (5–31), (7–19), and (7–22), we see that the difference in pressure coefficient across the wing is

$$C_{p_l} - C_{p_u} = 2\gamma. \tag{7–33}$$

Using the definition of C_p,

$$p_l - p_u = \rho_\infty U_\infty^2 \gamma. \tag{7–34}$$

Because the surface slopes are everywhere small, this is also essentially the load per unit plan area exerted on the wing in the positive z-direction. Since (7–34) is reminiscent of the Kutta formula (2–157), we note that γ can be interpreted as a circulation. As shown in Fig. 7–3, let the circulation about the positive y-direction be computed around a small rectangular box C of length dx in a chordwise cross section of the wing. Since the contributions of the vertical sides cancel, except for terms of higher order in dx, we find that the circulation around C is

$$U_\infty[1 + u_u]\, dx - U_\infty[1 + u_l]\, dx = U_\infty(u_u - u_l)\, dx = U_\infty \gamma\, dx. \tag{7–35}$$

Hence $U_\infty \gamma(x, y)$ is the spanwise component, per unit chordwise distance, of the circulation bound to the wing in the vicinity of point (x, y).

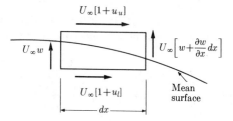

Fig. 7–3. Interpretation of γ by determining circulation around a circuit of length dx along the wing chord.

The lift per unit span at station y on the wing is

$$l(y) = \int_{\text{chord}} [p_l - p_u]\, dx = \rho_\infty U_\infty^2 \int_{\text{chord}} \gamma(x, y)\, dx = \rho_\infty U_\infty \Gamma(y), \tag{7–36}$$

Γ being the total bound circulation. If the wing tips are placed at $y = \pm b/2$, the total lift becomes

$$L = \int_{-b/2}^{b/2} l(y)\, dy = \rho_\infty U_\infty \int_{-b/2}^{b/2} \Gamma(y)\, dy. \tag{7–37}$$

The total pitching moment, pitching moment per unit span about an arbitrary axis, rolling moment, or any other desired quantity related to the loading may be constructed by an appropriate integration of (7–34).

Unlike a two-dimensional airfoil, the finite lifting wing does experience a downstream force (drag due to lift or "induced drag," D_i, sometimes also called "vortex drag," see Chapter 9) in a subsonic inviscid flow. An easy way to compute this resistance is by examining the wake at points remote behind the trailing edge. In fact, if we observe the wing moving at speed U_∞ through the fluid at rest, we note that an amount of mechanical work $D_i U_\infty$ is done on the fluid per unit time. Since the fluid is nondissipative and can store energy in kinetic form only, this work must ultimately show up as the value of T (cf. 2–11) contained in a length U_∞ of the distant wake. The nature of this wake we determine by finding its disturbance velocity potential. Over the wing region S, the φ-discontinuity is calculated, as in (7–26), to be

$$\Delta\varphi(x, y) \equiv \varphi(x, y, 0+) - \varphi(x, y, 0-)$$
$$= \int_{-\infty}^{x} [u_u(x_0, y, 0+) - u_l(x_0, y, 0-)] \, dx_0$$
$$= \int_{x_{\mathrm{LE}(y)}}^{x} \gamma(x_0, y) \, dx_0. \tag{7–38}$$

The last line here follows from the definition of γ, (7–22), the lower limit $-\infty$ being replaced by the coordinate x_{LE} of the local leading edge since there is no u-discontinuity ahead of this point. Beyond the trailing edge on the x, y-plane, $[u_u - u_l] = 0$ in view of the condition of continuity of pressure. Hence

$$\Delta\varphi_{\mathrm{wake}}(y) = \int_{\mathrm{chord}} \gamma(x_0, y) \, dx_0 = \frac{\Gamma(y)}{U_\infty}. \tag{7–39}$$

Since $\Delta\varphi_{\mathrm{wake}}$ is independent of x, the wake must consist of a sheet of trailing vortices parallel to x and having a circulation per unit spanwise distance

$$U_\infty \, \delta(y) = U_\infty \frac{\partial}{\partial y} (\Delta\varphi_{\mathrm{wake}}) = \frac{d\Gamma}{dy}. \tag{7–40}$$

The complete vortex sheet simulating the lifting surface, as seen from above, is sketched in Fig. 7–4. It is not difficult to show that the Kutta-Joukowsky condition of smooth flowoff is equivalent to the requirement that the vortex lines turn smoothly into the stream direction as they pass across the trailing edge. At points far downstream the motion produced by the trailing vortices becomes two-dimensional in x, z-planes (the so-called "Trefftz plane"). Although the wake is assumed to remain flat in accordance with the small-perturbation hypothesis, some rolling up and downward displacement in fact occurs [Spreiter and Sacks (1951)]. This rolling up can be shown to have influences on the loading that are only of

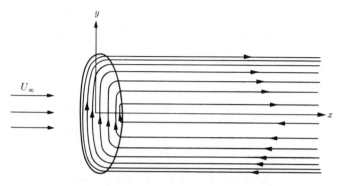

FIG. 7–4. Vortex sheet representing a loaded wing of finite span.

third order in θ and α. For the plane wake, we use (2–11) and the equality of work and kinetic-energy increment to obtain

$$D_i = -\frac{\rho_\infty U_\infty^2}{2} \iint_{S_{\text{wake}}} \varphi \frac{\partial \varphi}{\partial n} \, dS, \tag{7–41}$$

where S_{wake} comprises the upper and lower surfaces of unit length of the wake, as seen by an observer at rest in the fluid. Clearly, $dS = dy$, while

$$\frac{\partial \varphi}{\partial n} = \pm \varphi_z = \mp \frac{1}{2\pi U_\infty} \oint_{-b/2}^{b/2} \frac{d\Gamma}{dy_1} \frac{dy_1}{(y - y_1)} \tag{7–42}$$

from the properties of infinite vortex lines. Using (7–42) and (7–39) in (7–41),

$$D_i = -\frac{\rho_\infty U_\infty^2}{2} \int_{-b/2}^{b/2} \Delta\varphi_{\text{wake}}(y)\varphi_z(y) \, dy$$

$$= \frac{\rho_\infty}{4\pi} \oint_{-b/2}^{b/2} \oint_{-b/2}^{b/2} \Gamma(y) \frac{d\Gamma}{dy_1} \frac{dy_1 \, dy}{(y - y_1)} . \tag{7–43}$$

A more symmetrical form of (7–43) can be constructed by partial integration with respect to y. In the process, we make use of the Cauchy principal value operation at $y = y_1$ and note the fact that $\Gamma(\pm b/2) = 0$ in view of the continuous dropping of load to zero at the wing tips.

$$D_i = -\frac{\rho_\infty}{4\pi} \int_{-b/2}^{b/2} \int_{-b/2}^{b/2} \frac{d\Gamma}{dy} \frac{d\Gamma}{dy_1} \ln |y - y_1| \, dy_1 \, dy. \tag{7–44}$$

It is interesting that (7–37) and (7–44) imply the well-known result that minimum induced drag for a given lift is achieved, independently of the details of camber and planform shape, by elliptic spanwise load distribution

$$\Gamma = \Gamma(0)\sqrt{1 - y^2/(b/2)^2}. \tag{7–45a}$$

Thus, if we represent the circulation as a Fourier sine series

$$\Gamma = U_\infty b \sum_{n=1} A_n \sin n\theta, \tag{7–45b}$$

where

$$\frac{b}{2}\cos\theta = y, \tag{7–45c}$$

we find from (7–37)

$$L = \frac{\pi}{4}\rho_\infty U_\infty^2 b^2 A_1'. \tag{7–46}$$

However, (7–43) or (7–44) yields

$$D_i = \pi \frac{\rho_\infty U_\infty^2 b^2}{8} \sum_{n=1} n A_n^2. \tag{7–47}$$

Accordingly, the higher harmonics of the Fourier series add drag without increasing total lift. Hence an optimum is reached when only the A_1-term, corresponding to (7–45a), is present. One can then prove, for instance, that

$$C_{D_i} = \frac{C_L^2}{\pi A}, \tag{7–48}$$

where $A = b^2/S$.

7–4 Lifting-Line Theory

The first rational attempt at predicting loads on subsonic, three-dimensional wings was a method due to Prandtl and his collaborators, which was especially adapted in an approximate way to the large aspect-ratio, unswept planforms prevalent during the early twentieth century. Although our approach is not the classical one, we wish to demonstrate here how the lifting-line approximation follows naturally from an application of the method of matched asymptotic expansions. Let us consider a wing of the sort pictured in Fig. 7–4, and introduce a *second* small parameter ϵ_A, which is inversely proportional to the aspect ratio. Then if we write the local chord as

$$c(y) = \epsilon_A \bar{c}(y) \tag{7–49}$$

and examine the matched inner and outer solutions associated with the process $\epsilon_A \to 0$ at fixed span b, we shall be generating a consistent high aspect-ratio theory. As far as the thickness ratio, angle of attack, and camber are concerned, we assert that they are small at the outset and remain unchanged as $\epsilon_A \to 0$. Our starting point is therefore the problem embodied in (5-29)-(5-30), with $M = 0$; we study the consequences of superimposing a second limit on the situation which they describe.

Placing the wing as close as possible to the y-axis in the x, y-plane, we define new independent variables

$$\bar{x} = \frac{x}{\epsilon_A}, \qquad \bar{z} = \frac{z}{\epsilon_A} \tag{7-50}$$

to be held constant when performing the inner limit. The inner and outer series read

$$\varphi^i = \varphi_0^i(\bar{x}, y, \bar{z}) + \epsilon_A \varphi_1^i(\bar{x}, y, \bar{z}) + \cdots, \tag{7-51a}$$

$$\varphi^o = \varphi_0^o(x, y, z) + \epsilon_A \varphi_1^o(x, y, z) + \cdots \tag{7-51b}$$

[The significance of "zeroth-order" is, of course, different here from (5-6) and (5-10).] When ϵ_A vanishes, the entire bound vortex system of the wing (as seen by an "outer observer") contracts to a concentrated line between the wing tips $y = \pm b/2$ along the y-axis. There still remains a streamwise wake of circulation $d\Gamma/dy$ per unit width. According to (7-32) or (5-33), the velocity potential for this situation is

$$\varphi_1^o = \frac{1}{4\pi U_\infty} \int_{-b/2}^{b/2} \frac{\Gamma(y_1)z}{(y - y_1)^2 + z^2} \left[1 + \frac{x}{\sqrt{x^2 + (y - y_1)^2 + z^2}} \right] dy_1. \tag{7-52}$$

That this is indeed a first-order result will be confirmed by a later matching process.

To find the zeroth-order inner solution, we insert (7-50) into (5-29) and let $\epsilon_A \to 0$, whereupon

$$\frac{\partial^2 \varphi_0^i}{\partial \bar{x}^2} + \frac{\partial^2 \varphi_0^i}{\partial \bar{z}^2} = 0. \tag{7-53}$$

Since the error in (7-53) is $O(\epsilon_A^2)$, it is also true that

$$\frac{\partial^2 \varphi_1^i}{\partial \bar{x}^2} + \frac{\partial^2 \varphi_1^i}{\partial \bar{z}^2} = 0. \tag{7-54}$$

Equation (5-30) is a suitable statement of the inner boundary conditions and can be written

$$\frac{\partial \varphi^i}{\partial \bar{z}} = \epsilon_A \bar{F}_u(\bar{x}, y) \quad \text{at} \quad \bar{z} = 0+ \tag{7-55a}$$

$$\frac{\partial \varphi^i}{\partial \bar{z}} = \epsilon_A \bar{F}_l(\bar{x}, y) \quad \text{at} \quad \bar{z} = 0- \tag{7-55b}$$

for $(\bar{x}, y)$ on $\bar{S}$.

Here $\bar{S}$ is a planform projection whose chordwise dimensions remain finite as $\epsilon_A \to 0$. Inserting (7–51a) into (7–55), and equating corresponding orders in ϵ_A, we find

$$\frac{\partial \varphi_0^i}{\partial \bar{z}} = 0,$$

whence

$$\varphi_0^i = 0.$$

Also

$$\frac{\partial \varphi_1^i}{\partial \bar{z}} = \bar{F}_u(\bar{x}, y) \quad \text{at} \quad \bar{z} = 0+ \tag{7–56a}$$

$$\frac{\partial \varphi_1^i}{\partial \bar{z}} = \bar{F}_l(\bar{x}, y) \quad \text{at} \quad \bar{z} = 0- \tag{7–56b}$$

for $(\bar{x}, y)$ on $\bar{S}$.

A possible solution to (7–54) and (7–56) is the two-dimensional potential for a thin airfoil, which can be expressed as

$$\varphi_1^i = \frac{1}{2\pi} \int_{\text{chord}} \gamma_2(\bar{x}, y) \tan^{-1}\left(\frac{\bar{x} - \bar{x}_1}{\bar{z}}\right) d\bar{x}_1; \tag{7–57}$$

γ_2 can be determined by the solution process elaborated in Section 5–3; the correctness of (7–57) must be established *a posteriori*.

Proceeding to higher order in φ^i we find, in consequence of the vanishing of φ_0^i,

$$\frac{\partial^2 \varphi_2^i}{\partial \bar{x}^2} + \frac{\partial^2 \varphi_2^i}{\partial \bar{z}^2} = 0 \tag{7–58}$$

with

$$\frac{\partial \varphi_2^i}{\partial \bar{z}}(\bar{x}, y, 0) = 0, \quad \text{on} \quad \bar{S}. \tag{7–59}$$

With the matching process in view, we note that a possible solution of (7–58)–(7–59), with the necessary antisymmetry in $\bar{z}$, is

$$\varphi_2^i = \frac{1}{2\pi} \int_{\text{chord}} \Delta\gamma(\bar{x}_1, y) \tan^{-1}\left(\frac{\bar{x} - \bar{x}_1}{\bar{z}}\right) d\bar{x}_1 + \bar{z}f(y), \tag{7–60}$$

where $\Delta\gamma$ is to be found in terms of $f(y)$ by means of the boundary condition (7–59).

We now examine the matching. At the inner limit of the outer region $(x^2 + z^2) \to 0$, while at the outer limit of the inner region $(\bar{x}^2 + \bar{z}^2) \to \infty$. If correct, it must be possible to match the solutions along any ray $x/z = \bar{x}/\bar{z} = \text{const}$. We choose for convenience the ray $x/z = \bar{x}/\bar{z} = 0$ and assert that it can be proved that a similar matching along any other direction will yield the same results. We match the dimensionless velocity

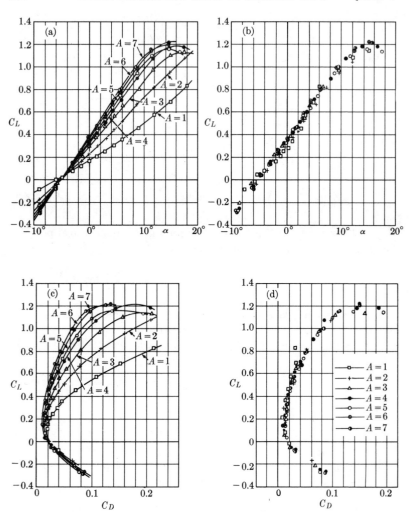

FIG. 7–5. Coefficients of lift plotted vs. angle of attack α and total drag coefficient C_D for several rectangular wings (values of A indicated on the figures). In parts (b) and (d), the same data are adjusted to a reference $A = 5$ by formulas based on lifting-line theory and elliptic loading. [Adapted from Prandtl, Wieselsberger, and Betz (1921).]

component w in the y, z-plane. With $x = 0$, after an integration by parts, (7–52) yields

$$w(0, y, z) = \epsilon_A \frac{\partial \varphi_1^o}{\partial z} = -\frac{\epsilon_A}{4\pi U_\infty} \int_{-b/2}^{b/2} \frac{(y - y_1) \, d\Gamma/dy_1}{(y - y_1)^2 + z^2} \, dy_1. \quad (7\text{–}61)$$

From the inner solution

$$w(0, y, \bar{z}) = \frac{1}{\epsilon_A} \frac{\partial \varphi^i}{\partial \bar{z}}$$

$$= \frac{1}{2\pi} \int_{\text{chord}} \frac{\bar{x}_1 \gamma_2(\bar{x}_1, y)}{\bar{x}_1^2 + \bar{z}^2} \, d\bar{x}_1 + \epsilon_A \left[\frac{1}{2\pi} \int_{\text{chord}} \frac{\bar{x}_1 \, \Delta\gamma(\bar{x}_1, y)}{\bar{x}_1^2 + \bar{z}^2} \, d\bar{x}_1 + f(y) \right]. \quad (7\text{–}62)$$

Comparing (7–61) and (7–62), we observe that both chordwise integrals in the latter vanish as $\bar{z} \to \infty$, whereupon the matching process yields

$$f(y) = -\frac{1}{4\pi U_\infty} \oint_{-b/2}^{b/2} \frac{d\Gamma}{dy_1} \frac{dy_1}{(y - y_1)}. \quad (7\text{–}63)$$

Finally,

$$\varphi_2^i = \frac{1}{2\pi} \int_{\text{chord}} \Delta\gamma(\bar{x}_1, y) \tan^{-1}\left(\frac{\bar{x} - \bar{x}_1}{\bar{z}} \right) d\bar{x}_1$$

$$- \frac{\bar{z}}{4\pi U_\infty} \oint_{-b/2}^{b/2} \frac{d\Gamma}{dy_1} \frac{dy_1}{(y - y_1)}, \quad (7\text{–}64)$$

and from (7–59),

$$\frac{U_\infty}{2\pi} \oint_{\text{chord}} \frac{\Delta\gamma(\bar{x}_1, y)}{(\bar{x} - \bar{x}_1)} \, d\bar{x}_1 = -\frac{1}{4\pi} \oint_{-b/2}^{b/2} \frac{d\Gamma}{dy_1} \frac{dy_1}{(y - y_1)} \equiv U_\infty \alpha_i(y) \quad (7\text{–}65)$$

Of course, Γ is the *total* bound circulation

$$\Gamma(y) = U_\infty \int_{\text{chord}} [\gamma_2(x, y) + \epsilon_A \, \Delta\gamma(x, y)] \, dx. \quad (7\text{–}66)$$

The foregoing results may be interpreted physically as follows: to first order in ϵ_A, the influence of finite span on the flow over a wing is to *reduce* the local incidence at any station y by an "induced angle of attack":

$$\alpha_i(y) = -\frac{1}{4\pi U_\infty} \oint_{-b/2}^{b/2} \frac{d\Gamma}{dy_1} \frac{dy_1}{(y - y_1)}. \quad (7\text{–}67)$$

For fixed y, the chordwise load distribution, etc., may be calculated on a two-dimensional basis if the geometrical incidence $[\alpha - \theta \, \partial\bar{h}/\partial x]$ is adjusted by the amount α_i. Since α_i is independent of x, this is equivalent to reducing only the flat-plate portion of the loading (cf. 5–83).

If one requires only the spanwise lift distribution $l = \rho_\infty U_\infty \Gamma$, a suitable equation can be derived by equating $\rho_\infty U_\infty \Gamma$ to a second expression for lift,

$$l(y) = 2\pi \frac{\rho_\infty}{2} U_\infty^2 c[\alpha_{\mathrm{ZL}} + \alpha_i], \qquad (7\text{–}68)$$

where $\alpha_{\mathrm{ZL}}(y)$ is the geometrical angle of attack at station y, measured from the zero-lift attitude. The prediction that the lift-curve slope of any two-dimensional airfoil is 2π has here been assumed, although in Prandtl's original development of (7–68) he replaced 2π by an experimentally determined value.

If we insert (7–67) into (7–68), we are led to the following integro-differential equation for Γ,

$$\Gamma = \pi U_\infty c \left[\alpha_{\mathrm{ZL}} - \frac{1}{4\pi U_\infty} \oint_{-b/2}^{b/2} \frac{d\Gamma}{dy_1} \frac{dy_1}{(y - y_1)} \right]. \qquad (7\text{–}69)$$

The substitutions (7–45b,c) result in the algebraic relation

$$\sum_{n=1} \left[1 + \frac{n\pi c}{2b \sin \theta} \right] A_n \sin n\theta = \pi \frac{c(\theta)}{b} \alpha_{\mathrm{ZL}}(\theta). \qquad (7\text{–}70)$$

This is usually solved by collocation at a number of spanwise stations equal to the number of constants A_n needed for satisfactory convergence.

Van Dyke (1963) has published a very thorough investigation of lifting-line theory as a singular perturbation problem. He has succeeded in replacing (7–69) by an explicit integral representation of the circulation and examined the influence of wing-tip shape on the validity of the results. It is also worth mentioning that a "lifting-line" procedure for swept wings can be derived by a somewhat more involved application of the foregoing procedure.

Perhaps the finest fundamental investigation of the validity of the lifting-line idealization remains that conducted by the Göttingen group [Prandtl, Wieselsberger, and Betz (1921), Section IV]. Their results, a few of which are summarized in Fig. 7–5, involve using formulas based on assumed elliptic loading to adjust measured lifts and induced drags to corresponding values on a reference wing with $A = 5$. By this means, the influences on the correlation due to profile drag and to deviations from 2π in the sectional lift-curve slopes are minimized. The figure reproduces curves of C_L vs. α and C_L vs. C_D as measured on a series of rectangular lifting surfaces, having aspect ratios between 1 and 7. In parts (b) and (c), the values adjusted to the effective $A = 5$ are seen to fall together within the experimental accuracy.

7–5 More Refined Theories of Lifting-Line Type

In view of the great mathematical simplification and the primary concern with spanwise load distribution that arises in connection with large aspect-ratio wings, several schemes have been proposed for reducing (7–28) to a single integral equation resembling (7–69). No other of these possesses the element of consistent rationality demonstrated in Section 7–4, but it is of interest to see how some of them are arrived at by approximation. The procedures which follow are suggested by the work of Reissner (1949).

Let us start with the lead term of (7–28b) and carry out the chordwise integration first.

$$\oiint_{S} \frac{\partial \gamma}{\partial y_1} \frac{dx_1 \, dy_1}{(y - y_1)} = \oint_{-b/2}^{b/2} \frac{1}{(y - y_1)} \left[\int_{\text{chord}} \frac{\partial \gamma}{\partial y_1} \, dx_1 \right] dy_1$$

$$= \oint_{-b/2}^{b/2} \frac{1}{(y - y_1)} \left[\frac{d}{dy_1} \int_{\text{chord}} \gamma \, dx_1 \right] dy_1$$

$$= \frac{1}{U_\infty} \oint_{-b/2}^{b/2} \frac{d\Gamma}{dy_1} \frac{dy_1}{(y - y_1)} . \tag{7–71}$$

Although the limits of chordwise integration may depend on y_1, we justify the interchange of integration and differentiation here by carrying our integration area an infinitesimal distance beyond the leading edge and observing that γ then vanishes at both limits. Thus we obtain the (still "exact") integral equation

$$w(x, y, 0) = - \frac{1}{4\pi U_\infty} \oint_{-b/2}^{b/2} \frac{d\Gamma}{dy_1} \frac{dy_1}{(y - y_1)}$$

$$- \frac{1}{4\pi} \oiint_{S} \frac{\partial \gamma}{\partial y_1} \frac{\sqrt{(x - x_1)^2 + (y - y_1)^2}}{(x - x_1)(y - y_1)} \, dx_1 \, dy_1. \tag{7–72}$$

We now examine the consequences of approximating the coefficient of $\partial \gamma / \partial y_1$ in various ways.

First, for large ratios b/c, we argue that $(y - y_1)^2 \gg (x - x_1)^2$ over most of the planform so that

$$\frac{\sqrt{(x - x_1)^2 + (y - y_1)^2}}{(x - x_1)(y - y_1)} = \frac{|y - y_1|}{(x - x_1)(y - y_1)} \left\{ 1 + O\left(\frac{(x - x_1)^2}{(y - y_1)^2} \right) \right\} . \tag{7–73}$$

Moreover, when we are integrating spanwise through the singularity at $y = y_1$,

$$\frac{\sqrt{(x - x_1)^2 + (y - y_1)^2}}{(x - x_1)(y - y_1)} \simeq \frac{|x - x_1|}{(x - x_1)(y - y_1)} \tag{7-74}$$

is antisymmetric near $y = y_1$, so that this contribution to the Cauchy integral tends to be self-canceling. Neglecting the error term from (7–73), we obtain

$$\oiint_S \frac{\partial\gamma}{\partial y_1} \frac{\sqrt{(x - x_1)^2 + (y - y_1)^2}}{(x - x_1)(y - y_1)} \, dx_1 \, dy_1$$

$$\simeq \oint_{\text{chord}} \frac{1}{(x - x_1)} \left[\int_{\substack{\text{local span} \\ \text{at } x}} \frac{|y - y_1|}{(y - y_1)} \frac{\partial\gamma}{\partial y_1} \, dy_1\right] dx_1$$

$$= \oint_{\text{chord}} \left[\int_{\text{left edge}}^y \frac{\partial\gamma}{\partial y_1} \, dy_1 - \int_y^{\text{right edge}} \frac{\partial\gamma}{\partial y_1} \, dy_1\right] \frac{dx_1}{(x - x_1)}$$

$$= 2\oint_{\text{chord}} \frac{\gamma(x_1, y)}{x - x_1} \, dx_1. \tag{7-75}$$

The integral across the chord in the last member is evidently to be taken at station y, since $\gamma(x_1, y) = 0$ ahead of the leading and behind the trailing edges. If (7–75) is inserted into (7–72), we recover lifting-line theory. In fact, it is not difficult to prove that a further multiplication* by $\sqrt{x/(c - x)}$ and integration with respect to x yields precisely (7–69).

Next, let us investigate the improved approximation of assigning $|x - x_1|$ a·rough average value $\frac{1}{2}c(y)$. That is, we write

$$\frac{\sqrt{(x - x_1)^2 + (y - y_1)^2}}{(x - x_1)(y - y_1)} \simeq \frac{\sqrt{[\frac{1}{2}c(y)]^2 + (y - y_1)^2}}{(x - x_1)(y - y_1)}. \tag{7-76}$$

We substitute (7–76) into (7–72), make a temporary local chordwise shift so that the leading and trailing edges are $x = 0$ and $c(y)$ at station y, multiply by $\sqrt{x/(c - x)}$ and integrate chordwise with respect to x. Thus we obtain

$$\pi\alpha_{\text{ZL}}(y) = \frac{1}{4U_\infty} \oint_{-b/2}^{b/2} \frac{d\Gamma}{dy_1} \frac{dy_1}{(y - y_1)}$$

$$+ \frac{1}{2U_\infty c(y)} \oint_{-b/2}^{b/2} \frac{d\Gamma}{dy_1} \frac{\sqrt{[\frac{1}{2}c(y)]^2 + (y - y_1)^2}}{(y - y_1)} \, dy_1. \tag{7-77}$$

---

* This assumes the leading and trailing edges have been shifted to $x = 0$ and $c(y)$.

In (7–77), we have identified the two-dimensional angle from zero lift α_{ZL} according to the relation

$$\frac{\pi}{2} c\alpha_{ZL} = -\int_0^c w(x, y, 0) \sqrt{\frac{x}{c - x}} \, dx, \qquad (7\text{–}78)$$

which is easily obtained from (5–84) by noting that

$$\Gamma = \pi U_\infty c\alpha_{ZL}.$$

Equation (7–77) is essentially the integro-differential equation of Weissinger's "L-Method" [Weissinger (1947)]. That author originally developed it by placing the loaded line of circulation $\Gamma(y)$ along the quarter-chord line (assumed unswept) and requiring the bound plus trailing vortex system to induce a flow parallel to the zero-lift line at the three-quarter chord point of each station y. His assumptions were justified by the well-known fact that putting the bound vortex at $c/4$ and satisfying the boundary condition at $3c/4$ yields the correct circulation for a flat plate or parabolic camber line in two-dimensional flow. For purposes of numerical solution, the singularity is eliminated from the second integral of (7–77) by adding and subtracting $c(y)/2$ in the numerator, whence

$$\pi\alpha_{ZL}(y) = \frac{1}{2U_\infty} \oint_{-b/2}^{b/2} \frac{d\Gamma}{dy_1} \frac{dy_1}{(y - y_1)}$$

$$+ \frac{1}{2U_\infty c(y)} \int_{-b/2}^{b/2} L\left(\frac{2}{c(y)}(y - y_1)\right) \frac{d\Gamma}{dy_1} \, dy_1, \qquad (7\text{–}79a)$$

where

$$L(q) \equiv \frac{\sqrt{1 + q^2} - 1}{q}. \qquad (7\text{–}79b)$$

For swept wings, Weissinger (1947) proposed following an exactly similar procedure, except that he employed loaded lines, inclined at angles $\pm\Lambda$ to the y-direction, to represent the right and left halves of the V-shaped planform. In this case, the function L of (7–79) is replaced by a much more complicated function of y, y_1, sweep angle, and local aspect ratio. The Weissinger techniques have been very successful for predicting spanwise load distribution on straight, swept-back, and swept-forward wings of moderate to large aspect ratio. Numerical solution of the integral equation is usually carried out by collocation, using the same Fourier-series substitution discussed in Section 7–4 but evaluating the nonsingular integral by quadrature. Most common in practice is an ingenious variant of the Fourier series [Multhopp (1938)], wherein the unknown coefficients become precisely the values of Γ at a preassigned set of stations across the wingspan. Many tables have been published to facilitate these calculations.

Reissner (1949) has suggested generalizing the technique which led to (7–77) as follows: He operates on (7–72) by applying various functions of x as weighting factors and integrating with respect to x across the chord. Simultaneously, in the nonsingular part of the kernel function, he approximates $\gamma(x_1, y_1)$ with a series of terms which contain suitable functions of x multiplied by the running lift $l(y)$, the pitching moment per unit span, and possible higher moments of the chordwise load distribution. The weighting factors for (7–72) are so chosen that the singular terms directly yield quantities proportional to $l(y)$, pitching moment, etc. Thus are obtained a set of simultaneous integrodifferential equations in the single independent variable y, the number being equal to the number of chordwise loading moments desired. Numerical solution can proceed by a generalization of the Fourier-series method for Γ, and all quadratures that must be done involve only nonsingular integrands.

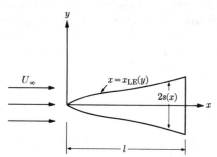

Fig. 7–6. Low-aspect-ratio wing in incompressible flow.

Quite a different adaptation of the lifting-line idea, but one worthy of mention here, is to focus primarily on the *chordwise* load distribution of an almost-plane wing with very low aspect ratio. For instance, let us consider a planform of the type illustrated in Fig. 7–6, with its trailing edge cut off normal to the flight direction.

Equation (7–72) may be written

$$w(x, y, 0) = -\frac{1}{4\pi} \oint_0^l \oint_{-s(x_1)}^{s(x_1)} \frac{\partial \gamma}{\partial y_1}$$
$$\times \left[\frac{1}{y - y_1} + \frac{\sqrt{(x - x_1)^2 + (y - y_1)^2}}{(x - x_1)(y - y_1)} \right] dx_1 \, dy_1. \qquad (7\text{–}80)$$

For reasons analogous to those which produced (7–75), we adopt the approximation

$$\sqrt{(x - x_1)^2 + (y - y_1)^2} \cong |x - x_1|. \qquad (7\text{–}81)$$

Noting that

$$\left[1 + \frac{|x - x_1|}{(x - x_1)} \right] = \begin{cases} 2, & x > x_1 \\ 0, & x < x_1, \end{cases} \qquad (7\text{–}82)$$

we manipulate the integral as follows:

$$w(x, y, 0) \cong -\frac{2}{4\pi} \int_0^x \oint_{-s(x_1)}^{s(x_1)} \frac{\partial\gamma}{\partial y_1} \frac{dx_1 dy_1}{(y - y_1)}$$

$$= -\frac{1}{2\pi} \oint_{-s(x)}^{s(x)} \frac{1}{(y - y_1)} \left[\int_{x_{\text{LE}}(y_1)}^x \frac{\partial\gamma}{\partial y_1} \, dx_1 \right] dy_1$$

$$= -\frac{1}{2\pi} \oint_{-s(x)}^{s(x)} \frac{1}{(y - y_1)} \frac{\partial}{\partial y_1} \int_{x_{\text{LE}}(y_1)}^x \gamma \, dx_1 \, dy_1$$

$$= -\frac{1}{2\pi} \oint_{-s(x)}^{s(x)} \frac{\partial}{\partial y_1} [\Delta\varphi(x, y_1)] \frac{dy_1}{(y - y_1)}. \qquad (7\text{–}83)$$

Here $x_{\text{LE}}(y)$ is the leading-edge coordinate defined in the figure, and we have interchanged differentiation and integration by allowing the integration area to extend a short distance ahead of the leading edge. No velocity-potential discontinuity is permissible ahead of this edge, so

$$\int_{x_{\text{LE}}(y_1)}^x \gamma(x_1, y_1) \, dx_1 = \int_{x_{\text{LE}}(y_1)}^x [u_u - u_l] \, dx_1 = \Delta\varphi(x, y_1). \qquad (7\text{–}84)$$

Clearly (7–83) is an integral equation for two-dimensional, antisymmetrical flow in planes normal to the flight direction. Thus, for wings in subsonic flight, the results of Sections 6–4 and 6–5 have been arrived at from another direction. It is of interest that (7–83) can be inverted to yield the velocity potential of the slender wing in terms of the known distribution of surface slope

$$\frac{\partial \Delta\varphi(x, y)}{\partial y} = \frac{2}{\pi\sqrt{s^2(x) - y^2}} \oint_{-s(x)}^{s(x)} \frac{w(x, y_1, 0)\sqrt{s^2(x) - y_1^2}}{(y - y_1)} \, dy_1. \qquad (7\text{–}85)$$

A condition of zero net circulation has been employed to make this solution unique. One important special case is that of no "spanwise camber," when w is independent of y. We are then led to

$$\frac{\partial \Delta\varphi}{\partial y} = \frac{2y}{\sqrt{s^2(x) - y^2}} w(x, 0), \qquad (7\text{–}86)$$

$$\Delta\varphi(x, y) = -2\sqrt{s^2(x) - y^2} \, w(x, 0). \qquad (7\text{–}87)$$

At the trailing edge,

$$\Delta\varphi(l, y) = \frac{\Gamma(y)}{U_\infty} = -2\sqrt{s^2(l) - y^2} \, w(l, 0), \qquad (7\text{–}88)$$

which will be recognized as a case of elliptic load distribution. Among other interesting results, we can replace $w(l, 0)$ by $(-\alpha)$ in (7–88) and calculate that the lift-curve slope for any such wing is $(\pi/2)A$, as was also derived in Section 6–7.

A "chordwise lifting-line theory" of the foregoing type but which applies higher up on the aspect-ratio scale has been put forth by Lawrence (1951). His scheme consists of integrating the exact equation (7–28b) spanwise, after applying the weighting factor $\sqrt{s^2(x) - y^2}$. He then makes the approximation

$$\sqrt{(x - x_1)^2 + (y - y_1)^2} \cong \tfrac{1}{2}[|x - x_1| + \sqrt{(x - x_1)^2 + s^2(x)}]$$

$$(7\text{–}89)$$

and thus arrives at a relatively tractable single integral equation in x.

7–6 Theories of Lifting-Surface Type

Numerous attempts have been published at devising an approximate solution for (7–28) which provides information on both spanwise and chordwise load distributions. These have involved multiple lifting lines, higher moments of the chordwise loading [e.g., Reissner (1949)], and various discrete-vortex schemes. Perhaps the best known and most widely used of the earlier techniques was that due to Falkner (1948), wherein a pattern of concentrated horseshoe vortices is sprinkled over the planform and their strengths are computed by applying the tangential-flow condition at a number of (carefully selected) stations equal to the number of undetermined circulation values.

In the era of the high-speed digital computer, there is an element of irrationality in clinging to any questionably consistent *approximation* that embodies the amount of numerical labor inherent in all of the traditional lifting-surface theories. One would prefer to make a direct attack on (7–28a) itself, employing a method of solution which converges uniformly to an exact result as greater refinement (i.e., larger numbers of unknowns) is introduced. Such are the approaches of Multhopp (1950) and Watkins *et al.* (1955, 1959, and others). We outline here the latter. The two have strong similarities, but Watkins' work has been programmed for widely available American computers and includes extensions for both compressible and unsteady flow (small simple harmonic wing oscillations, to be exact). Indeed, all the principal mathematical difficulties of these more general cases are present in the steady, incompressible reduction because the singularities in $(x - x_1)$ and $(y - y_1)$ remain unchanged. Hence our brief discussion is easily broadened to cover time-dependent, subsonic flow by introducing a more complicated "kernel function," or coefficient of the unknown $\gamma(x_1, y_1)$ in the integral equation.

Let (7–28a) first be written

$$w(x, y, 0) = \frac{1}{4\pi} \oint_{-b/2}^{b/2} \int_{x_{\text{LE}}(y)}^{x_{\text{TE}}(y)} \gamma(x_1, y_1) K \, dx_1 \, dy_1, \qquad (7\text{–}90)$$

where

$$K(x - x_1, y - y_1) \equiv \frac{1}{(y - y_1)^2}\left[1 + \frac{x - x_1}{\sqrt{(x - x_1)^2 + (y - y_1)^2}}\right]. \qquad (7\text{–}91)$$

We now define the dimensionless spanwise and chordwise variables, η and θ, according to

$$
\begin{aligned}
(y, y_1) &\equiv \frac{b}{2}\,(\eta, \eta_1) \\
x &\equiv \tfrac{1}{2}[x_{\text{TE}}(y) + x_{\text{LE}}(y)] - \tfrac{1}{2}[x_{\text{TE}}(y) - x_{\text{LE}}(y)]\cos\theta \\
x_1 &\equiv \tfrac{1}{2}[x_{\text{TE}}(y_1) + x_{\text{LE}}(y_1)] - \tfrac{1}{2}c(y_1)\cos\theta_1.
\end{aligned}
\qquad (7\text{–}92)
$$

Equations (7–92) have the effect of putting the wing tips at $\eta = \pm 1$, whereas the leading and trailing edges are plotted, respectively, along the lines $\theta = 0$ and π. The area element transforms thus:

$$dx_1 \, dy_1 = \frac{b}{4}\,c(\eta_1)\sin\theta_1 \, d\theta_1 \, d\eta_1. \qquad (7\text{–}93)$$

Substituting into (7–90)–(7–91) and extracting the singularity from the kernel function, we are led to the integral equation

$$w(\theta, \eta, 0) = \frac{1}{4\pi} \oint_{-1}^{1} \int_{0}^{\pi} \gamma(\theta_1, \eta_1)\,\frac{\overline{K}}{(\eta - \eta_1)^2}\,\frac{c(\eta_1)}{b}\sin\theta_1 \, d\theta_1 \, d\eta_1, \qquad (7\text{–}94)$$

where

$$\overline{K} = 1 + \frac{(2/b)\,(x - x_1)}{\sqrt{[(2/b)(x - x_1)]^2 + (\eta - \eta_1)^2}}. \qquad (7\text{–}95)$$

For brevity, $\overline{K}$ is expressed here directly in terms of the dimensionless chordwise distance $(2/b)(x - x_1)$; the latter is, of course, known when θ, θ_1, and the wing geometry have been specified.

Determination of load distribution can be reduced to a problem of numerical integrations and matrix inversion by choosing an appropriate series representation for the unknown. The form suggested in Watkins *et al.* (1959) is

$$\gamma(\theta_1, \eta_1) = \frac{4\pi b}{c(\eta_1)} \left\{ \sqrt{1 - \eta_1^2} \cot \frac{\theta_1}{2} \sum_{m=0} a_{0m} \eta_1^m \right.$$
$$\left. + \sqrt{1 - \eta_1^2} \sum_{m=0} \sum_{n=1} \frac{4a_{nm}}{2^{2n}} \eta_1^m \sin n\theta_1 \right\}. \qquad (7\text{-}96)$$

Equation (7-96) embodies the well-known series for chordwise loading, adopted from subsonic thin-airfoil theory, and a spanwise-load series closely related to (7-45b) for the lifting line. Thus the term $\sqrt{1 - \eta_1^2}$ contains the aforementioned elliptic distribution and shows the proper infinite slope at each wing tip. Here $\cot \theta_1/2$ is the two-dimensional, flat-plate distribution and has the necessary leading-edge singularity [cf. (5-83)]. All θ_1-terms satisfy the Kutta-Joukowsky condition.

For purposes of abbreviation, let us write

$$l_n(\theta_1) = \begin{cases} \cot \dfrac{\theta_1}{2}, & n = 0 \\[2mm] \dfrac{4}{2^{2n}} \sin n\theta_1, & n \geq 1. \end{cases} \qquad (7\text{-}97)$$

Then the substitution of (7-96)-(7-97) into (7-94) leads to the following integration problem:

$$w(\theta, \eta, 0) = \sum_{n=0} \sum_{m=0} a_{nm} \fint_{-1}^{1} \frac{\eta_1^m \sqrt{1 - \eta_1^2}}{(\eta - \eta_1)^2} \left\{ \int_0^{\pi} l_n \overline{K} \sin \theta_1 \, d\theta_1 \right\} d\eta_1. \qquad (7\text{-}98)$$

After the integrals here have been numerically evaluated for a number of (θ, η) combinations equal to the desired number of a_{nm} coefficients, the solution becomes a matter of algebra. Lift, pitching moment, induced drag, etc. are easily evaluated by appropriate operations on (7-96).

Watkins *et al.* (1959) go into considerable, necessary detail on the matter of programming the two numerical integrations called for in (7-98), but it does not seem worthwhile to provide a full account here. It should be noted that, for a given choice of n, the inner θ_1-integration contains no singularity. Care must be taken, however, because of the fact that when $\eta = \eta_1$, the value of $\overline{K}$ jumps from 0 to 2 as x_1 passes from greater to less than x. The outer η_1-integration is carried out according to Mangler's formula (7-30). In view of the rapid variations in the vicinity of $\eta = \eta_1$, Watkins *et al.* found it necessary to break the spanwise range into four distinct regions, the greatest concern being with a small region centered on $\eta = \eta_1$. Special steps must be taken for swept wings with pointed vertices and also for the vicinity of the integrable infinity of $l_0(\theta_1)$ at the leading edge $\theta_1 = 0$.

Sample computations by the foregoing procedure, for steady and unsteady flow, will be found in Watkins *et al.* (1959) and elsewhere [e.g., Hsu (1958), Cunningham and Woolston (1958)]. An ingenious refinement of the technique of solution, by which one is able to avoid completely the singularities of the kernel function at the price of working with a fixed set of stations over the planform, has been published by Hsu (1958).

As an example, we show in Fig. 7–7 an application of the kernel-function procedure to the incompressible flow around an uncambered delta wing of aspect ratio $A = 2.5$ at steady angle of attack. In the calculations, 9 loading modes were used (3 chordwise and 3 spanwise), and the upwash was satisfied approximately in the least-square sense at 16 control points.

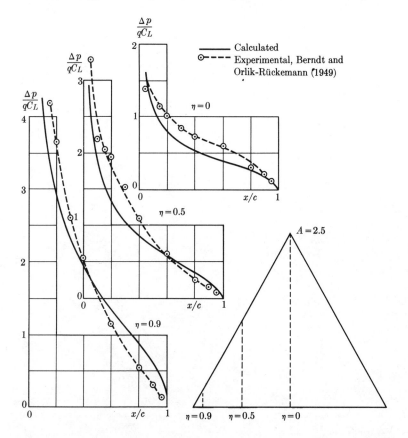

FIG. 7–7. Calculated and experimental lifting pressure distribution over a delta wing of $A = 2.5$. (Numerical lifting-surface theory, courtesy of Dr. Sheila Widnall.)

The agreement with the experimental lifting pressure distribution obtained by Berndt and Orlik-Rückemann (1949) is satisfactory but not as good as that for the total lift (cf. Fig. 6–10). This is mainly due to the fact that the thin-wing solution is basically an outer solution valid away from the neighborhood of the wing surface. Therefore, as application of the momentum theorem makes clear, the theory will give correct values of total lift and moment to first order in angle of attack whereas the lifting pressure distribution will contain first-order terms due to thickness which will be particularly important near the rounded leading edge.

Three-Dimensional Thin Wings
in Steady Supersonic Flow

8-1 Introduction

In this chapter we shall review some developments in the theory for supersonic flow around three-dimensional wings. The linearized theory for thin wings as formulated in Chapter 5 will be used throughout. Thus, the differential equation governing the perturbation velocity potential is given by (5–29), which for the present purpose will be written as follows:

$$-B^2\varphi_{xx} + \varphi_{yy} + \varphi_{zz} = 0, \qquad (8\text{--}1)$$

where $B = \sqrt{M^2 - 1}$. Associated with (8–1) is the boundary condition

$$\varphi_z = \begin{cases} \dfrac{\partial z_u}{\partial x} & \text{on } z = 0+ \\[2ex] \dfrac{\partial z_l}{\partial x} & \text{on } z = 0- \end{cases} \quad \text{for } x, y \text{ on } S, \qquad (5\text{--}30)$$

which results from the condition that the flow must be tangent to the wing surface. In addition one needs to prescribe that perturbations vanish ahead of the most upstream point of the wing. The Kutta condition need not be specified in supersonic flow unless the wing has a trailing edge with a high sweep so that the velocity component normal to the edge is subsonic. Such an edge will be said to be subsonic. A wing may have both supersonic and subsonic leading and trailing edges. A wing with only supersonic leading and trailing edges is commonly referred to as having a "simple planform." An example of such a wing is given in Fig. 8–1.

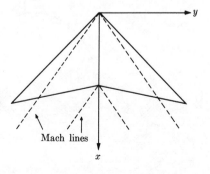

Fig. 8–1. Wing with simple planform.

8–2 Nonlifting Wings

For a nonlifting wing $z_l = -z_u$ in (5–30); hence

$$w(x, y, 0+) = -w(x, y, 0-) \equiv w_0(x, y), \qquad (8\text{–}2)$$

say, and the perturbation potential φ is thus symmetric in z. A convenient way to build up a symmetric potential is by covering the x,y-plane with a source distribution as was done in Chapter 7 for the incompressible-flow case. The solution for a source of strength $f(x_1, y_1)\, dx_1\, dy_1$ at the point $x_1, y_1, 0$ is (Eq. 5–38)

$$d\varphi = -\frac{1}{2\pi}\frac{f(x_1, y_1)\, dx_1\, dy_1}{\sqrt{(x - x_1)^2 - B^2[(y - y_1)^2 + z^2]}}. \qquad (8\text{–}3)$$

Hence by integrating, we obtain

$$\varphi(x, y, z) = -\frac{1}{2\pi}\iint_{\Sigma} \frac{f(x_1, y_1)\, dx_1\, dy_1}{\sqrt{(x - x_1)^2 - B^2[(y - y_1)^2 + z^2]}}, \qquad (8\text{–}4)$$

where the region of integration Σ is the portion of the x,y-plane intercepted by the upstream Mach cone from the field point x, y, z. This is illustrated in Fig. 8–2. Referring to Fig. 8–3 we see that the total volumetric outflow per unit area from the sources is $2U_\infty w_0(x, y)$, each side contributing half of this value. Hence the nondimensional source strength is

$$f(x, y) = 2w_0(x, y). \qquad (8\text{–}5)$$

By inserting (8–5) into (8–4) we obtain the following result, which was

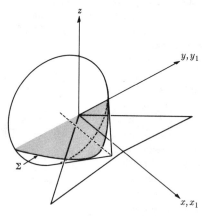

FIG. 8–2. Region of source distribution influencing the point (x, y, z).

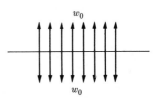

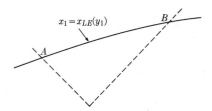

Fig. 8–3. Determination of source strength.

Fig. 8–4. Integration limits.

first derived by Puckett (1946):

$$\varphi = -\frac{1}{\pi} \iint_{\Sigma} \frac{w_0(x_1, y_1)\, dx_1\, dy_1}{\sqrt{(x - x_1)^2 - B^2[(y - y_1)^2 + z^2]}}. \qquad (8\text{–}6)$$

Next, the pressure coefficient is easily obtained from

$$C_p = -2\varphi_x. \qquad (5\text{–}31)$$

When evaluating C_p on the wing surface ($z = 0\pm$), it is convenient to integrate (8–6) first by parts with respect to x_1, and then perform the differentiation with respect to x. By using (6–37), or simply by noting the symmetry of the source solution in x and x_1, we then obtain

$$C_p(x, y, 0+) = \frac{2}{\pi}\left[\int_A^B \frac{w_0(x_{\mathrm{LE}}, y_1)\, dy_1}{\sqrt{(x - x_l)^2 - B^2(y - y_1)^2}}\right.$$
$$\left. + \iint_{\Sigma} \frac{w_{0x_1}(x_1, y_1)\, dx_1\, dy_1}{\sqrt{(x - x_1)^2 - B^2(y - y_1)^2}}\right], \qquad (8\text{–}7)$$

where A and B are the values of y_1 at the intersection of the upstream Mach lines from the point x, y with the leading edge of the wing (see Fig. 8–4). As a check of the above result we consider a two-dimensional wing for which w_0 is independent of y. Thus the integration over y_1 can be carried out directly. Using standard integral tables we find that

$$\int_{y-(x-x_1)/B}^{y+(x-x_1)/B} \frac{dy_1}{\sqrt{(x - x_1)^2 - B^2(y - y_1)^2}} = \frac{\pi}{B}. \qquad (8\text{–}8)$$

Hence (8–7) gives

$$C_p(x, y, 0+) = \frac{2}{B}\left[w_0(0) + \int_0^x w_{0x_1}\, dx_1\right] = \frac{2w_0}{B} \equiv \frac{2}{B}\frac{\partial z_u}{\partial x}, \qquad (8\text{–}9)$$

which is Ackeret's result for two-dimensional flow (see Section 5–4).

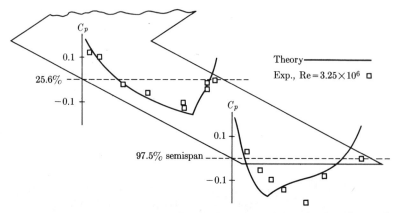

FIG. 8–5. Comparison of pressure distribution obtained from (8–8) with experiments for a 7% thick biconvex wing of 63° sweep at $M = 1.53$. (After Frick and Boyd, 1948. Courtesy of National Aeronautics and Space Administration.)

To give an indication of the accuracy of the theory we present in Fig. 8–5 a comparison with experimental results for a highly swept wing as obtained by Frick and Boyd (1948). The wing section is a biconvex circular arc profile of thickness ratio (measured in the flow direction) of 7%. As seen, the agreement is satisfactory except toward the wing tips where nonlinear disturbance accumulation apparently causes a shift in the pressure distribution downstream relative to the theoretical curve.

8–3 Lifting Wings of Simple Planform

For a wing with only supersonic edges, a wing with a "simple planform," there is no interaction between the flows on the upper and lower surfaces of the wing. Hence the flow above the wing, say, will be the same as on a symmetric wing with the same z_u, and (8–6) and (8–7) can therefore be directly applied. As an example, we give the result for an uncambered delta wing of sweep angle Λ (less than the sweep angle of the Mach lines) and angle of attack α. The lifting pressure difference between the lower and upper surfaces turns out to be [see, for example, Jones and Cohen (1960)]

$$\Delta C_p = \frac{4\alpha}{\pi B} \frac{m}{\sqrt{m^2 - 1}} \, \text{Re} \left\{ \cos^{-1} \frac{1 - m\tau}{m - \tau} + \cos^{-1} \frac{1 + m\tau}{m + \tau} \right\}, \quad (8\text{–}10)$$

where

$$m = B/\tan \Lambda, \quad (8\text{–}11)$$

$$\tau = By/x, \quad (8\text{–}12)$$

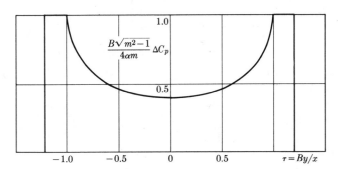

Fig. 8-6. Lifting pressure distribution on a delta wing with supersonic leading edges.

and the vertex is located at the origin. It is seen that ΔC_p is constant along rays through the wing vertex, as are all other physical properties of the flow. This is an example of so-called *conical flow* which will be discussed later. The pressure distribution given by (8–10) is shown in Fig. 8–6 for the case of $m = 1.2$. By integrating the lifting pressures over the wing surface one finds that the lift coefficient for a triangular wing becomes simply

$$C_L = 4\alpha/B, \tag{8-13}$$

that is, the lift is independent of the sweep when the leading edges are supersonic. This result is identical to that for two-dimensional flow, as may be determined from (8–9).

8–4 The Method of Evvard and Krasilshchikova*

Whenever the wing has an edge, or a part of one, that is subsonic, the problem of determining the flow around a lifting wing becomes much more complicated. The reason for this is apparent when we consider under what circumstances the simple formula (8–6) can be used. This solution requires that w_0 be known in the whole upstream region of influence from the field point. In the nonlifting case φ is symmetric in z; hence $\varphi_z = 0$ on the x,y-plane outside the wing. For the case of a simple planform no disturbance can propagate ahead of the leading edge and thus w_0 is zero ahead of the planform; that is, w_0 is known within the region of influence, regardless of whether the flow is symmetric or not. In case of a lifting wing with a subsonic leading edge, however, w_0 is not zero ahead of the edge; on the contrary it is always singular just ahead of the leading edge, except possibly at one value of α (see Fig. 8–7).

---

* Developed independently by Evvard (1950) and Krasilshchikova (1951).

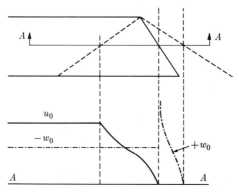

FIG. 8–7. Typical disturbance velocities u_0 and w_0 in the plane of a flat plate of planform shown.

However, if w_0 ahead of the subsonic edge could somehow be calculated it would be possible to use (8–6).

Let us consider the following problem: a lifting surface with an angle-of-attack distribution $\alpha(x, y)$ has a leading edge that is partially supersonic, partially subsonic (see Fig. 8–8). The slope may be discontinuous and the supersonic portion is assumed to end at O. The boundary conditions to be satisfied on the x,y-plane are

$$\varphi_z(x, y, 0) = w_0 = -\alpha(x, y), \qquad \text{on the wing,} \qquad (8\text{--}14)$$

$$\varphi(x, y, 0) = 0, \qquad \text{ahead of the wing.} \qquad (8\text{--}15)$$

The latter condition follows from the antisymmetry of φ.

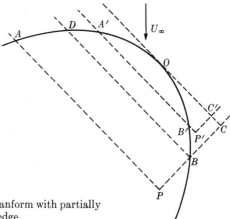

FIG. 8–8. Wing planform with partially subsonic leading edge.

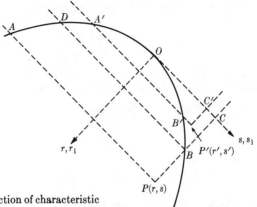

FIG. 8–9. Introduction of characteristic coordinates.

The potential at a point P on the wing can be expressed by aid of (8–6) as follows:

$$\varphi(x, y, 0) = -\frac{1}{\pi} \iint\limits_{OACP} \frac{w_0(x_1, y_1)\, dx_1\, dy_1}{\sqrt{(x - x_1)^2 - B^2(y - y_1)^2}}. \quad (8\text{–}16)$$

However, w_0 is not known in the region OCB. An integral equation for w_0 is obtained by use of (8–6) for a point in OCB. Then, because of (8–15),

$$0 = -\frac{1}{\pi} \iint\limits_{A'OC'B'} \frac{w_0(x_1, y_1)\, dx_1\, dy_1}{\sqrt{(x - x_1)^2 - B^2(y - y_1)^2}}. \quad (8\text{–}17)$$

To solve this integral equation it is convenient to introduce the characteristic coordinates r and s (see Fig. 8–9) defined by

$$\begin{aligned}
r &= (x - x_0) - B(y - y_0) \\
s &= (x - x_0) + B(y - y_0),
\end{aligned} \quad (8\text{–}18)$$

where x_0, y_0 are the coordinates of the point O. Similarly, as integration variables we introduce

$$\begin{aligned}
r_1 &= (x_1 - x_0) - B(y_1 - y_0) \\
s_1 &= (x_1 - x_0) + B(y_1 - y_0).
\end{aligned} \quad (8\text{–}19)$$

Thus

$$dx_1\, dy_1 = \left| \frac{\partial(x_1, y_1)}{\partial(r_1, s_1)} \right| dr_1\, ds_1 = \frac{1}{2B}\, dr_1\, ds_1, \quad (8\text{–}20)$$

and the integral equation (8–17) takes the following form:

$$\int_0^{r'} \int_{s_A(r_1)}^{s'} \frac{w_0(r_1, s_1)\, dr_1\, ds_1}{\sqrt{(r' - r_1)(s' - s_1)}} = 0, \qquad (8\text{–}21)$$

where $s_1 = s_A(r_1)$ is the equation of the supersonic portion of the leading edge (to the left of O). Equation (8–21) may be written in the following form:

$$\int_0^{r'} \frac{F(r_1)\, dr_1}{\sqrt{r' - r_1}} = 0, \qquad (8\text{–}22)$$

where

$$F(r_1) = \int_{s_A(r_1)}^{s'} \frac{w_0(r_1, s_1)\, ds_1}{\sqrt{s' - s_1}}. \qquad (8\text{–}23)$$

Equations (8–22) and (8–23) are of the form

$$f(t) = \int_0^t \frac{g(t_1)\, dt_1}{\sqrt{t - t_1}}. \qquad (8\text{–}24)$$

This is known as Abel's integral equation. Its solution, which is actually not needed for the present problem, is given in the Appendix to Section 8–4. It is sufficient to notice that, provided F has no singularities in the region, the solution to (8–22) is $F = 0$. Let us therefore write (8–23) as follows:

$$F(r_1) = \int_{s_A(r_1)}^{s_B(r_1)} \frac{w_0(r_1, s_1)\, ds_1}{\sqrt{s' - s_1}} + \int_{s_B(r_1)}^{s'} \frac{w_0(r_1, s_1)\, ds_1}{\sqrt{s' - s_1}}, \qquad (8\text{–}25)$$

where $s_1 = s_B(r_1)$ is the equation of the subsonic portion of the leading edge (to the right of O). Then, since $F = 0$,

$$\int_{s_A(r_1)}^{s_B(r_1)} \frac{w_0(r_1, s_1)\, ds_1}{\sqrt{s' - s_1}} = - \int_{s_B(r_1)}^{s'} \frac{w_0(r_1, s_1)\, ds_1}{\sqrt{s' - s_1}}. \qquad (8\text{–}26)$$

The integral on the left involves only values of w_0 on the planform which are known. Hence this integral can be calculated, and we are left with an integral equation of the form (8–24) which can be solved using the results given in the Appendix to Section 8–4. Sometimes it might be of interest to determine w_0 outside the leading edge; however, for most practical cases only the pressure on the wing is required, in which case it is not necessary to calculate w_0. In order to make the subsequent development easier to follow we will denote by $g(r, s)$ the unknown value of w_0 in the region outside the leading edge. Thus, from (8–26),

$$\int_{s_B(r_1)}^{s'} \frac{g(r_1, s_1)\, ds_1}{\sqrt{s' - s_1}} = - \int_{s_A(r_1)}^{s_B(r_1)} \frac{w_0(r_1, s_1)\, ds_1}{\sqrt{s' - s_1}}. \qquad (8\text{–}27)$$

Expressing (8–16) in the characteristic coordinates we obtain

$$\varphi(s, r) = - \frac{1}{2\pi B} \iint_{PBDA} \frac{w_0(s_1, r_1)\, ds_1\, dr_1}{\sqrt{(s - s_1)(r - r_1)}}$$

$$- \frac{1}{2\pi B} \iint_{BOD} \frac{w_0(s_1, r_1)\, ds_1\, dr_1}{\sqrt{(s - s_1)(r - r_1)}} - \frac{1}{2\pi B} \iint_{BCD} \frac{g(s_1, r_1)\, ds_1\, dr_1}{\sqrt{(s - s_1)(r - r_1)}} \, .$$

$$(8\text{–}28)$$

On combining the last two integrals and carrying out the integration over s_1, we first obtain, for each value of r_1, the following integral over s_1:

$$\int_{s_A}^{s_B} \frac{w_0(r_1, s_1)\, ds_1}{\sqrt{s - s_1}} + \int_{s_B}^{s} \frac{g(r_1, s_1)\, ds_1}{\sqrt{s - s_1}} \, . \qquad (8\text{–}29)$$

However, according to (8–27), this is zero for all values of s' in the region outside the edge, including $s' = s$. Hence the last two integrals in (8–28) cancel identically, and the final result, expressed in terms of Cartesian coordinates, becomes

$$\varphi = - \frac{1}{\pi} \iint_{PBDA} \frac{w_0(x_1, y_1)\, dx_1\, dy_1}{\sqrt{(x - x_1)^2 - B^2(y - y_1)^2}} \, . \qquad (8\text{–}30)$$

Notice that the potential is independent of the shape of the leading edge between D and B. The formula (8–30) may be extended also to cases where there is interaction between subsonic side edges (see Fig. 8–9), as long as there is some portion, however small, of the leading edge that is supersonic. The region of integration in (8–30) for the wing shown in Fig. 8–10 is indicated by the cross-hatched areas. This method has been used by Etkin and Woodward (1954) to obtain an approximate solution for a delta wing with subsonic leading edges in which the areas of integration are reflected back and forth right up to the pointed limit. By comparison with the exact solution (see the following section) it is found that the error incurred by neglecting the contributions after the third reflection is negligible.

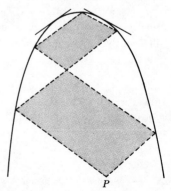

FIG. 8–10. Areas of interaction for interacting side edges.

8-5 Conical Flows

The Evvard-Krasilshchikova method fails to give an exact solution when no portion of the leading edge is supersonic, for example in the case of a delta wing with leading edges swept behind the Mach lines. In such cases one can often employ the methods of conical flow.

A conical flow is one for which the flow properties (velocity components, pressure, etc.) are constant along rays through one point. Examples of such flows are shown in Fig. 8-11. The wings in (a) and (c) are flat plates at a small angle of attack. In a conical flow there is no typical length. One can thus cut off any forward portion of the wing and the cone, and then, upon magnification, get back the original flow picture. What makes this possible is the fact that the position of the trailing edge does not affect the flow ahead of it. Conical flows are not realizable in a subsonic stream.

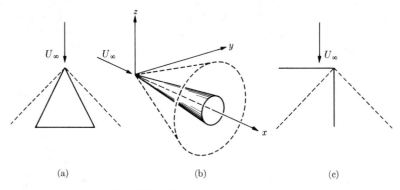

(a)   (b)   (c)

Fig. 8-11. Examples of conical flows.

We shall consider specifically the case (a) in Fig. 8-11*. The vertex of the wing is assumed to be located at the origin. To treat conical flows it is more convenient to work directly with the nondimensional perturbation velocity components u, v, w instead of with φ. Since the differential equation (8-1) for φ is linear, any of these components must also satisfy it. Thus

$$-B^2 u_{xx} + u_{yy} + u_{zz} = 0. \tag{8-31}$$

However, u is a function only of the conical variables $\bar{r}$, θ, which for convenience may be defined as follows:

$$\bar{r} \equiv \frac{B}{x} \sqrt{y^2 + z^2} = \sqrt{\bar{y}^2 + \bar{z}^2}, \qquad \theta \equiv \tan^{-1} y/z = \tan^{-1} \bar{y}/\bar{z}. \tag{8-32}$$

---

* The solution of this problem was first given by Stewart (1946).

Introducing (8–32) into (8–31), we obtain

$$(1 - \bar{r}^2)\bar{r}^2 u_{\bar{r}\bar{r}} + (1 - 2\bar{r}^2)\bar{r}u_{\bar{r}} + u_{\theta\theta} = 0, \qquad (8\text{–}33)$$

and similarly for the other components. In the y,z-plane, the Mach cone is represented by the unit circle and the wing by the slit along the y-axis of length $2m$, where m is defined by (8–11) (see Fig. 8–12).

The boundary condition for w is that $w = -\alpha$ on the slit. On the Mach cone, u, v, and w must all vanish. Furthermore, u and v must be antisymmetric in z whereas w must be symmetric.

The components are related through the irrotationality condition, which states

$$u_y = v_x; \qquad u_z = w_x; \qquad v_z = w_y. \tag{8–34}$$

It is possible to transform (8–33) into a more familiar form by introducing the new radial coordinate

$$\bar{\rho} = \frac{1 - \sqrt{1 - \bar{r}^2}}{\bar{r}};$$

$$\frac{d\bar{\rho}}{d\bar{r}} = \frac{1}{\sqrt{1 - \bar{r}^2}}\frac{\bar{\rho}}{\bar{r}}. \tag{8–35}$$

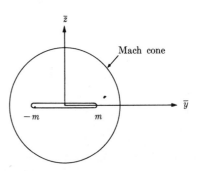

FIG. 8–12. Wing in conical variables.

This transformation is named after Tschaplygin. It has the property of transforming (8–33) into the two-dimensional Laplace equation

$$\bar{\rho}^2 u_{\bar{\rho}\bar{\rho}} + \bar{\rho}u_{\bar{\rho}} + u_{\theta\theta} = 0. \tag{8–36}$$

In the plane

$$\eta = \bar{\rho}\cos\theta \tag{8–37}$$
$$\zeta = \bar{\rho}\sin\theta,$$

the interior of the Mach cone again corresponds to the interior of the unit circle, but the wing is now represented by a slit between $\eta = m'$ and $\eta = -m'$ where

$$m' = \frac{1 - \sqrt{1 - m^2}}{m} \tag{8–38}$$

(see Fig. 8–13). The coordinates η and ζ are related to the original conical coordinates $\bar{y}$ and $\bar{z}$ through

$$\eta = \frac{\bar{y}}{1 + \sqrt{1 - \bar{y}^2 - \bar{z}^2}}$$

$$\zeta = \frac{\bar{z}}{1 + \sqrt{1 - \bar{y}^2 - \bar{z}^2}}. \tag{8–39}$$

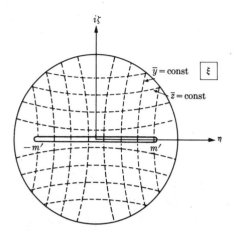

FIG. 8–13. Lines of constant y and z in ξ-plane.

Lines of constant $\bar{y}$ and $\bar{z}$ are indicated in Fig. 8–13. It shôuld be noted that the lines $\bar{y} = \mathrm{const}$ are normal to the η-axis for $\zeta = 0$, that is, the ζ-direction agrees with the $\bar{z}$-direction on $\zeta = 0$.

Since each of the velocity components is a solution of the two-dimensional Laplace equation in the η,ζ-plane, it is possible to use all the powerful techniques developed for incompressible flow. In particular, it is convenient to employ complex variables. Let

$$\xi = \eta + i\zeta. \tag{8–40}$$

Then we may define analytic functions $\hat{U}$, $\hat{V}$, $\hat{W}$ of ξ such that their real parts are identical to u, v, and w, respectively. The imaginary portions have no direct physical meaning. Thus let

$$\hat{U} = u + i\tilde{u}, \tag{8–41}$$

where $\tilde{u}$ is the artificially added imaginary part. We shall now consider the boundary conditions for u. On the wing, $w = \mathrm{const} = -\alpha$. Hence, from (8–34),

$$\frac{\partial u}{\partial z} = \frac{\partial w}{\partial x} = 0 \quad \text{on} \quad z = 0, \quad \left|\frac{y}{x}\right| < m. \tag{8–42}$$

Thus

$$\frac{\partial u}{\partial \zeta} = 0 \quad \text{on} \quad \zeta = 0, \quad |\eta| < m', \tag{8–43}$$

but according to the Cauchy-Riemann condition

$$\frac{\partial u}{\partial \zeta} = -\frac{\partial \tilde{u}}{\partial \eta} = 0. \tag{8–44}$$

Hence $\tilde{u}$ is constant on the slit $|\eta| < m'$, and this constant may be chosen equal to zero. Thus

$$\tilde{u} = 0, \qquad \zeta = 0, \qquad |\eta| < m'. \tag{8–45}$$

On the Mach cone the disturbances must vanish, hence

$$u = 0 \qquad \text{on} \qquad |\xi| = 1. \tag{8–46}$$

Because of the symmetry condition,

$$\frac{\partial u}{\partial \eta} = 0 \qquad \text{on} \qquad \eta = 0.$$

Hence, using the other Cauchy-Riemann relation,

$$\frac{\partial \tilde{u}}{\partial \zeta} = 0 \qquad \text{on} \qquad \eta = 0,$$

or

$$\tilde{u} = 0 \qquad \text{on} \qquad \eta = 0. \tag{8–47}$$

Also u must have singularities at the wing edges of the same type as a leading edge in a subsonic flow. To get a general idea how u, hence $\hat{U}$, may behave, we recall the result for a lifting slender wing in a sub- or supersonic flow. According to (6–66), the complex perturbation velocity potential for this case reads

$$\mathbb{W}(X) = \varphi + i\psi = -i\alpha[(X^2 - s^2)^{1/2} - X], \tag{8–48}$$

where s is the wing semispan and

$$X = y + iz.$$

Differentiation with respect to x gives $\hat{U}$, namely,

$$\hat{U} = \frac{\alpha s \, ds/dx}{(s^2 - X^2)^{1/2}}. \tag{8–49}$$

Hence on the y-axis ($z = 0+$)

$$\hat{U} = u = \frac{\alpha s \, ds/dx}{\sqrt{s^2 - y^2}}, \tag{8–50}$$

from which it is seen that u must have a square-root singularity at the leading edge. Equation (8–50) provides a very convenient starting point for finding a solution in the conical case. For a delta wing $s = x/\tan \Lambda$, and then (8–50) may be written

$$u = \frac{\alpha B}{\tan^2 \Lambda \sqrt{m^2 - \overline{y}^2}}. \tag{8–51}$$

When we introduce the expressions

$$\bar{y} = \frac{2\eta}{1 + \eta^2} \; ; \qquad m = \frac{2m'}{1 + m'^2} , \tag{8-52}$$

we obtain from (8–51) after some rearranging

$$u = \frac{2m'^2\alpha}{B(1 + m'^2)} \frac{(1 + \eta^2)}{\sqrt{(m'^2 - \eta^2)(1 - m'^2\eta^2)}} . \tag{8-53}$$

Hence by analytic continuation we find that

$$\hat{U} = \frac{2m'^2\alpha}{B(1 + m'^2)} \frac{(1 + \xi^2)}{[(m'^2 - \xi^2)(1 - m'^2\xi^2)]^{1/2}} \tag{8-54}$$

must be the limit of the function $\hat{U}$ as m or $m' \to 0$, since then the conical flow solution must approach the slender-body solution. It is easily seen that (8–54) is purely real on both the η-axis and the ζ-axis inside the unit circle. Also, $\hat{U}$ can be shown to be purely imaginary for $\xi = e^{i\theta}$ so that (8–46) is fulfilled. Consequently, (8–54) is a solution with the required properties. It is not the most general solution, however, since we may multiply it with any real constant because of the homogeneity of the boundary conditions. The proper solution is therefore

$$\hat{U} = \frac{2m'^2\alpha C(m')}{B(1 + m'^2)} \frac{(1 + \xi^2)}{[(m'^2 - \xi^2)(1 - m'^2\xi^2)]^{1/2}} , \tag{8-55}$$

where C remains to be evaluated. Clearly

$$C \to 1 \qquad \text{as} \qquad m' \to 0, \tag{8-56}$$

since the slender-body solution must be the proper limit for vanishing span of the delta wing; C must be determined such that

$$w = -\alpha \qquad \text{on the wing.} \tag{8-57}$$

Now, from the condition of irrotationality,

$$w = \int_{\text{Mach cone}}^{x} u_z(x_1, y, z) \, dx_1. \tag{8-58}$$

It is sufficient to determine w for one value of y/x on the wing, only. For convenience we shall choose $\bar{y} = 0$. Introducing as integration variable

$$\bar{z}_1 = Bz/x_1, \tag{8-59}$$

we obtain from (8–58)

$$w(0, 0) = B \int_0^1 \frac{d\bar{z}_1}{\bar{z}_1} u_{\bar{z}_1}(0, \bar{z}_1). \tag{8-60}$$

(The Mach cone corresponds to $\bar{z}_1 = 1$; whereas for $z = 0$ the limit $x_1 = x$ corresponds to $\bar{z}_1 = 0$.) We may in (8–60) make the further substitution of integration variable

$$\bar{z}_1 = \frac{2\zeta_1}{1 + \zeta_1^2} \qquad (8\text{–}61)$$

corresponding to the Tschaplygin transformation. Since $u_{\bar{z}_1}\, d\bar{z}_1 = u_{\zeta_1}\, d\zeta_1$, (8–60) becomes

$$w(0, 0) = B \int_0^1 \frac{1 + \zeta_1^2}{2\zeta_1} u_{\zeta_1}(0, \zeta_1)\, d\zeta_1. \qquad (8\text{–}62)$$

From (8–55) it follows, setting $\xi = i\zeta$, that

$$u(0, \zeta) = \frac{2m'^2 \alpha C(m')}{B(1 + m'^2)} \frac{1 - \zeta^2}{\sqrt{(m'^2 + \zeta^2)(1 + m'^2 \zeta^2)}}. \qquad (8\text{–}63)$$

Introducing this into (8–62) and setting $w(0, 0) = -\alpha$, we obtain the following expression for C:

$$\frac{1}{C(m')} = -\frac{m'^2}{(1 + m'^2)} \int_0^1 \frac{1 + \zeta_1^2}{\zeta_1} \frac{\partial}{\partial \zeta_1} \left[\frac{(1 - \zeta_1^2)}{\sqrt{(m'^2 + \zeta_1^2)(1 + m'^2 \zeta_1^2)}} \right] d\zeta_1. \qquad (8\text{–}64)$$

Hence

$$\frac{1}{C(m')} = m'^2(1 + m'^2) \int_0^1 \frac{(1 + \zeta_1^2)^2\, d\zeta_1}{[(m'^2 + \zeta_1^2)(1 + m'^2 \zeta_1^2)]^{3/2}}. \qquad (8\text{–}65)$$

By making the substitution $t = \zeta_1/m'$, it is easily seen that $C(0) = 1$, as it should. The evaluation of (8–65) is facilitated by the substitutions

$$\zeta_1^2 = \frac{1 - \sin\theta}{1 + \sin\theta}, \qquad k = \frac{1 - m'^2}{1 + m'^2} = \sqrt{1 - m^2}. \qquad (8\text{–}66)$$

After considerable manipulation, this transformation gives

$$C = 1/E(k), \qquad (8\text{–}67)$$

(see Fig. 8–14) where

$$\begin{aligned}
E(k) &= (1 - k^2) \int_0^{\pi/2} \frac{d\theta}{(1 - k^2 \sin^2\theta)^{3/2}} \\
&= \int_0^{\pi/2} \sqrt{1 - k^2 \sin^2\theta}\, d\theta
\end{aligned} \qquad (8\text{–}68)$$

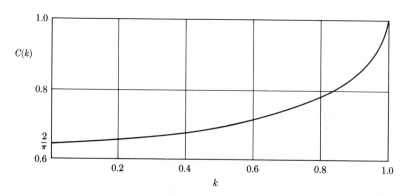

Fig. 8–14. The function $C(k) = 1/E(k)$.

is the complete elliptic integral of second kind and modulus k. The identity of the two integrals in (8–68) can be demonstrated as follows:

$$\int_0^{\pi/2} \left[\frac{1 - k^2}{(1 - k^2 \sin^2 \theta)^{3/2}} - \sqrt{1 - k^2 \sin^2 \theta} \right] d\theta$$

$$= -k^2 \int_0^{\pi/2} \frac{\partial}{\partial \theta} \left(\frac{\sin \theta \cos \theta}{\sqrt{1 - k^2 \sin^2 \theta}} \right) d\theta = 0.$$

The velocity distribution is given by (8–51) multiplied by C. Thus for the lifting pressure we obtain

$$\Delta C_p = \frac{4\alpha B C}{\tan^2 \Lambda \sqrt{m^2 - y^2}}. \tag{8–69}$$

The lift coefficient is found by integration to be

$$C_L = \frac{2\pi\alpha C}{\tan \Lambda} = \alpha \frac{\pi C}{2} A. \tag{8–70}$$

This result checks with slender-body theory value $[\alpha(\pi/2)A]$ for $m = 0$ ($k = 1$) and the value according to supersonic-edge theory $(4\alpha/B)$ for $m = 1$ ($k = 0$). The theoretical result (8–70) agrees well with experiments for low α ($<5°$) as shown in Fig. 8–15.

The calculation of drag is complicated by the fact that a proper treatment requires that the singularity in the velocity distribution at the leading edge be taken into account. For an uncambered wing with supersonic edges, the drag is simply the component in the flow direction of the normal force. Hence, to lowest order in α,

$$D = \alpha L. \tag{8–71}$$

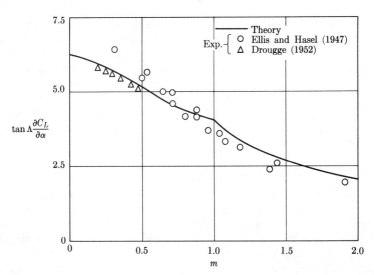

Fig. 8–15. Lift-curve slope for delta wings at supersonic speeds. Comparison of theory and experiment.

However, when the leading edge is subsonic, the high velocities near the edge create a finite suction force, i.e., a thrust, which reduces the drag from the value given by (8–71). A familiar case is that of a two-dimensional flat plate in a subsonic flow for which the leading-edge thrust exactly balances the drag given by (8–71). By considering a separate inner flow region near the leading edge, one realizes that the flow in this immediate neighborhood must be approximately two-dimensional in planes normal to the leading edge with a "free-stream" Mach number given by the normal velocity component. Calculations performed in this way show that (see Jones and Cohen, 1960, p. 201) the total leading-edge thrust T for a delta wing of semispan s is

$$T = \frac{\pi}{2} \rho_\infty U_\infty^2 \alpha^2 \frac{s^2}{[E(k)]^2} \sqrt{1 - m^2}. \qquad (8\text{--}72)$$

Hence with

$$D = \alpha L - T,$$

we obtain

$$C_D = \frac{2C_L^2}{\pi A} [E(k) - \tfrac{1}{2}\sqrt{1 - m^2}]. \qquad (8\text{--}73)$$

Experiments show that, even for wings with smoothly rounded leading edges, only a fraction of the theoretical leading-edge suction force can be realized. Wings with sharp leading edges show no suction force, as would be expected.

The specific example considered was one in which both leading edges were subsonic so that only the flow inside the Mach cone was of interest. For the region outside the Mach cone one can find a transformation similar to Tschaplygin's (8–35) that transforms (8–33) to the two-dimensional wave equation for $r_1 > 1$. In a conical problem with one or both leading edges supersonic (e.g., the case of a side edge of a rectangular wing) one would then need to join the solutions in the hyperbolic and elliptic regions in such a manner that the velocity components become continuous across the Mach cone.

8–6 Numerical Integration Schemes·

In many practical cases the functional dependence of w_0 on x and y is too complicated to permit analytical integration of (8–6), and one then has to resort to numerical methods. The usual procedure in such cases is to subdivide the area of integration into suitably selected elements, and then assume w_0 constant within each elementary area. This was the approach tried in an early paper by Linnaluoto (1951). Later developments of this kind have primarily been made with the oscillating-wing problem in mind. We may here cite the papers of Zartarian and Hsu (1955), and Pines *et al.* (1955). In both these a "Mach-box" scheme is employed in which the elementary areas consist of rectangles having the Mach lines as diagonals (Fig. 8–16a). The most accurate development to date seems to be a recent paper by Stark (1964) in which he uses rhombic boxes bounded by Mach lines (Fig. 8–16b) and takes great care in accounting for upwash singularities at swept leading edges.

In these methods the integration over each elementary area can be carried out in advance and a table prepared giving the value of the integral for unit w_0 as a function of the position of the box relative to the point x,y on the wing.

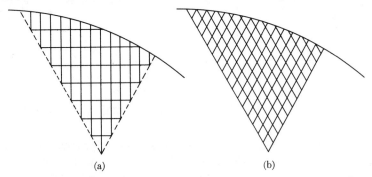

(a)   (b)

Fig. 8–16. Subdivision of integration area in numerical schemes. (a) "Mach box." (b) "Characteristic box."

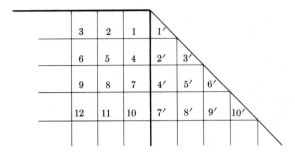

FIG. 8–17. Use of supersonic box method for a side edge.

It is also possible to use the numerical scheme in the case of a subsonic leading edge without making direct use of the Evvard-Krasilshchikova method. Consider for simplicity the case shown in Fig. 8–17. For all the boxes denoted by 1, 2, 3, 4, etc. w_0 is known. For the boxes denoted $1'$, $2'$, $3'$, $4'$, etc. w_0 is initially unknown. The boundary condition in the primed region is that $\varphi = 0$. Hence, considering first the potential in the center of box $1'$ we obtain directly that $w_0 = 0$ in $1'$. If we consider next the potential in the center of box $2'$, the integral then involves the boxes $2'$, $1'$, and 1. In all of these, except box $2'$, w_0 is known. Setting $\varphi = 0$ in $2'$ gives a linear equation for w_0 in $2'$ that can be easily solved. In this manner one can proceed step by step downstream in the primed region and find w_0, whereupon φ on the wing may be calculated.

Related applications to nonplanar problems are considered in Chapter 11.

Appendix to Section 8–4

Take the Laplace transform of each side of (8–24). Since

$$\int_0^\infty \frac{e^{-st}}{\sqrt{t}}\, dt = \sqrt{\frac{\pi}{2}}\, s^{-1/2}, \tag{A–1}$$

we obtain by use of the convolution theorem

$$F(s) = \sqrt{\frac{\pi}{2}}\, s^{-1/2} G(s), \tag{A–2}$$

where $F(s)$ and $G(s)$ denote the Laplace transforms of $f(t)$ and $g(t)$. Thus

$$G(s) = \sqrt{\frac{2}{\pi}}\, s^{1/2} F(s) \equiv \sqrt{\frac{2}{\pi}}\, s^{-1/2}[sF(s)], \tag{A–3}$$

in which $sF(s)$ is recognized as the transform of $f'(t)$ provided $f(0) = 0$. By inverting (A–3) using the convolution theorem and (A–1), we obtain

$$g(t) = \frac{2}{\pi} \int_0^t \frac{f'(t_1)}{\sqrt{t - t_1}} \, dt_1. \tag{A–4}$$

This solution is valid for the case $f(t)$ continuous and $f(0) = 0$.

9

Drag at Supersonic Speeds

9–1 Introduction

In supersonic flow there are basically three different mechanisms whereby drag is created. The first two are essentially the same as in subsonic flow, namely through the action of viscosity in the boundary layer and through the release of vorticity that accompanies production of lift. The determination of the skin friction drag involves calculation of the boundary layer in a manner very similar to what was considered in incompressible flow and will not be described further in this book. However, it should be pointed out that skin friction may amount to a considerable fraction of the total drag (a typical figure is 30%) and must of course be included in any realistic drag analysis. Thus many of the results that will be given later must be interpreted with this in mind.

The vortex drag arises from the momentum, and hence kinetic energy, left in the fluid as a lifting vehicle travels through it. Since the vorticity remains essentially stationary with the fluid, there is no fundamental difference between subsonic and supersonic speeds. In fact, the vortex drag can be calculated by use of the formula (7–44) for the induced drag of a lifting three-dimensional wing in incompressible flow. In the supersonic case, however, the lift will induce an additional drag component, namely wave drag.

The wave drag is an aerodynamic phenomenon unique to supersonic flow and is associated with the energy radiated away from the vehicle in the form of pressure waves in much the same way as a fast-moving ship causes waves on the water surface. The present chapter will be concerned mostly with wave drag. For a planar wing one can distinguish between wave drag due to thickness and wave drag due to lift. The sum of wave drag and vortex drag is often called the pressure drag, since it is manifested by the pressure times the chordwise slope of the wing or body surface.

9–2 Calculation of Supersonic Drag by Use of Momentum Theory

The distinction between wave drag and vortex drag will be clearer when we attempt to calculate the drag from momentum considerations. We shall use the same type of control surface as in the case of a slender body treated in Section 6–6 except that now the dimensions of the cylinder will be chosen differently. See Fig. 9–1.

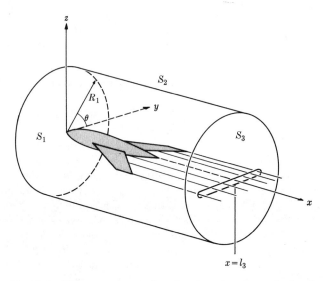

Fig. 9–1. Momentum control surface for wave drag calculated.

As in the slender-body case the radius R_1 of the cylindrical surface will be chosen as several times that of the span of the configuration; however, since the configuration is not necessarily slender, this also means that R_1 is several times its length. In the slender-body case, configurations with a blunt base were included, and hence it was necessary to place the rear disk S_3 at the base section. In the present case we will restrict the analysis to bodies that are pointed in the rear; S_3 may then be placed in an arbitrary location behind the body. The trailing vortices will pass through S_3. It is convenient to locate S_3 so far behind that the flow at S_3 is essentially two-dimensional in the crossflow plane (cf. the "Trefftz plane" discussed in Section 7–3). This requires that all Mach lines from the body cross S_2 far upstream of S_3.

The flow of momentum across the surfaces S_1, S_2, and S_3 was considered in Section 6–6, and the same general analysis applies in the present case. Thus from (6–71) with zero base drag we have

$$D = -\rho_\infty U_\infty^2 \iint\limits_{S_2} \varphi_x \varphi_r \, dS_2 \; - \; \iint\limits_{S_3} (p - p_\infty) \, dS_3 \; - \; \rho_\infty U_\infty^2 \iint\limits_{S_3} \varphi_x (1 + \varphi_x) \, dS_3.$$

$$(9\text{–}1)$$

On S_3 the flow is two-dimensional in the y,z-plane; hence the last term vanishes and

$$p - p_\infty = -\tfrac{1}{2}\rho_\infty U_\infty^2 (\varphi_y^2 + \varphi_z^2).$$

$$(9\text{–}2)$$

Thus the expression for the pressure drag becomes

$$D = -\rho_\infty U_\infty^2 \iint\limits_{S_2} \varphi_x \varphi_r \, dS_2 + \tfrac{1}{2}\rho_\infty U_\infty^2 \iint\limits_{S_3} (\varphi_y^2 + \varphi_z^2) \, dS_3. \qquad (9\text{–}3)$$

The second integral gives the vortex drag, which is identical to the induced drag for subsonic flow considered in Section 7–3. Hence, according to (7–44),

$$D_v = \tfrac{1}{2}\rho_\infty U_\infty^2 \iint\limits_{S_3} (\varphi_y^2 + \varphi_z^2) \, dS_3$$

$$= -\frac{\rho_\infty}{4\pi} \int_{-b/2}^{b/2} \int_{-b/2}^{b/2} \frac{d\Gamma}{dy} \frac{d\Gamma}{dy_1} \ln |y - y_1| \, dy \, dy_1, \qquad (9\text{–}4)$$

where b is the total span and $\Gamma(y)$ the circulation distribution in the wake. The first term in (9–3) represents the wave drag:

$$D_w = -\rho_\infty U_\infty^2 \iint\limits_{S_2} \varphi_x \varphi_r \, dS_2, \qquad (9\text{–}5)$$

and will be the subject of further study.

9–3 Drag of a Lineal Source Distribution

It has been demonstrated in several earlier examples how flow around a body can be built up by superimposing elementary source solutions. It is therefore of interest to determine the drag of a line of sources, a lineal source distribution. With the momentum method it is not necessary to know the particular body the sources represent, since the drag can be computed directly from the knowledge of source strength. The perturbation velocity potential due to a lineal source distribution of strength $U_\infty f(x)$ is

$$\varphi = -\frac{1}{2\pi} \int_0^{x-Br} \frac{f(x_1) \, dx_1}{\sqrt{(x - x_1)^2 - B^2 r^2}}, \qquad (6\text{–}25)$$

where

$$r = \sqrt{y^2 + z^2}.$$

We shall assume that f is continuous and that $f(0) = f(l) = 0$. The derivatives φ_x and φ_r are obtained by first integrating (6–25) by parts

$$\varphi = \frac{1}{2\pi} f(x - Br) \ln Br$$

$$-\frac{1}{2\pi} \int_0^{x-Br} f'(x_1) \ln \left[x - x_1 + \sqrt{(x - x_1)^2 - B^2 r^2} \right] dx_1. \qquad (9\text{–}6)$$

Thus on $r = R_1$, we have

$$
\varphi_r = \frac{f(x - BR_1)}{2\pi R_1} + \frac{1}{2\pi} \int_0^{x - BR_1} \frac{B^2 R_1 f'(x_1)}{\sqrt{(x - x_1)^2 - B^2 R_1^2}}
$$

$$
\times \frac{dx_1}{\left[x - x_1 + \sqrt{(x - x_1)^2 - B^2 R_1^2} \right]}
$$

$$
= \frac{1}{2\pi} \int_0^{x - BR_1} \frac{(x - x_1) f'(x_1) \, dx_1}{R_1 \sqrt{(x - x_1)^2 - B^2 R_1^2}} . \tag{9-7}
$$

The second form of (9–7) is arrived at by multiplying numerator and denominator in the first integral by $x - x_1 - \sqrt{(x - x_1)^2 - B^2 R_1^2}$. Also

$$
\varphi_x = -\frac{1}{2\pi} \int_0^{x - BR_1} \frac{f'(x_1) \, dx_1}{\sqrt{(x - x_1)^2 - B^2 R_1^2}} . \tag{9-8}
$$

Setting $dS_2 = 2\pi R_1 \, dx$ in (9–5), the total wave drag is thus

$$
D_w = \frac{\rho_\infty U_\infty^2}{2\pi} \int_{BR_1}^{l_3} dx \int_0^{x - BR_1} \frac{(x - x_1) f'(x_1) \, dx_1}{\sqrt{(x - x_1)^2 - B^2 R_1^2}}
$$

$$
\times \int_0^{x - BR_1} \frac{f'(x_2) \, dx_2}{\sqrt{(x - x_2)^2 - B^2 R_1^2}} . \tag{9-9}
$$

For $x - BR_1 > l$ the upper limit in the inner integrals may be replaced by l. The lower limit in the integral over x follows from the fact that $\varphi = 0$ for $x < Br$. We shall now evaluate the integral in the limit as $R_1 \to \infty$. It should then be remembered that, according to the way the rear disk S_3 was chosen, l_3 must be much larger at all times than R_1. Hence l_3 goes to infinity as $R_1 \to \infty$ in such a manner that $BR_1/l_3 \ll 1$.

Let $x' = x - BR_1$. Then (9–9) becomes

$$
D_w = \frac{\rho_\infty U_\infty^2}{2\pi} \int_0^{l - BR_1} dx' \left[\int_0^{x', l} \frac{(x' + BR_1 - x_1) f'(x_1) \, dx_1}{\sqrt{(x' - x_1)(x' - x_1 + 2BR_1)}} \right.
$$

$$
\left. \times \int_0^{x', l} \frac{f'(x_2) \, dx_2}{\sqrt{(x' - x_2)(x' - x_2 + 2BR_1)}} \right], \tag{9-10}
$$

the upper limit in the inner integrals being x' or l depending on whether x' is less or greater than l.

Now consider the factors containing R_1:

$$\frac{x' + BR_1 - x_1}{\sqrt{(x' - x_1 + 2BR_1)(x' - x_2 + 2BR_1)}}. \qquad (9\text{-}11)$$

As $BR_1 \to \infty$, but x' remains finite, (9–11) will tend to $\frac{1}{2}$. For $x' \gg BR_1$, which is the case for part of the integration region, the ratio (9–11) tends to unity. It is therefore convenient to split the interval of integration over x up into two parts, the first from 0 to a, and the second from a to $(l_3 - BR_1)$. The dividing point $x = a$ will be chosen so that $a \gg l$ but that a remains finite in the limit $BR_1 \to \infty$, hence $a/BR_1 \to 0$. In the first region we can in the limit replace (9–11) by $\frac{1}{2}$, whereas in the second region we may neglect x_1 and x_2 compared to x'. Thus in the limit of large BR_1, (9–10) becomes

$$D_w = \frac{\rho_\infty U_\infty^2}{4\pi} \int_0^a dx' \int_0^{x',l} \frac{f'(x_1)\, dx_1}{\sqrt{x' - x_1}} \int_0^{x',l} \frac{f'(x_2)\, dx_2}{\sqrt{x' - x_2}}$$

$$+ \frac{\rho_\infty U_\infty^2}{2\pi} \int_a^{l_3-BR_1} dx' \int_0^l \frac{(x' + BR_1)f'(x_1)\, dx_1}{\sqrt{x'(x' + 2BR_1)}} \int_0^l \frac{f'(x_2)\, dx_2}{\sqrt{x'(x' + 2BR_1)}}.$$
$$(9\text{-}12)$$

In the second part we can directly carry out the integrals over x_1 and x_2. Since $f(0) = f(l)$ this part vanishes. Hence

$$D_w = \frac{\rho_\infty U_\infty^2}{4\pi}\, I, \qquad (9\text{-}13)$$

where

$$I = \int_0^a dx' \int_0^{x',l} \frac{f'(x_1)\, dx_1}{\sqrt{x' - x_1}} \int_0^{x',l} \frac{f'(x_2)\, dx_2}{\sqrt{x' - x_2}}, \qquad (9\text{-}14)$$

and $a \gg l$. The obvious way to proceed is to interchange order of integration in (9–14). When the integral over x' is carried out first, the lower limit is either x_1 or x_2 depending on which is the larger. Now

$$\int \frac{dx'}{\sqrt{(x' - x_1)(x' - x_2)}} = \ln\left[2x' - x_1 - x_2 + 2\sqrt{(x' - x_1)(x' - x_2)}\right].$$
$$(9\text{-}15)$$

Hence the integral over x' becomes

$$\ln\left[2a - x_1 - x_2 + 2\sqrt{(a - x_1)(a - x_2)}\right] - \ln|x_1 - x_2|. \qquad (9\text{-}16)$$

However, x_1 and x_2 are at most l. Thus for $a \gg l$, we have

$$\ln\left[2a - x_1 - x_2 + 2\sqrt{(a - x_1)(a - x_2)}\right] \simeq \ln 4a - 2(x_1 + x_2)/a,$$

and, consequently,

$$I = -\int_0^l \int_0^l f'(x_1)f'(x_2) \ln |x_2 - x_1| \, dx_1 \, dx_2$$

$$+ \ln 4a \int_0^l \int_0^l f'(x_1)f'(x_2) \, dx_1 \, dx_2 + O(l/a). \qquad (9\text{--}17)$$

The second integral vanishes identically because of the condition $f(0) = f(l) = 0$. Hence, in the limit of $a \to \infty$,

$$I = -\int_0^l \int_0^l f'(x_1)f'(x_2) \ln |x_1 - x_2| \, dx_1 \, dx_2. \qquad (9\text{--}18)$$

Thus, the wave drag of a lineal source distribution is

$$D_w = -\frac{\rho_\infty U_\infty^2}{4\pi} \int_0^l \int_0^l f'(x_1)f'(x_2) \ln |x_1 - x_2| \, dx_1 \, dx_2. \qquad (9\text{--}19)$$

This result may be compared to that obtained from slender-body theory, (6–82). It was shown in Chapter 6 that for a slender body

$$f(x) = S'(x). \qquad (9\text{--}20)$$

Thus, with $S'(l) = 0$,

$$D_w = -\frac{\rho_\infty U_\infty^2}{2\pi} \int_0^l f'(x) \, dx \int_0^x f'(x_1) \ln (x - x_1) \, dx_1. \qquad (9\text{--}21)$$

It is left to the reader to prove that (9–19) and (9–21) are equal.

A very interesting property of (9–19) is that, upon setting $\Gamma = U_\infty f$, it becomes identical to the formula (9–4) for the vortex drag of a lifting wing. This property was first discovered by von Kármán (1936). Equation (9–19) also shows that the drag coefficient of a slender body in supersonic flow is independent of Mach number, provided the body has a pointed nose and ends in either a point or a cylindrical portion, as was already noted in Section 6–6.

9–4 Optimum Shape of a Slender Body of Revolution

It is of both theoretical and practical interest to find area distributions of slender bodies which, for given length, volume, or some other constraint, give the lowest possible wave drag. Application of (9–19) with (9–20) then gives rise to a variational problem that can be treated with standard methods. The similarity with the vortex drag problem treated in Section 7–3 suggests another approach, however, which leads more directly to the desired results. Thus, in a similar fashion we introduce the Glauert variable θ, defined by

$$x = \frac{l}{2}(1 + \cos\theta). \tag{9-22}$$

The nose is represented by $\theta = \pi$ and the base by $\theta = 0$. Expressing f as a Fourier sine series

$$f = l \sum_{n=1}^{\infty} A_n \sin n\theta, \tag{9-23}$$

and carrying out the double integration in (9–19), we obtain (cf. 7–47)

$$D_w = \frac{\pi\rho_\infty U_\infty^2 l^2}{8} \sum_{n=1}^{\infty} n A_n^2. \tag{9-24}$$

The corresponding area distribution can be found by integrating over x, since $f = S'$. Hence

$$S(\theta) = \frac{l^2}{4}\left\{ A_1\left(\pi - \theta + \frac{\sin 2\theta}{2}\right) \right.$$
$$\left. + \sum_{n=2}^{\infty} A_n\left[\frac{\sin(n+1)\theta}{n+1} - \frac{\sin(n-1)\theta}{n-1}\right]\right\}. \tag{9-25}$$

A further integration gives that the total volume is

$$V = \frac{\pi l^3}{8}(A_1 - \tfrac{1}{2}A_2). \tag{9-26}$$

Consider first the case of given base area. From (9–25) it follows that only A_1 contributes to the base area ($\theta = 0$) and hence

$$A_1 = \frac{4S(l)}{\pi l^2}. \tag{9-27}$$

However, all components contribute to the drag, and the drag is thus a minimum for $A_n = 0$, $n \neq 1$, and has the value

$$D_w = \frac{2\rho_\infty U_\infty^2}{\pi} \frac{[S(l)]^2}{l^2} \tag{9-28}$$

or, expressed in terms of the drag coefficient based on the base area,

$$C_{Dw} = \frac{4}{\pi}\frac{S(l)}{l^2}. \tag{9-29}$$

The area distribution of this body

$$S = \frac{S(l)}{\pi}(\pi - \theta + \tfrac{1}{2}\sin 2\theta), \tag{9-30}$$

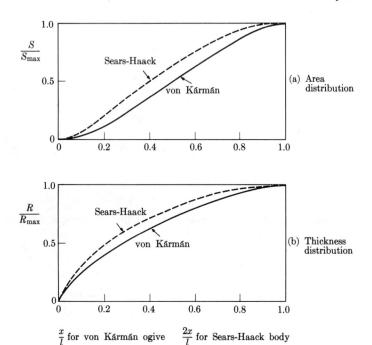

FIG. 9–2. Von Kármán ogive and Sears-Haack body.

and its associated thickness distribution are shown in Fig. 9–2. This body is called the *von Kármán ogive* after its discoverer.

We shall consider the case when the body is pointed at both ends and the volume is given. Then, since $A_1 = 0$, we obtain from (9–26)

$$A_2 = -\frac{16V}{\pi l^3}. \tag{9–31}$$

Again, for minimum drag, all the other coefficients must be zero. This gives for the drag of the *Sears-Haack body*

$$D_w = \frac{64V^2}{\pi l^4}\rho_\infty U_\infty^2 \tag{9–32}$$

or, expressed as a drag coefficient based on the maximum cross-sectional area

$$C_{Dw} = \frac{24V}{l^3}. \tag{9–33}$$

The area distribution of the Sears-Haack body

$$S(\theta) = \frac{4V}{\pi l} (\sin \theta - \tfrac{1}{3} \sin 3\theta), \qquad (9\text{–}34)$$

and its thickness distribution are shown in Fig. 9–2. This shape was discovered independently by Sears (1947) and Haack (1947).

It is interesting to note that both the von Kármán ogive and the Sears-Haack body are slightly blunted. Although linearized theory cannot be expected to hold close to a blunted nose or tail, it will give reasonable accuracy away from these regions provided the bluntness is not "excessive."

Cases for which $S'(l) \neq 0$ were discussed by Harder and Rennemann (1957). They found that the slender-body result (6–86) could not be used in that case since it did not give a proper drag minimum.

Generally, the drag is not very sensitive to small departures from the optimum shapes, *provided the area distribution is smooth.*

9–5 Drag of a General Source Distribution. Hayes' Method

Consider next a spatial distribution of sources inside a volume V representing the flow around some particular body whose shape we need not know for the present purpose. The strength of this distribution per unit volume will be denoted by $U_\infty \tilde{f}(x, y, z)$, which may be discontinuous in the y- and z-directions, but not in the x-direction. Thus, the lineal distribution considered in Section 9–4 is of the form

$$\tilde{f} = f(x)\ \delta(y)\ \delta(z), \qquad (9\text{–}35)$$

where δ is the Dirac delta function.

We shall now proceed to calculate the wave drag of the body using (9–5). Let the drag contribution of a strip of the cylinder S_2 between $\theta = \theta_0$ and $\theta = \theta_0 + \Delta\theta$ be ΔD_w. We define

$$\frac{dD_w}{d\theta} = \frac{\text{drag contribution}}{\text{unit angle}} \equiv \lim_{\Delta\theta \to 0} \left(\frac{\Delta D_w}{\Delta\theta} \right), \qquad (9\text{–}36)$$

$$D_w = \int_0^{2\pi} \frac{dD_w}{d\theta}\, d\theta. \qquad (9\text{–}37)$$

Consider a fixed meridian plane $\theta = \theta_0$, and a point $P(x_0, R_1, \theta_0)$ on the cylinder in the region that receives perturbations from the source distribution (see Fig. 9–3). The potential φ in this point depends on the contributions from all the sources within the upstream Mach cone from P. The contribution from a source located at $Q = (x_1, y_1, z_1)$ is proportional to the source strength $\tilde{f}(Q)$ and inversely proportional to the hyperbolic

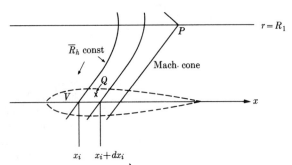

FIG. 9–3. Contribution to the potential at P from source at Q.

radius $\overline{R}_h(P, Q)$ between P and Q which is

$$\overline{R}_h = \sqrt{(x_0 - x_1)^2 - B^2[(y_0 - y_1)^2 + (z_0 - z_1)^2]}$$
$$= \sqrt{(x_0 - x_1)^2 - B^2[(R_1 \cos \theta_0 - y_1)^2 + (R_1 \sin \theta_0 - z_1)^2]}.$$
$$(9\text{–}38)$$

[For example, see (6–25).] This hyperbolic radius is constant on hyperboloids of revolution with $r = R_1$, $\theta = \theta_0$ as axis. Consider now the sources between two such hyperboloids, which intersect the x-axis at $x = x_i$ and $x = x_i + dx_i$. To evaluate the contribution to $\varphi(P)$ of these sources, one may transfer their total source strength to the axis. In this way the distribution in V is replaced by an equivalent lineal distribution, i.e., by an equivalent body of revolution. So far, this lineal distribution depends on x_0 and R_1 as well as θ_0.

Consider now, still for a fixed $\theta = \theta_0$, the limit as $R_1 \rightarrow \infty$. Then the hyperboloids of revolution may be replaced by planes that cut the meridian plane $\theta = \theta_0$ orthogonally along the Mach lines $x - Br =$ const. In place of letting $R_1 \rightarrow \infty$, one can perhaps observe this limiting behavior more clearly by causing the body width and thickness to shrink to zero, whereupon the curvature of the hyperboloids of revolution becomes unimportant inside the body. The equation for these planes, the "Mach planes," is

$$x_i = x - By \cos \theta_0 - Bz \sin \theta_0,$$
$$(9\text{–}39)$$

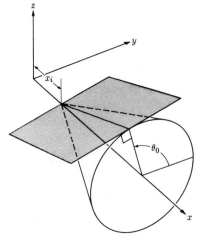

FIG. 9–4. Mach plane.

where x_i is the intercept by the plane of the x-axis. The Mach planes are tangent to the downstream Mach cones from the point $x = x_i$ (see Fig. 9–4).

The total source strength between two such adjacent planes thus can be moved to the x-axis. The equivalent body of revolution in the limit $R_1 \to \infty$ will depend at most on θ_0, that is, it will be independent of R_1 and x_0. The corresponding lineal source strength distribution will be denoted by $f(x; \theta_0)$. A consequence of the independence of x_0 and R_1 is that $f(x; \theta_0)$ may be used for computing φ_r and φ_x at P as well as φ. In general it may not be used to compute φ_θ. Evidently, φ_θ is zero for a lineal source distribution since this gives the flow around a nonlifting body of revolution, but the original source distribution does not give zero φ_θ in general. On the other hand φ_θ is not needed for evaluating the drag.

Since φ_r and φ_x may be computed from the equivalent lineal source distribution for fixed θ, it follows that $dD_w/d\theta$ may be computed in exactly the same way as the drag of a lineal source distribution was computed. The result will differ from (9–21) only by a factor $\frac{1}{2}\pi$. Hence we have proved the following: The drag D_w of a spatial distribution of sources of strength $U_\infty \tilde{f}(x_1, y_1, z_1)$ per unit volume is given by

$$D_w = \int_0^{2\pi} \frac{dD_w}{d\theta}\, d\theta, \tag{9–40}$$

$$\frac{dD_w}{d\theta} = -\frac{\rho_\infty U_\infty^2}{8\pi^2} \int_0^l \int_0^l f'(x_1; \theta) f'(x_2; \theta) \ln |x_1 - x_2|\, dx_1\, dx_2, \tag{9–41}$$

$$f(x_i; \theta)\, dx_i = \iiint_{V(x_i;\theta)} \tilde{f}(Q)\, dV, \tag{9–42}$$

where $V(x_i; \theta)$ is the region contained between two Mach planes perpendicular to $\theta_0 = \theta$ and intersecting the x-axis at $x = x_i$ and $x = x_i + dx_i$.

Equation (9–40) is valid for spatial distributions $\tilde{f}$ such that the equivalent linear distribution $f(x; \theta)$ is continuous in x, and $f(0; \theta) = f(l; \theta) = 0$, for each θ.

This result was first given by Hayes (1947). The derivation presented above is essentially that of Graham et al. (1957).

9–6 Extension to Include Lift and Side Force Elements

For a configuration that experiences lift and side forces one cannot represent the flow by a distribution of sources alone. The sources represent essentially volume elements. Lift and side forces may be represented by suitably oriented doublets, or, more conveniently, by elementary horseshoe vortices (see Section 5–2). Lift and side force elements were included in Hayes' original analysis. We state here his final result without proof.

Define a function $\tilde{h}$ such that

$$\tilde{h} = \tilde{f} - B(\tilde{l} \sin \theta + \tilde{s} \cos \theta), \tag{9-43}$$

where

$$\rho_\infty U_\infty^2 \tilde{l}(x, y, z) = \text{lift/unit volume,}$$

$$\rho_\infty U_\infty^2 \tilde{s}(x, y, z) = \text{side force/unit volume.}$$

The distributions of lift and side force may have singularities similar to (9–35). A continuous spatial distribution of forces is, of course, impossible to realize in any physical situation, as is also a continuous volumetric source distribution, but in principle it could be thought of as obtained from an infinitely dense cascade of infinitesimal wings, each holding an elementary horseshoe vortex.

Define further, as in (9–42),

$$h(x_i; \theta) = \iiint\limits_{V(x_i; \theta)} \tilde{h}(Q) \, dV, \tag{9-44}$$

where $V(x_i; \theta)$ is, as before, the region between two adjacent Mach planes perpendicular to θ. Then the drag will be given by

$$\frac{dD_w}{d\theta} = -\frac{\rho_\infty U_\infty^2}{8\pi^2} \int_0^l \int_0^l h'(x_1; \theta) h'(x_2; \theta) \ln |x_2 - x_1| \, dx_1 \, dx_2. \tag{9-45}$$

The term $B(\tilde{l} \sin \theta + \tilde{s} \cos \theta)$ is proportional to the component of force in the direction of θ, and the result thus indicates that this is the only lateral force component contributing to the wave drag.

Equation (9–45) makes it possible to determine the wave drag of an arbitrary spatial system containing thickness and carrying both lift and side forces. In order to determine the total pressure drag of the system it is necessary to evaluate the vortex drag produced by the lift and side force. This calculation will involve the pressure forces acting on the rear control surface S_3, as discussed before in Section 9–2. For a lifting planar wing the appropriate result is that given by (9–4).

9–7 The Supersonic Area Rule

Hayes' result can be directly applied for the calculation of wave drag of any body satisfying the "closure condition" $f(0; \theta) = f(l; \theta) = 0$, provided we know how the source strength is related to the shape of the body. Such a relationship is known for the case of thin wings and slender bodies of revolution, where the source strength is directly proportional to the rate of change of the area in the flow direction.

Let us first consider a thin nonlifting thin wing for which (5–30), (8–2), and (8–5) together with (5–1) give

$$\tilde{f}(x, y, z) = 2\tau \frac{\partial \bar{g}}{\partial x} \delta(z), \qquad (9\text{–}46)$$

where $\delta(z)$ is the Dirac delta function and $2\tau\bar{g}$ is the thickness distribution. The intersections of two neighboring Mach planes with the x,y-plane are shown in Fig. 9–5. The surface element for the strip of area between the two intersections is $dx_i \, dy$ and the total equivalent source strength is thus

$$f(x_i; \theta) \, dx_i = 2\tau \int \frac{\partial \bar{g} \, (x_i + By \cos \theta; y)}{\partial x} \, dy \, dx_i.$$

That is, provided $\bar{g}$ is continuous,

$$f(x_i; \theta) = \frac{\partial}{\partial x} \int 2\tau\bar{g}(x_i + By \cos \theta; y) \, dy. \qquad (9\text{–}47)$$

This integral can be given a simple geometrical interpretation. It represents the area cut off by the Mach plane and projected on a plane normal to the x-axis (see Fig. 9–5). Let this area be $S_w(x_i; \theta)$. Thus from (9–47)

$$f(x_i; \theta) = S'_w(x_i; \theta), \qquad (9\text{–}48)$$

where a prime denotes differentiation with respect to x.

Consider next a slender body of revolution. We already know that the flow around such a body can be represented by sources along the x-axis of strength $U_\infty S'_f(x)$, so that

$$f(x_i) = S'_f(x_i), \qquad (9\text{–}49)$$

independent of θ. Since for a very slender body of revolution the projected area of the oblique cut will not differ appreciably from the cut normal to the x-axis, we may thus use the geometrical design shown in Fig. 9–5 in this case also.

It follows from the equivalence rule, Section 6–5, that the geometrical method will give a good approximation to the actual equivalent source strength for any slender fuselage whose surface slope is everywhere small. Thus, it should hold approximately for any nonlifting configuration con-

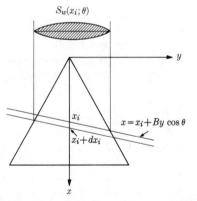

Fig. 9–5. Mach plane cut of a thin wing.

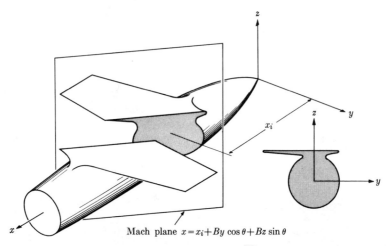

Mach plane $x = x_i + By \cos\theta + Bz \sin\theta$

FIG. 9–6. Mach plane cut of a wing-fuselage combination. (From Lomax and Heaslet, 1956. Courtesy of the American Institute of Aeronautics and Astronautics.)

sisting of a slender fuselage and a thin wing so that, using (9–48) and the equation for the wave drag (9–41), we obtain

$$D_w = -\frac{\rho_\infty U_\infty^2}{8\pi^2} \int_0^{2\pi} d\theta \int_0^l \int_0^l S''(x_1; \theta)S''(x_2; \theta) \ln|x_1 - x_2| \, dx_1 \, dx_2,$$

$$(9\text{--}50)$$

where $S'(x; \theta)$ is obtained by cutting the configuration by a series of Mach planes as shown in Fig. 9–6.

For a winged body of revolution the area distribution of the equivalent body of revolution is thus

$$S(x; \theta) = S_f(x) + S_w(x; \theta), \qquad (9\text{--}51)$$

where S_f is the fuselage cross-sectional area and S_w follows from (9–46) and (9–47). This corresponds to a simplified evaluation of $S(x; \theta)$ as sketched in Fig. 9–7. That is, the variation with z is neglected when the oblique plane cuts the wing, and the variation with both z and y is neglected when the oblique plane cuts the body.

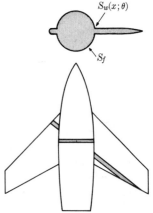

FIG. 9–7. Simplified evaluation of Mach plane cut. (From Lomax and Heaslet, 1956. Courtesy of the American Institute of Aeronautics and Astronautics.)

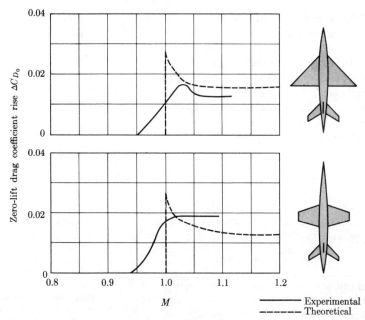

FIG. 9–8. Comparison of theoretical and experimental wave drag. (From Holdaway, 1953. Courtesy of National Aeronautics and Space Administration.)

An example of the use of (9–50) together with the simplified evaluation of $S(x; \theta)$ is shown in Fig. 9–8, taken from Holdaway (1953).

We shall now show how (9–50) and (9–51) can be used to determine shapes with low drag. Let us first expand $S'_w(x; \theta)$ in a Fourier series in θ. Such a series will, because of the symmetry of the wing, contain only terms of the form $\cos 2n\theta$. Thus

$$S'_w(x; \theta) = \sum_{n=0}^{\infty} A_{2n}(x) \cos 2n\theta. \qquad (9\text{–}52)$$

Now this series may be introduced into (9–50) and the integration over θ carried out. This gives

$$D_w = - \frac{\rho_\infty U_\infty^2}{4\pi} \int_0^l \int_0^l [S''_f(x_1) + A'_0(x_1)][S''_f(x_2) + A'_0(x_2)]$$

$$\times \ln |x_1 - x_2| \, dx_1 \, dx_2 - \frac{\rho_\infty U_\infty^2}{8\pi} \sum_{n=1}^{\infty} \int_0^l \int_0^l A'_{2n}(x_1) A'_{2n}(x_2)$$

$$\times \ln |x_1 - x_2| \, dx_1 \, dx_2. \qquad (9\text{–}53)$$

It follows from (9–53) that the interaction of fuselage and the wing is given by the first term of the series (9–52) only, and that the drag of the higher components is determined solely by the wing design. The complete effect of wing-body interference, therefore, appears in the sum of A_0 and the fuselage area distribution determined by planes normal to the free stream. Now

$$A_0(x) = \frac{1}{2\pi} \int_0^{2\pi} S_w'(x; \theta) \, d\theta, \quad (9\text{–}54)$$

and the minimum drag, for a given wing design, is obtained by selecting a fuselage shape such that

$$S_f'(x) + A_0(x) \qquad 9(\text{–}55)$$

corresponds to a body of optimum shape. The optimum slender body for a given volume and length is the Sears-Haack body, thus the fuselage should be chosen such that (9–55) corresponds to a Sears-Haack body as given by (9–34). The fuselage selected in this way will generally be indented in the region around the wing junction.

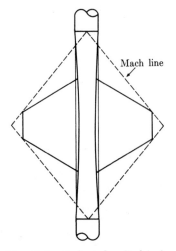

Fig. 9–9. Indented cylindrical fuselage for minimum wave drag at a supersonic Mach number. (From Lomax and Heaslet, 1956. Courtesy of the American Institute of Aeronautics and Astronautics.)

A simple illustration of this is the case when the body, instead of being finite, is an infinite cylinder. The optimum body indentation is then given by

$$\Delta S_f'(x) + A_0(x) = 0, \qquad (9\text{–}56)$$

which might give a shape like that illustrated in Fig. 9–9. One can show that in this case the total drag of the configuration is equal to the drag of the exposed wing alone, minus the drag of the indentation alone. For $M = 1$ the coefficients A_{2n} for $n \geq 1$ become zero and the total wave drag would then be zero. However, the linearized theory will, of course, break down for transonic speeds as is further discussed in Chapter 12.

The procedure described above is the *supersonic area rule* as presented by Lomax and Heaslet (1956). In the limit of $M = 1$ it goes over into the transonic area rule, since then the Mach planes become normal to the x-axis. To give an example of the drag reductions that can be realized by use of the supersonic area rule we reproduce in Fig. 9–10 some results given by Holdaway (1953). As seen from the curve, the modification gives a total drag reduction of about 20%, and the drag values predicted by the

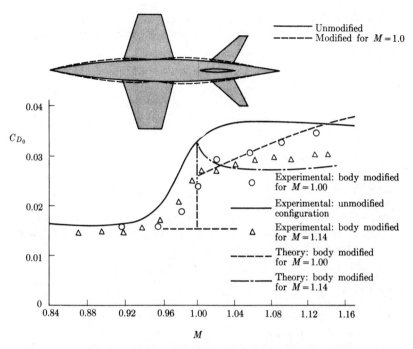

FIG. 9–10. Theoretical and experimental zero-lift wave drag for a wing-body-tail combination with modified bodies. (Adapted from Holdaway, 1954. Courtesy of National Aeronautics and Space Administration.)

theory are in good agreement with experimental values when the skin friction drag has been added ($\simeq$ the drag at subsonic speeds). The only exception is the immediate region around $M = 1$, particularly for the lower transonic region ($M < 1$), for which linearized theory predicts zero wave drag.

9–8 Wave Drag Due to Lift

Consider first the wave drag due to lift on a lifting wing of zero thickness located in the x,y-plane. By setting $\bar{f} = \bar{s} = 0$ in (9–43) we find from (9–44) and (9–45) that

$$D_w = - \frac{\rho_\infty U_\infty^2 B^2}{8\pi^2} \int_0^{2\pi} \sin^2 \theta \, d\theta \int_0^l \int_0^l l'(x_1; \theta) l'(x_2; \theta)$$

$$\times \ln |x_1 - x_2| \, dx_1 \, dx_2, \tag{9–57}$$

where $l(x_i; \theta) \, dx_i$ is the total lift (divided by $\rho_\infty U_\infty^2$) on the portion of the wing cut off by two Mach planes intersecting the x-axis at $x = x_i$ and $x = x_i + dx_i$:

$$
\begin{aligned}
l(x_i; \theta) \, dx_i &= \iint\limits_{V(x_i; \theta)} \bar{l}(Q) \, dV \\
&= \frac{dx_i}{\rho_\infty U_\infty^2} \int \Delta p(x_i + By \cos \theta, y) \, dy. \quad (9\text{--}58)
\end{aligned}
$$

From (9–57) we can draw some important conclusions. First, the wave drag due to lift is proportional to B^2; consequently it vanishes as $M \to 1$. Secondly, by analogy with the result for the vortex drag, minimum wave drag is obtained if, for each θ, the resulting loading $l(x; \theta)$ is elliptic in x. This can be realized, for example, by an elliptic planform with constant loading. This result was first given by Jones (1952). The wave drag of such a wing can be shown to be

$$
D_w = \frac{2L^2}{\pi \rho_\infty U_\infty^2 b^2} \left[\sqrt{1 + \frac{b^2 B^2}{c^2}} - 1 \right], \quad (9\text{--}59)
$$

where L is the total lift, b is the span of the wing, and c is the maximum chord. In order to get a low wave drag, both the chord and span should thus be large, i.e., the lift should be spread out as much as possible both in the spanwise and chordwise directions. The vortex drag is also minimum for an elliptic loading and is found from (7–46) and (7–47) to be

$$
D_v = \frac{2L^2}{\pi \rho_\infty U_\infty^2 b^2}. \quad (9\text{--}60)
$$

Thus the total drag due to lift for the elliptic wing of constant loading is

$$
D_i = D_w + D_v = \frac{2L^2}{\pi \rho_\infty U_\infty^2 b^2} \sqrt{1 + \frac{b^2 B^2}{c^2}} \quad (9\text{--}61)
$$

or, in terms of C_L and C_{Di}

$$
C_{Di} = \frac{B C_L^2}{4} \sqrt{1 + \left(\frac{1}{\pi B A} \right)^2}, \quad (9\text{--}62)
$$

showing that the drag due to lift for a straight elliptic wing is always greater than in two-dimensional flow ($C_{Di} = B C_L^2/4$). Lower values can be obtained by yawing the elliptic wing behind the Mach cone as shown by Jones (1952).

The wave drag interference between lift and thickness is zero for a midwing configuration. This can be seen by introducing (9–49) together with the Fourier series (9–52) into (9–45). Then the wave drag becomes

$$
D_w = -\frac{\rho_\infty U_\infty^2}{8\pi^2} \int_0^{2\pi} d\theta \int_0^l \int_0^l
$$

$$
\times \left[S_f''(x_1) + \sum_0^\infty A_{2n}'(x_1) \cos 2n\theta - B \sin \theta l'(x_1; \theta) \right]
$$

$$
\times \left[S_f''(x_2) + \sum_0^\infty A_{2n}'(x_2) \cos 2n\theta - B \sin \theta l'(x_2; \theta) \right]
$$

$$
\times \ln |x_1 - x_2| \, dx_1 \, dx_2. \tag{9–63}
$$

If $l(x; \theta)$ is also expanded in a Fourier series it will only contain $\cos n\theta$-terms since it must be an even function of θ. Hence, since

$$
\sin \theta \cos m\theta = \tfrac{1}{2}[\sin (m + 1)\theta - \sin (m - 1)\theta],
$$

any term containing the cross product between the coefficients for the two Fourier series will vanish in the integration over θ, and the interference drag will be zero.

Although for a planar wing there is no drag interaction between the thickness and lift distributions, an interaction between body thickness and wing lift is obtained if the fuselage is not in the plane of the wing. This is evident from the fact that, if the wing is not located on the x,y-plane, $l(x; \theta)$ will not be an even function of θ and the Fourier series for l will thus also contain $\sin n\theta$-terms. A favorable interference can often be obtained with a high-wing configuration as shown, for example, by Lomax and Heaslet (1956). This possibility can be easily demonstrated without any calculations by the aid of Fig. 9–11.

The wing is assumed to be a flat plate at zero angle of attack located above the fuselage in the position shown. The upwash created by the forward portion of the fuselage causes a lift on the wing. Also the reflected overpressure region from the wing will impinge on the rear portion of the fuselage where the surface slopes are negative. Hence the wing will have a lift without drag, and the drag of the fuselage will be lower due to interference from the wing. Through the interference a lift has thus been generated with a decrease of total pressure drag. Further developments may be found in a paper by Ferri *et al.* (1957).

FIG. 9–11. Interference between thickness and lift.

10

Use of Flow-Reversal Theorems in Drag Minimization Problems

10–1 Introduction

In the previous chapter we considered the important problem of how to determine shapes of lowest possible drag at supersonic speed. Generally, this problem is quite difficult. Hayes' drag formula, (9–45), however, allows us to treat a considerably simpler problem, namely that of finding distributions of singularities like sources and force elements having the smallest possible drag. It remains as a separate problem to find the shape corresponding to the singularity distribution. This may sometimes be more difficult than to obtain the singularity distribution itself. Nevertheless, certain overall properties, like total lift, side force, or volume, may be calculated directly from the singularity distribution without knowledge of the body shape. Thus one can determine the minimum pressure drag for, say, a given total lift under some general constraints such as given wing planform. Obviously, such information is useful in assessing the quality of a particular configuration since it provides an absolute minimum for purposes of comparison.

For drag minimization problems it has been found useful to employ certain general theorems that relate the drag of lift, side force, and volume distributions in forward and reverse flows. Such theorems were first put forward by von Kármán (1947a) and Hayes (1947). Ursell and Ward (1950) showed that the theorems hold under quite general conditions. Munk (1950) introduced the very important concept of a "combined flow field," obtained by superimposing the disturbance velocities in forward and reverse flows. Using this idea, Jones (1951, 1952) was able to derive criteria for identifying configurations of minimum drag. However, Jones' method does not give any clue to how to design an optimum shape, nor does it give the actual value of the minimum drag.

If we are primarily interested in the attainable minimum drag, and not in the first place in the actual shape of the body producing the drag, it would be desirable to go one step further in simplification and determine the minimum drag without having to calculate the actual optimum singularity distribution representing the body. That this is indeed possible was first stated by Nikolsky (1956) reporting on Russian work in this field (see later publications by Kogan, 1957, and Zhilin, 1957). Independent proofs confirming this possibility were given by Ward (1956), E. W. Graham (1956), and Germain (1957a, b). We shall here present Graham's formulation, which is based on the combined flow field concept.

10–2 Drag of a General Singularity Distribution from a "Close" Viewpoint

Previously, we have calculated the drag by considering the flow of momentum through a control surface at large distances from the body. An alternative approach is to take the "close" viewpoint, namely to integrate pressures times slope of the surface:

$$D = -\iint p(\mathbf{i} \cdot \mathbf{n}) \, dS, \tag{10–1}$$

where $\mathbf{n}$ is the outward normal to the body.

Instead of fixing attention on a certain body, we shall, as in the derivation of Hayes' drag formula, consider a general spatial distribution of sources of strength $U_\infty \tilde{f}$, and lift and side-force elements of strength $\rho_\infty U_\infty^2 \tilde{l}$ and $\rho_\infty U_\infty^2 \tilde{s}$ respectively, all per unit volume. These will all be considered continuous in the present section. The lift and side-force distributions are associated with a distribution of elementary horseshoe vortices. The source distribution may be given a geometrical interpretation. Consider a lineal distribution of strength f, for which slender-body theory gives (see equation 6–26)

$$f = \frac{dS}{dx} \equiv \lim_{\Delta x \to 0} \frac{\Delta S}{\Delta x}.$$

Hence from (9–35) we obtain

$$f \, \Delta x = \tilde{f} \, \Delta x \, \Delta y \, \Delta z = \Delta S, \tag{10–2}$$

so that $\tilde{f}$ may be interpreted as a distribution of *frontal area elements*. In problems with fixed total body volume it is convenient, instead of $\tilde{f}$, to work with a distribution $\tilde{v}$, defined as

$$\frac{\partial \tilde{v}}{\partial x} = \tilde{f}. \tag{10–3}$$

From (10–2) it follows that $\tilde{v}$ represents *volume elements*.

The volume or frontal area elements, and the force elements, will cause perturbation velocities in the flow with components $U_\infty u$, $U_\infty v$, and $U_\infty w$. For a continuous distribution the pressure induced is given to first order by

$$p - p_\infty = -\rho_\infty U_\infty^2 u. \tag{10–4}$$

This expression holds for any steady nonviscous small-perturbation flow (potential or nonpotential) as follows from the x-momentum equation.

Consider now the drag of one small lift element

$$\Delta l = \rho_\infty U_\infty^2 \tilde{l} \, \Delta x \, \Delta y \, \Delta z.$$

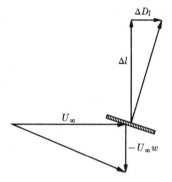

Physically, it can be represented by a small piece of a zero-thickness wing aligned with the local streamline and carrying the lift Δl. In a frictionless flow within the small-perturbation approximation (see Fig. 10–1), its drag will be

$$\Delta D_l = -\Delta l w, \qquad (10\text{–}5)$$

Fig. 10–1. Drag on a lift element.

where $U_\infty w$ is the upwash at the point where the element is introduced. Similarly, the drag of a small side-force element

$$\Delta s = \rho_\infty U_\infty^2 \tilde{s} \, \Delta x \, \Delta y \, \Delta z$$

becomes (side force defined as positive in the positive y-direction)

$$\Delta D_s = -\Delta s v. \qquad (10\text{–}6)$$

The drag of a frontal area element $\Delta S = \tilde{f} \, \Delta x \, \Delta y \, \Delta z$ is

$$\Delta D_f = \Delta S(p - p_\infty) = -\rho_\infty U_\infty^2 u \, \Delta S. \qquad (10\text{–}7)$$

Summing up all contributions from infinitesimal elements we obtain for the total drag

$$\frac{D}{\rho_\infty U_\infty^2} = -\iiint (\tilde{l} w + \tilde{s} v + \tilde{f} u) \, dV, \qquad (10\text{–}8)$$

where the integral is to be taken over the entire volume V in which singularities are present. Alternatively, if volume elements are used instead of frontal area elements to represent the thickness distribution of the body, one finds that the force on each volume element $\Delta V = \tilde{\nu} \, \Delta x \, \Delta y \, \Delta z$ is a buoyancy force proportional to the local pressure gradient in the x-direction and volume strength:

$$\Delta D_\nu = -\Delta V \frac{\partial p}{\partial x} = \rho_\infty U_\infty^2 \frac{\partial u}{\partial x} \Delta V, \qquad (10\text{–}9)$$

so that (10–8) may also be written

$$\frac{D}{\rho_\infty U_\infty^2} = -\iiint (\tilde{l} w + \tilde{s} v) \, dV + \iiint \tilde{\nu} \frac{\partial u}{\partial x} \, dV. \qquad (10\text{–}10)$$

This could also be obtained from (10–8) by integrating the last term by parts.

Obviously, whenever an additional elementary singularity is introduced into the field, it will cause a change in the disturbance velocity field proportional to the strength of the singularity. This in turn will change the drag of the previous elements by an amount of the same order of magnitude as the drag of the additional element itself. Thus there is an interference drag between the original distribution and the added element.

The disturbance velocity field in (10–8) and (10–10) is caused by all the lift, side-force, and thickness elements themselves. It should be emphasized that these equations are strictly valid only when the distributions l, $\bar{s}$, and $\bar{f}$ as well as the perturbation velocities u, v, and w are nonsingular. Hence, any leading edge suction is excluded, as well as lineal source distributions, since these give infinite perturbation velocities along the source line. Otherwise, (10–8) and (10–10) are valid regardless of whether the flow is sub- or supersonic. The expressions for drag due to lift and side force also hold for unsteady flow.

When using the present "close viewpoint" there is no way of distinguishing between vortex and wave drag; only the total pressure drag is obtained. If desired, one can compute the vortex drag by use of the Trefftz plane and then separate it from the total drag to obtain the wave drag in supersonic flight.

10–3 The Drag Due to Lift in Forward and Reverse Flows

We shall first consider the drag of a lift distribution alone. See Fig. 10–2.

Each element of drag involves the product of the lift of one element Δl_1 and the angle of attack induced at its position by another element Δl_2. This angle may be written

$$w_{12} = W_{12}\, \Delta l_2, \qquad (10\text{--}11)$$

where W_{12} is the "influence function" (Green's function for the problem, i.e., the upwash angle induced at position 1 due to a unit lift at position 2). The element of drag is then proportional to

$$-\Delta l_1 W_{12}\, \Delta l_2. \qquad (10\text{--}12)$$

Now consider the case for which the direction of the free stream of velocity U_∞ is reversed but the distribution of lift is kept the same. To do this, it is generally necessary to change the angle

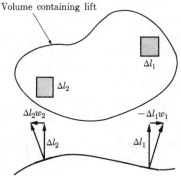

Fig. 10–2. Mutual drag of two lifting elements.

of attack of each lifting element so that the disturbance velocities will be different from those in forward flow. In the reverse flow, element 2 will lie in the same relation to element 1 as was formerly occupied by Δl_1 relative to Δl_2. Hence the influence function W_{21} in reverse flow will be equal to W_{12} in forward flow and the element of drag in reverse flow will thus be

$$-\Delta l_2 W_{21} \, \Delta l_1 \, = \, -\Delta l_1 W_{12} \, \Delta l_2. \tag{10–13}$$

Consequently, for every element of drag in forward flow there will be an equal element in reverse flow. The total drag, which is obtained by summing the contribution from each pair of lifting elements within the volume, will thus be the same in forward and reverse flow. Hence we have proved that:

> *The drag of a given distribution of lift is unchanged by a reversal of the flow direction.*

10–4 The Drag Due to Thickness in Forward and Reverse Flows

A completely analogous approach can be used for a given distribution of sources or frontal area elements. Consider two frontal area elements at points 1 and 2 of strength ΔS_1 and ΔS_2, respectively. Referring to (10–7), the drag on the element 1 due to the element 2 is proportional to the frontal area or source strength at 1 times the pressure (u-perturbation) induced at 1 due to the area element at 2. This pressure disturbance may be written

$$-\rho_\infty U_\infty^2 u_{12} \, = \, -\rho_\infty U_\infty^2 U_{12} \, \Delta S_2, \tag{10–14}$$

where U_{12} denotes the u-perturbation at point 1 due to an element of unit strength at 2. The element of drag is thus

$$-\rho_\infty U_\infty^2 U_{12} \, \Delta S_2 \, \Delta S_1. \tag{10–15}$$

If the direction of flow is reversed, the sign of the frontal area elements of sources must be reversed in order that the volume remain positive. Otherwise the argument is exactly identical with the one employed in the preceding section, and one thus can directly extend the result obtained there and state that:

> *The drag of a given volume or thickness distribution is the same in forward and reverse flow.*

10–5 The Drag of a General Distribution of Singularities in Steady Supersonic Forward or Reverse Flows

Using Hayes' method, we can obtain an alternative derivation of the two preceding theorems, valid for supersonic steady flow only, but which does allow leading edge pressure singularities to be present as well as lineal source distributions.

Consider first a lineal source distribution with $f(0) = f(l) = 0$. The drag of this in forward flow is given by (9–19):

$$D_F = - \frac{\rho_8 U_\infty^2}{4\pi} \int_0^l \int_0^l f_F'(x_1) f_F'(x_2) \ln |x_1 - x_2| \, dx_1 \, dx_2, \qquad (10\text{–}16)$$

where index F denotes forward flow. The reversal of the flow direction is equivalent to replacing x_1 and x_2 by $-x_1$ and $-x_2$. In order to preserve the sign of the frontal area or volume distribution, the source strength must change sign in reverse flow. Thus the drag in reverse flow is

$$D_R = - \frac{\rho_\infty U_\infty^2}{4\pi} \int_{-l}^0 \int_{-l}^0 f_R'(-x_1) f_R'(-x_2) \ln |x_2 - x_1| \, dx_1 \, dx_2, \qquad (10\text{–}17)$$

which is seen to be identical to (10–16) with $f_R = -f_F$. Hence:

The drag of a lineal source distribution is equal in forward and reverse flight.

For a general distribution of singularities we have, according to (9–45), in forward flow

$$D_F = - \frac{\rho_\infty U_\infty^2}{8\pi^2} \int_0^{2\pi} d\theta \int_0^l \int_0^l h_F'(x_1; \theta) h_F'(x_2; \theta) \ln |x_1 - x_2| \, dx_1 \, dx_2, \qquad (10\text{–}18)$$

where

$$h_F(x_i; \theta) = f_F(x_i; \theta) - B[l_F(x_i; \theta) \sin \theta + s_F(x_i; \theta) \cos \theta]. \qquad (10\text{–}19)$$

Here l_F and s_F are the lumped lift and side-force elements along the Mach planes as defined for l in (9–58). Consider next the reverse flow case.

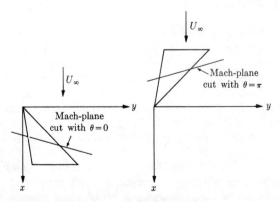

Fig. 10–3. Mach-plane cuts in forward and reverse flows.

Each Mach-plane cut at angle θ in reverse flow is identical to that for $\theta + \pi$ in forward flow with x replaced by $-x$ (see Fig. 10–3). Hence

$$
\begin{aligned}
h_R(x; \theta) &= -f_F(-x; \theta + \pi) \\
&\quad -B[l_F(-x; \theta + \pi) \sin \theta + s_F(-x; \theta + \pi) \cos \theta] \\
&= -h_F(-x; \theta + \pi),
\end{aligned}
\tag{10–20}
$$

where the sign of the source strength has been reversed. Introducing (10–20) into the drag formula, we then find that the wave drag of a general distribution of singularities is the same in forward and reverse flows provided the sign of the source strength is reversed. The vortex drag depends only on the spanwise distribution of lift and side forces and must therefore be the same in forward and reverse flow. Hence:

> *The drag of a general distribution of thickness, lift, and side-force elements in steady supersonic flow is the same in forward and reverse flight.*

10–6 The Combined Flow Field

The idea of the combined flow field was first introduced by Munk (1950) and later employed by Jones (1951, 1952). The combined flow is the (completely artificial) flow field obtained by superimposing half the disturbance velocities in forward and reverse flow of a given distribution of lift, side force, and thickness. Thus the disturbance velocities $\bar{u}$, $\bar{v}$, and $\bar{w}$ in the combined flow field are

$$
\begin{aligned}
\bar{u} &= \tfrac{1}{2}(u_F + u_R) \\
\bar{v} &= \tfrac{1}{2}(v_F + v_R) \\
\bar{w} &= \tfrac{1}{2}(w_F + w_R).
\end{aligned}
\tag{10–21}
$$

(The factor $\tfrac{1}{2}$ is introduced for convenience.) The reverse-flow quantities are measured in the same coordinate system, but with the free stream velocity U_∞ in the negative x-direction. In the combined flow field, discontinuities in the u-component are canceled, and so are discontinuities in w and v due to source distributions. Discontinuities in the v- and w-components due to lift and side-force surface distributions will remain, however (see, for example, Graham *et al.*, 1957). As a simple example of a combined field, we may consider the subsonic flow around an elementary horseshoe vortex, i.e., around a single lifting element. This is illustrated in Fig. 10–4. In the combined field, the infinitesimal vortex joining the two streamwise vortices is canceled, leaving two infinite streamwise vortices. The combined flow around any arrangement of lifting elements in subsonic

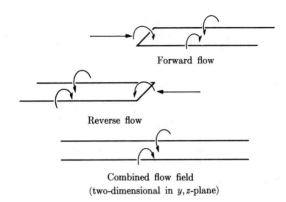

Forward flow

Reverse flow

Combined flow field
(two-dimensional in y, z-plane)

FIG. 10–4. Lifting element in subsonic flow.

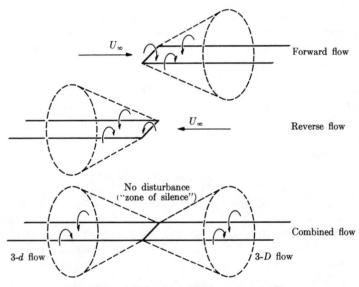

FIG. 10–5. A lifting element in supersonic flow.

flight will hence result in an infinite strip of vortices, and the flow will be entirely two-dimensional in the crossflow (y,z-) plane. In supersonic flight, however, the combined field will generally be three-dimensional (see Fig. 10–5).

By use of (10–8) or (10–10) and noting that the drag is equal in forward and reverse flight so that $D = D_F = D_R$, we may express the drag in the combined flow in the following way:

$$\frac{D}{\rho_\infty U_\infty^2} = \frac{D_F + D_R}{2\rho_\infty U_\infty^2} = -\iiint (\bar{l}\bar{w} + \bar{s}\bar{v})\, dV - \iiint \bar{j}\bar{u}\, dV. \qquad (10\text{--}22)$$

Alternatively, the last term in (10–22) may be given as

$$\iiint \bar{v}\frac{\partial \bar{u}}{\partial x}\, dV. \qquad (10\text{--}23)$$

Because of the cancellation of discontinuities in u-velocities as well as of (some of) the v- and w-discontinuities in the combined field, (10–22) is less restricted than (10–8) and (10–10). Thus it holds when leading-edge suction is present, as well as for a lineal distribution.

10–7 Use of the Combined Flow Field to Identify Minimum Drag Conditions

The usefulness of the combined flow field can be demonstrated as follows. Consider a distribution of singularities consisting of lift elements alone. We want to study the change in drag that occurs when a small element of lift δl is introduced at one point. This drag change will consist of the drag of the added element alone plus the interference drag δD_i between δl and l. The drag of the element alone is of order $(\delta l)^2$, whereas the interference drag is of order δl. In order to calculate the interference drag we divide the basic distribution l into two portions l_1 and l_2, of which l_1 is unaware of δl in forward flow and l_2 is unaware of δl in reverse flow (see Fig. 10–6). Such a division is, of course, only possible in a supersonic flow. The interference drag in forward flow may be written:

$$\delta D_{iF}(\delta l, l) = \delta D_{iF}(\delta l, l_1) + \delta D_{iF}(\delta l, l_2). \qquad (10\text{--}24)$$

Now, $\delta D_{iF}(\delta l, l_1)$ can only be $-\rho_\infty U_\infty^2 \delta l w_F$, where $U_\infty w_F$ is the upwash velocity at δl due to l_1 in forward flight, since the law of forbidden signals allows no w-change at l_1 due to δl. The interference drag between δl and l_2 must be the same in forward as in reverse flow, so that (10–24) may consequently be written

$$\delta D_i(\delta l, l) = \delta D_{iF}(\delta l, l_1) + \delta D_{iR}(\delta l, l_2). \qquad (10\text{--}25)$$

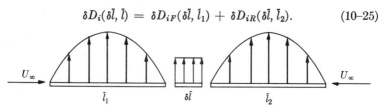

FIG. 10–6. Division of basic lift distribution.

Hence, if $U_\infty w_R$ is the upwash at $\delta \bar{l}$ due to $\bar{l}_2$ in reverse flow,

$$\frac{\delta D_i}{\rho_\infty U_\infty^2} = -\delta \bar{l} w_F - \delta \bar{l} w_R = -2\delta \bar{l}\bar{w}. \tag{10–26}$$

This result holds for either forward or reverse flow.

Assume now that we want to investigate whether a particular lift distribution gives a minimum drag. The total lift and the wing planform (or the volume containing the lift) are assumed to be given. We can test whether the distribution is minimum by removing a small lift element $\delta \bar{l}$ from one point P_1 and adding another of equal strength at a point P_2. This operation does not change the total lift, but the drag will get an increment which is equal to the sum of:

(1) The drag of each element isolated [of order $(\delta \bar{l})^2$].

(2) The drag due to interference between the two elements [of order $(\delta \bar{l})^2$].

(3) The drag due to interference between the elements and the original distribution (of order $\delta \bar{l}$).

If $\delta \bar{l}$ is made sufficiently small the third term will dominate. This term may be calculated by use of (10–26). Hence the change in drag to first order will be

$$\frac{\delta D_i}{\rho_\infty U_\infty^2} = 2\delta \bar{l}(\bar{w}_1 - \bar{w}_2). \tag{10–27}$$

If $\bar{w}_2 > \bar{w}_1$, then the drag can evidently be lowered by transferring some lift from P_1 to P_2. On the other hand, if $\bar{w}_1 > \bar{w}_2$, the drag will be decreased by making $\delta \bar{l}$ negative, i.e., transferring lift from P_2 to P_1. There will be no first-order change in drag if $\bar{w}_1 = \bar{w}_2$. Repeating this argument for every pair of points on the planform, we thus find that to first order the drag will be invariant to small arbitrary changes in the lift distribution, i.e., will have an extremum, if $\bar{w}$ is constant over the planform. This extremum must be a minimum, because the second variation given by the terms proportional to $(\delta \bar{l})^2$ must be positive, since it represents the drag of the two lifting elements acting alone. Hence:

(a) *If a distribution of lift is to have minimum drag for a given planform and given total lift, $\bar{w}$ must be constant over the planform.*

In a similar way a necessary condition for minimum drag for a thickness distribution of given length and base area may be derived. Such a body must be imagined as having a downstream parallel extension of infinite length. It is easily seen that the condition is that $\bar{u} = $ const. since, if it is not, area elements can be transferred within the space occupied by the source distribution so as to obtain a drag reduction. Hence:

(b) *If the drag of a nonlifting body with given length and base area is to be minimum, the pressure in the combined flow field must be constant in the region occupied by the thickness distribution.*

It can be shown that this condition is fulfilled for a von Kármán ogive.

Finally, for a thickness distribution of given total length and volume to have minimum drag, $\partial \bar{u}/\partial x$ must be constant since otherwise volume

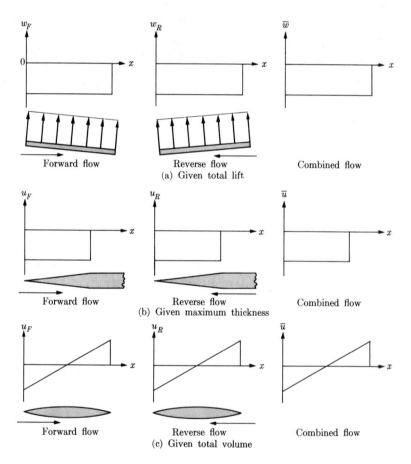

Fig. 10–7. Configurations having minimum drag in two-dimensional supersonic flow under various constraints.

elements can be transferred so as to obtain a drag reduction. Hence:

(c) *In order for a thickness distribution of given length and volume to have minimum drag, the pressure gradient in the combined flow field must be constant in the region occupied by the thickness distribution.*

It can be shown that this condition holds for a Sears-Haack body.

Finally, let us consider the case where the total lift is given over a prescribed region of space, but total volume and side force within the region are not prescribed. From (a) it follows that $\bar{w}$ must be constant over the region *containing the lift elements.* Furthermore, $\bar{v}$ must be zero, otherwise side-force elements may be added to reduce the drag. In addition $\partial\bar{u}/\partial x$, and $\bar{u}$, must be zero, since otherwise volume or thickness distributions of nonzero net strength may be added to reduce the drag. Hence:

(d) *In order for a distribution of singularities within a given space to have minimum drag for fixed total lift, the combined flow field must have $\bar{u} = \bar{v} = 0$ and $\bar{w} = $ const within the region.*

It should be noticed that this condition does not necessarily lead to zero total side force and volume within the region.

Simple examples of the application of these criteria can be given for two-dimensional supersonic flow. A two-dimensional uniform loading has minimum drag for given total lift, since both in forward and reverse flow the corresponding upwash is constant and hence $\bar{w}$ is constant. A single wedge gives minimum drag for a given chord and frontal area, since u_F and u_R, and hence $\bar{u}$, are constant over the wedge. A nonlifting airfoil made up of two parabolic arcs has minimum drag for given total volume and chord, since both in forward and reverse flow $\partial u/\partial x$ is constant over the airfoil, and hence also $\partial\bar{u}/\partial x$. These cases are illustrated in Fig. 10–7.

10–8 The Calculation of Minimum Drag by Solution of an Equivalent Two-Dimensional Potential Problem

Although the criteria derived in the preceding section provide relatively simple methods to determine whether a given singularity distribution is optimum by considering the properties of the combined flow field, the problem of determining the optimum singularity distribution corresponding to a given combined flow field can be produced by individual forward and reverse flows in an infinite number of ways. This implies that there are singularity distributions which give zero perturbation velocities in the combined flow field. Such a simple distribution in two-dimensional flow is the thickness distribution corresponding to Busemann's biplane (see Fig. 10–8).

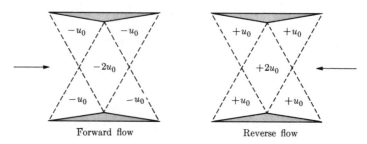

Fig. 10–8.   Example of two-dimensional thickness distribution producing $\bar{u} = \bar{w} = 0$.

However, as shown by Graham (1956), it may not be necessary for some purposes to calculate the actual distribution of singularities, since the desired information can be obtained directly from the combined flow field. For example, the total drag and lift, as well as the spanwise lift distribution, and thus the vortex drag, can be obtained directly from the combined flow field. Hence one may calculate the absolute minimum of drag without knowing what detailed distributions of lift, side force, and thickness produce the minimum drag. In some cases these distributions will probably be highly singular and impracticable. However, for preliminary design purposes it may nevertheless be useful to know the absolute limits that can be achieved.

Before proceeding, two definitions will be useful. The *Mach envelope* of a system is the bounding surface of the region in space which is affected by the system in forward and reverse flow. Thus, for a slender body pointed at both ends the Mach envelope is a double Mach cone (see Fig. 10–9). The *Mach envelope rim* is the closed line which marks the greatest lateral extent of the Mach cone; i.e., the rim is the intersection of the envelope of Mach waves caused by the system in forward flow with

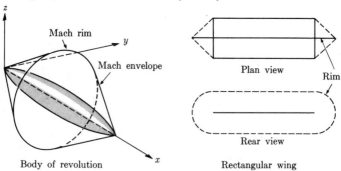

Fig. 10–9.   Mach rim and envelope for two simple bodies.

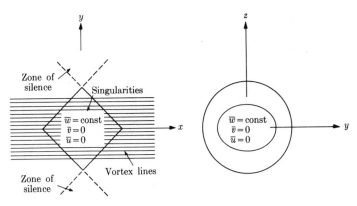

FIG. 10–10. Combined flow field for minimum drag.

the envelope of Mach waves caused by the system in reverse flow. For the case of a pointed slender body, for example, the rim is a circle.

Consider now the problem of finding the minimum drag for a given total lift but with no constraint on side force and volume, i.e., case (d) in the preceding section. Assume for simplicity that the Mach envelope is a double Mach cone. For minimum drag it is required that $\bar{u} = \bar{v} = 0$ and $\bar{w} = $ const in the region carrying lift. The combined flow field is illustrated in Fig. 10–10.

In the zone of silence no disturbance is felt in either forward or reverse flow; hence there $\bar{u} = \bar{v} = \bar{w} = 0$. Inside the Mach envelope we have $\bar{u} = \partial \bar{u} / \partial x = 0$, and the flow is consequently entirely two-dimensional in the crossflow plane that cuts the Mach envelope rim plane. For the plane through the Mach rim we thus have the following boundary value problem for the combined perturbation velocity potential $\bar{\varphi}$ for the region between the inner region and the Mach envelope rim:

(1) $\bar{\varphi}_{yy} + \bar{\varphi}_{zz} = (M^2 - 1)\bar{\varphi}_{xx} = 0.$

$$(10\text{–}28)$$

(2) On the boundary of the inner region

$$\bar{\varphi}_z = \text{const} = \bar{w}_0. \quad (10\text{–}29)$$

Alternatively, this condition may be written

$$\bar{\varphi}_n = \bar{w}_0 \sin \gamma, \quad (10\text{–}30)$$

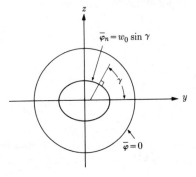

FIG. 10–11. Boundary-value problem for combined velocity potential.

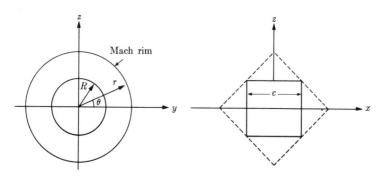

FIG. 10–12. Ring wing in supersonic flow.

where n is the unit outward normal to the inner boundary and γ the angle of the normal to the inner boundary with the y-axis (see Fig. 10–11).

(3) Outside the Mach envelope rim there must be no disturbance. This can be obtained by setting

$$\bar{\varphi} = 0 \qquad (10\text{–}31)$$

on the Mach envelope rim. Let the total lift $L = L_1 + L_2$, where L_1 is the lift carried to the left of a plane through the Mach rim and L_2 the lift to the right of this plane. Then, considering the forward flow alone and using a momentum control surface having its rear disk located at the plane through the Mach rim, we find, to the small perturbation approximation, that

$$L_1 = -\rho_\infty U_\infty^2 \iint\limits_{-\infty}^{\infty} w_F \, dy \, dz. \qquad (10\text{–}32)$$

In a similar fashion L_2 may be obtained by considering the reverse flow alone. Thus for the total lift*

$$L = -\rho_\infty U_\infty^2 \iint\limits_{-\infty}^{\infty} (w_F + w_R) \, dy \, dz = -2\rho_\infty U_\infty^2 \iint\limits_{-\infty}^{\infty} \overline{w} \, dy \, dz.$$

By introducing the potential for the combined flow field this may be simplified to

$$L = 2\rho_\infty U_\infty^2 \int \Delta\bar{\varphi} \, dy - 2\rho_\infty U_\infty^2 \overline{w}_0 S, \qquad (10\text{–}33)$$

where $\Delta\bar{\varphi}$ is the difference in $\bar{\varphi}$ between the upper and lower boundary of the lift-carrying region and S its cross-sectional area.

From (10–22) the drag of the distribution is found to be

$$D = -\overline{w}_0 L. \qquad (10\text{–}34)$$

---

* This derivation is due to Graham (1956) who also considered a more general case for which the Mach rim was not normal to the free stream.

As a simple example we shall calculate the minimum drag of a ring wing, a problem considered by Graham (1956) and in more detail by Beane (1960). For simplicity, only the case of large chord will be treated so that there is only an outer circular rim (see Fig. 10–12). The radius of the ring wing is set equal to R and its chord equal to c. The radius of the Mach envelope rim is easily found to be

$$R_r = R + \frac{c}{2B} = R(1 + c^*), \tag{10–35}$$

where

$$c^* = \frac{c}{2BR}.$$

In polar coordinates the boundary value problem reads

$$\bar{\varphi}_r = \bar{w}_0 \sin \theta \quad \text{for } r = R, \tag{10–36}$$

$$\bar{\varphi} = 0 \quad \text{for } r = R(1 + c^*). \tag{10–37}$$

It is easily verified that for $R \leq r \leq R(1 + c^*)$ the solution is

$$\bar{\varphi} = \frac{\bar{w}_0 \sin \theta}{r} \left[\frac{r^2 - (1 + c^*)^2 R^2}{1 + (1 + c^*)^2} \right]. \tag{10–38}$$

Inside the wing the solution is

$$\bar{\varphi} = \bar{w}_0 z = \bar{w}_0 r \sin \theta. \tag{10–39}$$

Hence the potential jump through the wing is

$$\Delta\bar{\varphi} = -2R\bar{w}_0 \sin \theta \frac{(1 + c^*)^2}{1 + (1 + c^*)^2}. \tag{10–40}$$

Application of (10–33) gives for the total lift (the cross-sectional area $S = 0$ in this case)

$$L = -4\rho_\infty U_\infty^2 \pi R^2 \bar{w}_0 \frac{(1 + c^*)^2}{1 + (1 + c^*)^2}, \tag{10–41}$$

and for the drag (10–34) gives

$$D = -\bar{w}_0 L = 4\rho_\infty U_\infty^2 \pi R^2 \bar{w}_0^2 \frac{(1 + c^*)^2}{1 + (1 + c^*)^2}. \tag{10–42}$$

Hence

$$\frac{C_D}{C_L^2} = \frac{1 + (1 + c^*)^2}{8(1 + c^*)^2} \to \frac{5}{32} \quad \text{for } c^* = 1, \tag{10–43}$$

where C_D and C_L are based on πR^2.

A thorough theoretical and experimental investigation of low-drag ring-wing configurations in which nonlinear effects were taken into account has been reported by Browand, et al. (1962).

11

Interference and Nonplanar
Lifting-Surface Theories

11-1 Introduction

For a general discussion of interference problems and linearized theoretical methods for analyzing them, the reader is referred to Ferrari (1957).
His review contains a comprehensive list of references, and although it
was editorially closed in 1955 only a few articles of fundamental importance
seem to have been published since that date.

The motivation for interference or interaction studies arises from the
fact that a flight vehicle is a collection of bodies, wings, and tail surfaces,
whereas most aerodynamic theory deals with individual lifting surfaces, or
other components in isolation. Ideally, one would like to have theoretical
methods of comparable accuracy which solve for the entire combined flow
field, satisfying all the various boundary conditions simultaneously.
Except for a few special situations like cascades and slender wing-body
configurations, this has proved impossible in practice. One has therefore
been forced to more approximate procedures, all of which pretty much
boil down to the following: first the disturbance flow field generated by
one element along the mean line or center surface of a second element is
calculated; then the angle-of-attack distribution and hence the loading
of the second element are modified in such a way as to cancel this "interference flow field" due to the first. Such interference effects are worked
out for each pair of elements in the vehicle which can be expected to inter-
act significantly. Since the theories are linear, the various increments can
be added to yield the total loading.

There are some pairs of elements for which interference is unidirectional.
Thus a supersonic wing can induce loading on a horizontal stabilizer
behind it, whereas the law of forbidden signals usually prevents the
stabilizer from influencing the wing. In such cases, the aforementioned
procedure yields the exactly correct interference loading within the limits
imposed by linearization. When the interaction is strong and mutual, as
in the case of an intersecting wing and fuselage, the correct combined flow
can be worked out only by an iteration process, a process which seems
usually to be stopped after the first step.

Interference problems can be categorized by the types of elements involved. The most common combinations are listed below.

(1) Wing and tail surfaces.

(2) Pairs or collections of wings (biplanes or cascades).

(3) Nonplanar lifting surfaces (T- and V-tails, hydrofoil-strut combinations).

(4) Wing or tail and fuselage or nacelle.

(5) Lifting surface and propulsion system, especially wing and propeller.

(6) Tunnel boundary, ground and free-surface effects.

It is also convenient to distinguish between subsonic and supersonic steady flight, since the flow fields are so different in·the two conditions.

In the present discussion, only the first three items are treated, and even within this limitation a number of effects are omitted. Regarding item 4, wing-fuselage interference, however, a few comments are worth making. Following Ferrari, one can roughly separate such problems into those with large aspect-ratio, relatively unswept wings and those with highly-swept, low aspect-ratio wings. Both at subsonic and (not too high) supersonic speeds, the latter can be analyzed by slender-body methods along the lines described in Chapter 6. The wings of wider span need different approaches, depending on the Mach number [cf. Sections C, 6–11 and C, 35–50 of Ferrari (1957)]. For instance, subsonically it appears to be satisfactory to replace the fuselage with an infinite cylinder and work with two-dimensional crossflow methods in the Trefftz plane.* At supersonic speeds, however, the bow wave from the pointed body may have a major influence in modifying the spanwise load distribution.

11–2 Interfering or Nonplanar Lifting Surfaces in Subsonic Flow

A unified theory of interference for three-dimensional lifting surfaces in a subsonic main stream can be built up around the concept of pressure or acceleration-potential doublets. We begin by appealing to the Prandtl-Glauert-Göthert law, described in Section 7–1, which permits us to restrict ourselves to incompressible fluids. Granted the availability of high-speed computing equipment, it then proves possible to represent the loading distribution on an arbitrary collection of surfaces (biplane, multiplane, T-tail, V-tail, wing-stabilizer combination, etc.) by distributing appropriately oriented doublets over all of them and numerically satisfying the flow-tangency boundary condition at a large enough set of control points. The procedure is essentially an extension of the one for planar wings that is sketched in Section 7–6.

---

* This scheme is associated with the names of Multhopp (1941) and Vandrey (1938).

Two observations are in order about the method described below. First it overlooks two sometimes significant phenomena that occur when applied to a pair of lifting surfaces aligned streamwise (e.g., wing and tail). These are the rolling up of the wake vortex sheet and finite thickness or reduced dynamic pressure in the wake due to stalling. They are reviewed at some length in Sections C,2 and C,4 of Ferrari (1957).

The second remark concerns thickness. In what follows, we represent the lifting surfaces solely with doublets, which amounts to assuming negligible thickness ratio. When two surfaces do not lie in the same plane, however, the flow due to the thickness of one of them can induce interference loads on the other, as indicated in Fig. 11–1. The presence of this thickness and the disturbance velocities produced at remote points thereby may be represented by source sheets in extension of the ideas set forth in Section 7–2. Since the procedure turns out to be fairly straightforward, it is not described in detail here.

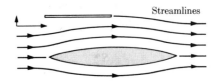

FIG. 11–1. Illustrating how streamlines due to thickness of one wing can modify the angle of attack of a second wing out of the plane of the first.

The necessary ideas for analyzing most subsonic interference loadings of the type listed under items 1, 2, and 3 can be developed by reference to the thin, slightly inclined, nonplanar lifting surface illustrated in Fig. 11–2. We use a curvilinear system of coordinates x, s to describe the surface of S, and the normal direction n is positive in the sense indicated. The small camber and angle of attack, described by the vertical deflection $\Delta z(x, y)$ or corresponding small normal displacement $\Delta n(x, s)$, are superimposed on the basic surface $z_0(y)$. The latter is cylindrical, with generators in the free-stream x-direction. To describe the local surface slope in y,z-planes, we use

$$\psi(y) = \tan^{-1}\left(\frac{dz_0}{dy}\right). \tag{11-1}$$

The flow tangency boundary condition can be written in terms of the dimensionless normal-velocity perturbation,

$$v_n[x, y, z_0(y)] = \frac{\partial}{\partial x}(\Delta n), \qquad \text{for } (x, y) \text{ on } S. \tag{11-2}$$

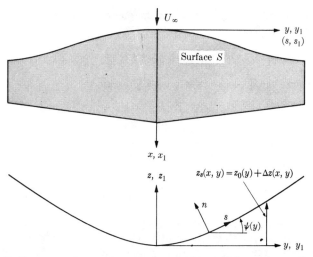

FIG. 11-2. Geometry of a nonplanar surface, without thickness, in a uniform flow.

As mentioned previously, a lifting flow with the desired properties can be constructed from a layer of pressure or acceleration-potential doublets all over S, having their axes pointed in the local n-direction. We start by deducing from (7–32) or (5–35) that, for such a doublet placed at x_1, y_1, 0 with its axis along z, the disturbance potential is

$$\varphi^\Gamma(x, y, z) = \frac{\gamma(x_1, y_1)}{4\pi} \frac{z}{[(y - y_1)^2 + z^2]}\left[1 + \frac{x - x_1}{r}\right]$$

$$= \frac{\gamma(x_1, y_1)}{4\pi}\left[\frac{z}{r[r - (x - x_1)]}\right]. \qquad (11\text{-}3)$$

Here we have

$$r = \sqrt{(x - x_1)^2 + (y - y_1)^2 + z^2}, \qquad (11\text{-}4)$$

and $\gamma = \Delta p/\rho_\infty U_\infty^2$ is proportional to the lifting pressure difference. We can reposition this doublet to x_1, s_1 on the curved surface S by making the following three substitutions in the second form of (11–3):

(1) Replace $\gamma(x_1, y_1)$ by $\Delta p(x_1, s_1)/\rho_\infty U_\infty^2$.
(2) $r = \sqrt{(x - x_1)^2 + (y - y_1)^2 + [z - z_0(y_1)]^2}$.
(3) Replace z by the normal distance $\mathbf{n}_1 \cdot \mathbf{r}$, $1\mathbf{n}_1$ being a unit vector in the n-direction at x_1, y_1, $z_0(y_1)$.

Thus we are led to the disturbance potential describing the complete surface:

$$\varphi(x, y, z) = \frac{1}{4\pi\rho_\infty U_\infty^2} \iint\limits_S \frac{\Delta p(x_1, s_1)[\mathbf{n}_1 \cdot \mathbf{r}]}{r[r - (x - x_1)]} \, dx_1 \, ds_1. \qquad (11\text{-}5)$$

To compute $v_n[x, y, z_0(y)] \equiv v_n(x, s)$ at points on the surface S, we need to perform the operation

$$v_n(x, s)\Big|_{\text{on } S} = \frac{\partial \varphi(x, y, z)}{\partial n}\Big|_{\text{on } S}, \qquad (11\text{-}6)$$

where, as is obvious from examining the figure,

$$\frac{\partial}{\partial n} = \cos \psi(y) \frac{\partial}{\partial z} - \sin \psi(y) \frac{\partial}{\partial y}.$$

The operation of taking $\partial/\partial n$ can be interchanged with the integration in the formula for the potential. Thus we find

$$v_n(x, s)\Big|_{\text{on } S} = \frac{1}{4\pi\rho_\infty U_\infty^2} \lim_{z \to z_0(y)} \left\{ \iint_S \Delta p \mathbf{n}_1 \cdot \frac{\partial}{\partial n} \left[\frac{\mathbf{r}}{r[r - (x - x_1)]} \right] dx_1 \, ds_1 \right\}.$$
$$(11\text{-}7)$$

When carrying out these differentiations, we use the above formula for $\partial/\partial n$ and also the fact $\mathbf{n}_1 = -\mathbf{j} \sin \psi(y_1) + \mathbf{k} \cos \psi(y_1)$. This permits us to overlook the $\mathbf{i}$-term in the vector $\mathbf{r}$ and replace it as follows:

$$\mathbf{r} \to (y - y_1)\mathbf{j} + [z - z_0(y_1)]\mathbf{k}.$$

During the differentiation process we discover that the singularity in the coefficient of $\Delta p(x_1, s_1)$ has exactly the same form as that in the planar integral equation (7-28a) and can, in fact, be isolated by factoring a $(y - y_1)^{-2}$ out in front of the whole kernel function. Thereupon, the cofactor has no worse than a discontinuous, finite jump where it passes through the point $y = y_1$, $x = x_1$ (or $x_1 = x$, $s = s_1$). Clearly, finite-part integration in Mangler's sense, (7-30), is indicated, and this interpretation can be rigorously proved.

After carrying out a couple of pages of straightforward manipulation, we are led finally to the integral equation of the nonplanar lifting surface,

$$v_n(x, s) = \frac{1}{4\pi\rho_\infty U_\infty^2} ⨏\!\!\!\int_S \Delta p(x_1, s_1) K_{\text{NP}}(x_0, y_0, z_{00}) \, dx_1 \, ds_1, \qquad (11\text{-}8)$$

where

$$K_{\text{NP}} = \frac{1}{y_0^2} \left\{ \sin[\psi(y) + \psi(y_1)] \frac{y_0^3 z_{00}}{r_1^4} \left[2 + \frac{2x_0^3 + 3x_0 r_1^2}{r_0^3} \right] \right.$$

$$- \cos \psi(y) \cos \psi(y_1) \frac{y_0^2 z_{00}^2}{r_1^4} \left[1 + \frac{x_0^3 + 2x_0 r_1^2}{r_0^3} - \frac{y_0^2}{z_{00}^2} \left(1 + \frac{x_0}{r_0} \right) \right]$$
$$(11\text{-}9a)$$

$$\left. - \sin \psi(y) \sin \psi(y_1) \frac{y_0^4}{r_1^4} \left[1 + \frac{x_0^3 + 2x_0 r_1^2}{r_0^3} - \frac{z_{00}^2}{y_0^2} \left(1 + \frac{x_0}{r_0} \right) \right] \right\}.$$

The auxiliary notation defined here consists of

$$\left. \begin{array}{l} x_0 = x - x_1, \qquad y_0 = y - y_1, \qquad z_{00} = z_0(y) - z_0(y_1) \\ \qquad r_1 = \sqrt{y_0^2 + z_{00}^2} \\ \qquad r_0 = \sqrt{x_0^2 + y_0^2 + z_{00}^2} \end{array} \right\} . \qquad (11\text{–}9b)$$

It is a simple matter to confirm that (11–8)–(11–9) reduce to the planar-wing system (7–91)–(7–92) when $\psi(y) = \psi(y_1) = 0$. There are fairly obvious reductions in other simplified cases of practical interest.

Regarding the numerical solution of the integral equation, it can be made dimensionless and solved by exactly the same procedure [Watkins *et al.* (1959)] that is outlined for the planar case in Section 7–6. It is preferable to adopt x_1 and s_1 as integration variables, because they remain single-valued over any imaginable surface shape, whereas y, y_1 may be multiple-valued on something like a ring wing. For separate but interfering surfaces, the integration must be carried out over two distinct pieces of S, as is clarified in succeeding examples. In the computing program, tables or subroutines must be stored which relate z_0 to y and s to y.

11–3 Special Cases and Numerical Solution

To illustrate the application of the integral equation (11–8) and possible simplifications of its kernel function (11–9a), let us look at four special cases in increasing order of complexity.

1. The Single Plane Surface. This example, which was dealt with in Section 7–6, is obtained by a reduction in which we set

$$\left. \begin{array}{l} \psi(y) = \psi(y_1) = z_0(y) = z_0(y_1) = 0 \\ v_n(x, s) = w(x, y, 0) \end{array} \right\} . \qquad (11\text{–}10)$$

The limits of integration on x_1 become $x_{\mathrm{LE}}(y_1)$ and $x_{\mathrm{TE}}(y_1)$, while y_1 ranges between the wingtips at $\pm b/2$. Thus we get

$$w(x, y, 0) = \frac{1}{4\pi\rho_\infty U_\infty^2} \oint_{-b/2}^{b/2} \int_{x_{\mathrm{LE}}(y_1)}^{x_{\mathrm{TE}}(y_1)} \frac{\Delta p(x_1, y_1)}{y_0^2} \left[1 + \frac{x_0}{\sqrt{x_0^2 + y_0^2}} \right] dx_1\, dy_1 , \qquad (11\text{–}11)$$

which corresponds to (7–90)–(7–91).

The process of solution involves the substitutions (7–92) and (7–96). It evidently leads to a system of simultaneous algebraic equations that can be abbreviated, in matrix notation,

$$\{w\} = [K]\{a_{nm}\} . \qquad (11\text{–}12)$$

Once the square matrix of coefficients $[K]$ is computed numerically, taking suitable account of the finite part at $y = y_1$, a standard inversion yields the coefficients a_{nm} of the pressure series from the known distribution of angle of attack or upwash over the mean surface.

2. The Single Nonplanar Surface. This case corresponds exactly to the integral equation in the form (11–8). The leading and trailing edges can be described by functions $x_{LE}(s)$ and $x_{TE}(s)$. By centering the coordinate systems the tips may be located at $s = \pm s_{TIP}$, but explicit relations between y and s, and z_0 and s, must also be available.

Provided there are no discontinuities in the surface slope (that is, $s(y)$ and $v_n(x, s)$ continuous functions), there is no objection to a solution procedure paralleling that in Section 7–6. We simply define the convenient auxiliary variables θ, σ according to

$$\left.\begin{array}{l} x = \tfrac{1}{2}[x_{TE}(s) + x_{LE}(s)] - \tfrac{1}{2}c(s) \cos \theta \\ x_1 = \tfrac{1}{2}[x_{TE}(s_1) + x_{LE}(s_1)] - \tfrac{1}{2}c(s_1) \cos \theta_1 \\ (s, s_1) = s_{TIP}[\sigma, \sigma_1] \end{array}\right\}, \quad (11\text{–}13)$$

$c \equiv (x_{TE} - x_{LE})$ being the chord. Then a series of the form

$$\frac{\Delta p(\theta_1, \sigma_1)}{\rho_\infty U_\infty^2} = 4\pi \frac{s_{TIP}}{c(\sigma_1)} \left\{ \sqrt{1 - \sigma_1^2} \cot \frac{\theta_1}{2} \sum_{m=0} a_{0m}\sigma_1^m \right.$$

$$\left. + \sqrt{1 - \sigma_1^2} \sum_{m=0} \sum_{n=1} \frac{4a_{nm}}{2^{2n}} \sigma_1^m \sin n\theta_1 \right\} \quad (11\text{–}14)$$

reduces the general equation to

$$v_n(\theta, \sigma) = \sum_{n=0} \sum_{m=0} a_{nm} \fint_{-1}^{1} \frac{\sigma_1^m \sqrt{1 - \sigma_1^2}}{y_0^2} \left\{ \int_0^\pi l_n(\theta_1)\overline{K}_{NP} \sin \theta_1 \, d\theta_1 \right\} d\sigma_1. \quad (11\text{–}15)$$

Here $\overline{K}_{NP}$ is deduced from K_{NP} by an obvious factorization of the singularity at $y = y_1$, whereas $l_n(\theta_1)$ are the same functions defined in (7–97). Special care must be observed with reentrant shapes, like ring wings, where $y(s)$ is not single-valued and y_0 can vanish between pairs of points that are not at the same spanwise station. With this reservation, we note that a procedure very similar to Watkins' (1959) converts the foregoing results into the matrix equation

$$\{v_n\} = [\overline{K}_{NP}]\{a_{nm}\}. \quad (11\text{–}16)$$

The adaptation of procedures developed for the planar wing has been shown to be quite straightforward. For instance, multiplying and dividing the integrand in (11–15) by $(\sigma - \sigma_1)^2$ permits the outer integral to be recast as

$$\fint_{-1}^{1} \frac{\sigma_1^m \sqrt{1 - \sigma_1^2}}{(\sigma - \sigma_1)^2} \{\cdots\} \, d\sigma_1,$$

while the inner portion remains well-behaved, so that the scheme already worked out for dealing with the singular portions of the numerical integration can be carried over without modification.

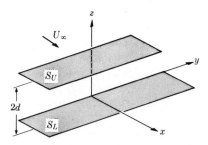

FIG. 11–3. Schematic of a biplane configuration.

3. The Biplane. As an example of an interference situation, consider a pair of almost-plane parallel lifting surfaces. See Fig. 11–3. The lower, S_L, might be located adjacent to the x,y-plane, while the upper, S_U, could be near $z = 2d$, d being the vertical half-spacing between them. In specializing (11–8), it is convenient to separate upwash contributions from S_L and S_U. The angle $\psi = 0$ always, and $z_{00} = 0$ when looking at one wing's effect on itself, while $z_{00} = 2d$ when the doublets of one are affecting the other. After a little manipulation, we find the following expression for the upwash induced at S_L:

$$w_L(x, y) = \frac{1}{4\pi\rho_\infty U_\infty^2} \fint\!\!\int_{S_L} \frac{\Delta p_L(x_1, y_1)}{y_0^2} \left[1 + \frac{x_0}{\sqrt{x_0^2 + y_0^2}}\right] dx_1 \, dy_1$$

$$+ \frac{1}{4\pi\rho_\infty U_\infty^2} \iint_{S_U} \frac{(2d)^2 \, \Delta p_U}{[y_0^2 + (2d)^2]} \tag{11–17}$$

$$\times \left\{\frac{y_0^2}{(2d)^2}\left[1 + \frac{x_0}{r_0}\right] - 1 - \frac{x_0^3 + 2x_0[y_0^2 + (2d)^2]}{r_0}\right\} dx_1 \, dy_1,$$

where (here only)

$$r_0^2 = x_0^2 + y_0^2 + (2d)^2.$$

A similar formula is readily constructed for w_U. We now propose to introduce dimensionless variables (θ^L, η^L) and (θ^U, η^U) to describe S_L and S_U, respectively. Series like (7–96), giving Δp_L and Δp_U in terms of sets of coefficients a_{nm}^L and a_{nm}^U are then inserted into the four integrals in the formulas for w_L and w_U, leading ultimately to a pair of matrix equations:

$$\{w_L\} = [K_{LL}]\{a_{nm}^L\} + [K_{LU}]\{a_{nm}^U\}, \qquad (11\text{–}18\text{a})$$
$$\underset{(N)}{} \quad \underset{(N\times N)}{} \underset{(N)}{} \quad \underset{(N\times M)}{} \underset{(M)}{}$$

$$\{w_U\} = [K_{UL}]\{a_{nm}^L\} + [K_{UU}]\{a_{nm}^U\}. \qquad (11\text{–}18\text{b})$$
$$\underset{(M)}{} \quad \underset{(M\times N)}{} \underset{(N)}{} \quad \underset{(M\times M)}{} \underset{(M)}{}$$

The elements of square matrices $[K_{LL}]$ and $[K_{UU}]$ are identical with those of $[K]$ for the individual planar wings. The cross-terms involve a little more complicated integrations but now contain no singularities. For a complete solution, one would solve simultaneously the double system of equations for the $(M + N)$ unknowns a_{nm}^L and a_{nm}^U. It might also be possible to devise an iteration process starting from solution of the individual uncoupled sets of equations, then feeding the previous iteration's results into the cross-terms prior to carrying out the next iteration.

Two subcases have special engineering interest.

(a) *Free water surface effect at high Froude number.* Suppose that S_L is a planar hydrofoil running at a depth d below a parallel water surface. If $U/\sqrt{gl} \gg 1$, g being the gravitational constant, wavemaking on the free surface is not important and a boundary condition $\varphi = 0$ at $z = d$ provides a satisfactory representation. If thickness effects are neglected, this boundary condition is met by the biplane configuration with S_U and S_L identical surfaces and $\Delta p_U = \Delta p_L$. Indeed, this is just a special case of the general method of images, which is also employed below to establish a ground plane.

Since the pressure series are identical, the loading on the actual hydrofoil can be calculated by contracting the first of (11–18a) into

$$\{w_L\} = [K_{LL} + K_{LU}]\{a_{nm}^L\}. \qquad (11\text{–}19\text{a})$$
$$\underset{(N)}{} \quad \underset{(N\times N)}{} \quad \underset{(N)}{}$$

(b) *Ground effect.* The presence of a planar ground surface at $z = d$ calls for the boundary condition $w = \varphi_z = 0$ there. If then we regard S_U as a single surface flying at height d, S_L provides a suitable image if it is identical and $\Delta p_L = -\Delta p_U$. In this case the matrix equivalent of the integral equation may be contracted to

$$\{w_U\} = [K_{UU} - K_{UL}]\{a_{nm}^U\}. \qquad (11\text{–}19\text{b})$$
$$\underset{(M)}{} \quad \underset{(M\times M)}{} \quad \underset{(M)}{}$$

4. An Intersecting Nonplanar Arrangement: The T-Tail.

Consider an arrangement which, in rear view, looks like Fig. 11–4.

The "normal" washes, pressure loadings, and chords are w_H, v_V, $\Delta p_H(x, y)$, $\Delta p_V(x, z)$, $c_H(y)$, $c_V(z)$ on the horizontal and vertical surfaces, respectively. For the horizontal surface $\psi(s) = \psi(y) = 0$ and for the vertical $\psi(s) = \psi(z) = -90°$.

Looking at the general kernel function and trying to satisfy the boundary condition at S_H, we are led to something of the following sort:

$$w_H(x, y) = \frac{1}{4\pi\rho_\infty U_\infty^2} \iint_{S_V} \Delta p_V(x_1, z_1) K_{HV} \, dx_1 \, dy_1$$

$$+ \frac{1}{4\pi\rho_\infty U_\infty^2} \oiint_{S_H} \frac{\Delta p_H(x_1, y_1)}{(y - y_1)^2} \left[1 + \frac{x - x_1}{\sqrt{(x - x_1)^2 + (y - y_1)^2}} \right] dx_1 \, dy_1,$$

(11–20a)

where

$$K_{HV} = \sin(-90°) \frac{y_0 z_{00}}{r_1^4} \left[2 + \frac{2x_0^3 + 3x_0 r_1^2}{r_0^3} \right]. \qquad (11\text{–}20b)$$

Since

$$\left.\begin{array}{l} y_0 = y - y_1 = y - 0 \\ z_{00} = z_0(y) - z_0(y_1) = 0 - z_1 \\ r_1 = \sqrt{y^2 + z^2} \\ r_0 = \sqrt{x_0^2 + y^2 + z^2} \end{array}\right\}, \qquad (11\text{–}21)$$

we get

$$K_{HV} = \frac{y z_1}{[y^2 + z_1^2]^2} \left[2 + \frac{2x_0^3 + 3x_0[y^2 + z_1^2]}{\sqrt{x_0^2 + y^2 + z_1^2}} \right]. \qquad (11\text{–}22)$$

Thus we are obviously led to a pair of integral equations, which might be approximated in matrix form:

$$\{w_H\} = [K'_{HH}] \left\{ \frac{\Delta p_H}{\rho_\infty U_\infty^2} \right\} + [K'_{HV}] \left\{ \frac{\Delta p_V}{\rho_\infty U_\infty^2} \right\},$$

(11–23a)

$$\{w_V\} = [K'_{VH}] \left\{ \frac{\Delta p_H}{\rho_\infty U_\infty^2} \right\} + [K'_{VV}] \left\{ \frac{\Delta p_V}{\rho_\infty U_\infty^2} \right\}.$$

(11–23b)

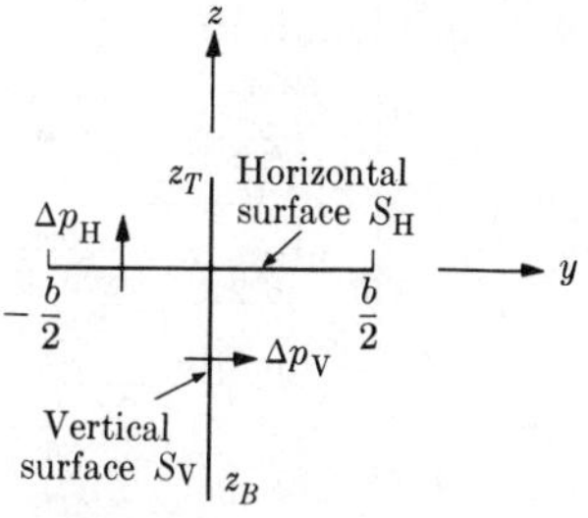

Fig. 11–4. Geometry and directions of pressure loading on an idealized T-tail.

When approaching practical problems, we note that they can be separated into cases of symmetrical and antisymmetrical loading with respect to the x,y-plane. In the former situation, $\Delta p_V = 0$, so the problem corresponds to one of a single planar surface.

From among the several examples that have now been worked out of incompressible flow loading on nonplanar and interfering surfaces, we choose to reproduce from Saunders (1963) the three cases shown in Fig. 11–5. More details can be found in the reference and the paper by Ashley, Widnall, and Landahl (1965), but the method of computation is essentially that described in the present chapter.

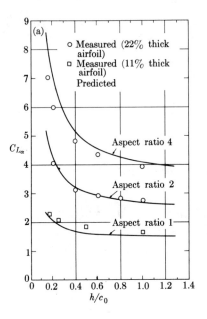

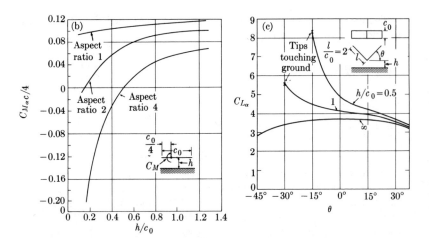

Fig. 11–5. (a) Steady-state lift-curve slopes of three rectangular wings in constant-density flow, as influenced by height h/C_0 chords above a parallel ground plane. [Taken from Saunders (1963).] (b) Theoretical slopes of pitching moment about a quarter-chord axis for the same three wings in ground effect. (c) Combined influences of dihedral angle and ground proximity on the lift-curve slope of a constant-chord, zero-thickness surface in constant-density flow.

11–4 Loads on Interfering Surfaces in Subsonic Flow

Any generalized force on the curved surface of Fig. 11–2 can be written in the form

$$Q = \iint_S \Delta p(x, s) f(x, s) \, dx \, ds$$

$$= \tfrac{1}{2} s_{\text{TIP}} \int_{-1}^{1} \int_{0}^{\pi} \Delta p(\theta, \sigma) f(\theta, \sigma) \, c(\sigma) \, \sin \theta \, d\theta \, d\sigma. \qquad (11\text{–}24)$$

For example, the lift is given by taking $f = \cos \psi(\sigma)$, whereas the nose-up pitching moment about the y-axis would result from

$$f = -\cos \psi(\sigma) x(\theta)$$
$$= \tfrac{1}{2} c(\sigma) \, \cos \theta \cos \psi(\sigma) - \tfrac{1}{2}[x_{\text{TE}}(\sigma) + x_{\text{LE}}(\sigma)] \cos \psi(\sigma).$$

Running loads per unit s-distance are obtained by eliminating the s- or σ-integration.

If the series (11–14) is inserted for $\Delta p(\theta, \sigma)$ the resulting computation can often be carried out in closed form, especially if $f(\theta, \sigma)$ can be expressed as a Fourier series in θ and power series in σ.

For instance, the lift works out as follows:

$$L = 4\pi s_{\text{TIP}}^2 \rho_\infty U_\infty^2 \left\{ \sum_{m=0} a_{0m} \int_{-1}^{1} \sqrt{1 - \sigma^2} \, \sigma^m \cos \psi(\sigma) \right.$$

$$\times \left[\tfrac{1}{2} \int_0^{\pi} \sin \theta \cot \frac{\theta}{2} \, d\theta \right] d\sigma + \sum_{m=0} \sum_{n=1} a_{nm} \int_{-1}^{1} \sqrt{1 - \sigma^2} \, \sigma^m \cos \psi(\sigma)$$

$$\times \left. \left[\tfrac{1}{2} \int_0^{\pi} \sin \theta \sin n\theta \, d\theta \right] d\sigma \right\} = 2\pi^2 s_{\text{TIP}}^2 \rho_\infty U_\infty^2$$

$$\times \left[\sum_{m=0} \left(a_{0m} + \frac{a_{1m}}{2} \right) \int_{-1}^{1} \sqrt{1 - \sigma^2} \, \sigma^m \cos \psi(\sigma) \, d\sigma \right], \qquad (11\text{–}25)$$

since

$$\int_0^{\pi} \sin n\theta \sin \theta \, d\theta = \begin{cases} \pi/2, & n = 1 \\ 0, & n > 1 \end{cases}$$

$$\int_0^{\pi} \sin \theta \cot \frac{\theta}{2} \, d\theta = \pi.$$

Another case of special interest is the induced drag. By Trefftz-plane considerations analogous to those that produced (7–43) we can work out

$$D_i = -\frac{\rho_\infty U_\infty^2}{2} \int_{-s_{\text{TIP}}}^{s_{\text{TIP}}} \Gamma(s) v_{n_\infty}(s) \, ds, \qquad (11\text{–}26)$$

where $\Gamma(s) = U_\infty \, \Delta\varphi_{TE}(s)$ is the circulation bound to the wing at station s. Now $\Gamma(s)$ can be calculated from the normal force N per unit s-distance and the pressure as

$$\Gamma = \frac{N}{\rho_\infty U_\infty} = \frac{1}{\rho_\infty U_\infty} \int_{x_{LE}}^{x_{TE}} \Delta p(x, s) \, dx = \cdots$$

Also $v_{n_\infty}(s)$ is the normal-wash in the wake as $x \to \infty$. From the original kernel-function formulation with $x \to \infty$ it works out to be

$$v_{n_\infty}(s) = -\frac{1}{4\pi\rho_\infty U_\infty^2} \oint_{-s_{TIP}}^{s_{TIP}} N(\sigma) K_{w_\infty} \, d\sigma, \qquad (11\text{--}27)$$

where

$$K_{w_\infty} = \frac{1}{y_0^2 r_1^4} \left\{ 4y_0^3 z_{00} \sin\left[\psi(y) + \psi(y_1)\right] + 2y_0^2[y_0^2 - z_{00}^2] \right.$$
$$\left. \times \left[\cos\psi(y) \cos\psi(y_1) + \frac{y_0^2}{z_{00}^2} \sin\psi(y) \sin\psi(y_1) \right] \right\}. \qquad (11\text{--}28)$$

Other reductions and simplifications are possible. An interesting discussion of drag minimization for single nonplanar surfaces has been published by Cone (1962).

11–5 Nonplanar Lifting Surfaces in Supersonic Flow

For analyzing the corresponding problem by linearized theory at supersonic Mach numbers, we rely entirely on the technique of supersonic aerodynamic influence coefficients (AIC's) and present only the general outlines of a feasible computational approach. Their only special virtue over other techniques is that they have been practically and successfully mechanized for a variety of steady and unsteady problems.

It should be pointed out first that there exists a formula for a supersonic pressure doublet analogous to the one that has been worked with in the preceding section [see Watkins and Berman (1956)]. This is quite difficult for manipulation, however, because of the law of forbidden signals and other wavelike discontinuities which occur. As a result, even the case of a planar lifting surface at supersonic speed has not been worked through completely and in all generality using pressure doublets. Rather, the problem has been handled by the sorts of special techniques described in Chapter 8. Examples of numerical generalization for arbitrary distributions of incidence will be found in Etkin (1955), Beane *et al.* (1963), and Pines *et al.* (1955).

For nonplanar wings, the method of AIC's has been developed and mechanized for the high-speed computer, but only in cases where two or more individually-plane surfaces intersect or otherwise interfere. Thus a three-dimensional biplane, T-tail or V-tail can be handled, but ring or channel wings remain to be studied.

In preparation for the interference problem, let us describe more thoroughly the procedure for a single surface that was outlined in Section 8–5. For simplicity, let the trailing edges be supersonic; but in the case of a subsonic leading edge the forward disturbed region of the x,y-plane is assumed to be extended by a diaphragm, where a condition of zero pressure (or potential) discontinuity must be enforced. This situation is illustrated in Fig. 11–6.

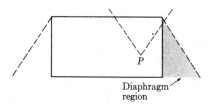

Diaphragm region

FIG. 11–6. Plane wing in supersonic flow, illustrating the right wingtip diaphragm.

For the disturbance potential anywhere in $z \geq 0+$, (8–6) gives

$$\varphi(x, y, z) = -\frac{1}{\pi} \iint_{\Sigma} \frac{w(x_1, y_1) \, dx_1 \, dy_1}{\sqrt{(x - x_1)^2 - B^2[(y - y_1)^2 + z^2]}}. \quad (11\text{--}29)$$

The region Σ is the portion of the wing plus diaphragm area intercepted by the upstream Mach cone from (x, y, z). When $z \to 0+$ so that φ is being calculated for a point P on the upper wing surface, Σ reduces to the area between the two forward Mach lines (see the figure).

Let us restrict ourselves to the lifting problem and let the wing have zero thickness. We know $w(x_1, y_1)$ from the given mean-surface slope over that portion of Σ that does not consist of diaphragm; on the latter, w is unknown but a boundary condition $\varphi = 0$ applies. As illustrated in Figs. 8–15 and 8–16, let the wing and diaphragm be overlaid to the closest possible approximation with rectangular elementary areas ("boxes") having a chordwise dimension b_1 and spanwise dimension b_1/B. In (11–29), introduce the transformation [cf. Zartarian and Hsu (1955)]

$$(\xi, \xi_1) = \frac{(x, x_1)}{b_1}$$

$$(\eta, \eta_1, \zeta) = \frac{(y, y_1, z)}{b_1/B}. \quad (11\text{--}30)$$

One thus obtains something of the form

$$\varphi(\xi, \eta, \zeta) = - \frac{b_1}{\pi B} \iint_{\Sigma} \frac{w(\xi_1, \eta_1)\, d\xi_1\, d\eta_1}{\sqrt{(\xi - \xi_1)^2 - [(\eta - \eta_1)^2 + \zeta^2]}}. \qquad (11\text{--}31)$$

This simultaneously employs dimensionless independent variables and converts all supersonic flows to equivalent cases at $M = \sqrt{2}$. The Mach lines now lie at 45° to the flight direction. The elementary areas, which had their diagonals parallel to the Mach lines in the x,y-plane, are thus deformed into squares.

Next let it be assumed that w is a constant over each area element and equal to the value $w_{\nu,\mu}$ at the center. Both ν and μ are integers used to count the positions of these areas rearward and to the right from $\nu = 0$ and $\mu = 0$ at the origin of coordinates. With this further approximation, the potential may be written

$$\varphi(\xi, \eta, \zeta) = - \frac{b_1}{B} \sum_{\nu, \mu} w_{\nu,\mu} \Phi_{\nu,\mu}(\xi, \eta, \zeta), \qquad (11\text{--}32)$$

where the summation extends over all boxes and portions of boxes in the forward Mach cone. The definition of $\Phi_{\nu,\mu}$ as an integral over an area element is fairly obvious. For example, for a complete box

$$\Phi_{\nu,\mu} = \int_{(\mu-1/2)}^{(\mu+1/2)} \int_{(\nu-1/2)}^{(\nu+1/2)} \frac{d\xi_1\, d\eta_1}{\sqrt{(\xi - \xi_1)^2 - [(\eta - \eta_1)^2 + \zeta^2]}}. \qquad (11\text{--}33)$$

This can easily be worked out in closed form. The computation is mechanized by choosing ξ, η (and possibly ζ) to be integers, corresponding to the centers of "receiving boxes." Thus, if we choose $\xi = n$, $\eta = m$, $\zeta = l$, we get

$$\begin{aligned}
\Phi_{\nu,\mu}(n, m, l) &= \int_{(\mu-1/2)}^{(\mu+1/2)} \int_{(\nu-1/2)}^{(\nu+1/2)} \frac{d\xi_1\, d\eta_1}{\sqrt{(n - \xi_1)^2 - [(m - \eta_1)^2 + l^2]}} \\
&= \int_{(\bar{\mu}-1/2)}^{(\bar{\mu}+1/2)} \int_{(\bar{\nu}-1/2)}^{(\bar{\nu}+1/2)} \frac{d\bar{\xi}_1\, d\bar{\eta}_1}{\sqrt{\bar{\xi}_1^2 - \bar{\eta}_1^2 - l^2}} \equiv \Phi_{\bar{\nu},\bar{\mu},l}, \qquad (11\text{--}34)
\end{aligned}$$

where $\bar{\mu} = m - \mu$, $\bar{\nu} = n - \nu$. (Special forms apply for combinations of ν, μ, l on the upstream Mach-cone boundary, which may be determined by taking the real part of the integral. Also $\Phi_{\bar{\nu},\bar{\mu},l} = 0$ when

$$\sqrt{\bar{\xi}_1^2 - \bar{\eta}_1^2 - l^2}$$

is imaginary throughout the range of integration.)

In a similar way, we can work out, for the vertical and horizontal velocity components in the field at a point (n, m, l), expressions of the following forms:

$$v(n, m, l) = \sum_{\nu,\mu} w_{\nu,\mu} V_{\bar{\nu},\bar{\mu},l}$$

$$w(n, m, l) = \sum_{\nu,\mu} w_{\nu,\mu} W_{\bar{\nu},\bar{\mu},l}. \tag{11–35}$$

The AIC's V and W involve differentiation of the $\Phi_{\bar{\nu},\bar{\mu},l}$ formula with respect to η and ζ, respectively, but they can be worked out without difficulty.

Now to find the load distribution on a single plane surface, we set $l = 0$ and order the elementary areas from front to back in a suitable way. The values of disturbance potentials at the centers of all these areas can then be expressed in the matrix form

$$\{\varphi_{n,m}\} = -\frac{b_1}{B}\,[\Phi_{\bar{\nu},\bar{\mu},0}]\{w_{\nu,\mu}\}. \tag{11–36}$$

A suitable ordering of the areas consists of making use of the law of forbidden signals to assure that all numbers in $\Phi_{\bar{\nu},\bar{\mu},0}$ are zero to the left and below the principal diagonal. It is known that $\varphi_{n,m} = 0$ at all box centers on the diaphragms, whereas $w_{\nu,\mu}$ is given at all points on the wing. The former information can be used to solve successively for the values of $w_{\nu,\mu}$ at the diaphragm, the computation being progressive and never requiring the inversion of a matrix.

Once $w_{\nu,\mu}$ is known for all centers on the wing and diaphragm, the complete distribution of φ may be determined. From this, we can calculate the pressure distribution (which is antisymmetrical top to bottom) by the relatively inaccurate process of numerical differentiation. If only lift, moment, pressure drag, or some other generalized forces are needed, however, we find that these can be expressed entirely in terms of the potential discontinuity over the surface and along the trailing edge. Hence, the differentiation step can be avoided. Zartarian (1956) and Zartarian and Hsu (1955) provide many details.

Turning to the interfering surfaces, we illustrate the method by two examples.

1. The Biplane. Let us consider two supersonic wings, with associated diaphragm regions, separated by a distance d in the ζ-direction (Fig. 11–7). Because there is a certain artificiality in the use of *sources* to represent the flow over the upper surface of each of these wings, some care must be

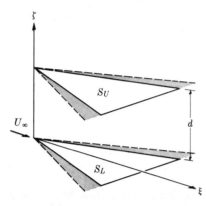

Fig. 11-7. Two interfering-plane supersonic wings with attached diaphragms. (Mach lines are at 45° in ξ-, η-, ρ-coordinates.)

exercised in setting up the interference problem. We have found that the best way to avoid paradoxes is to focus on the two conditions:

(a) The streamlines must be parallel to the mean surface over the area of each wing.

(b) $\Delta\varphi = 0$ must be enforced over each diaphragm ($\Delta p = 0$ on wake diaphragms).

These can best be handled in the biplane case by placing additional sources over each wing-diaphragm combination, whose purpose is to *cancel* the upwash induced over one particular wing area due to the presence of the other wing. There is no need to be concerned with interfering upwash over the diaphragms, since the diaphragm is not a physical barrier and the interference upwash there does not cause any *discontinuity* of potential.

Having placed suitable patterns of square area elements over S_U and S_L, we can write for the upwash induced at wing boxes on S_U due to the presence of S_L:

$$w_{UL}(n, m, d = l) = \sum_{\nu,\mu} (w_L)_{\nu,\mu} W_{\bar{\nu},\bar{\mu},d}. \qquad (11\text{-}37)$$

The summation here extends over all wing and diaphragm boxes on S_L that can influence point (n, m, d). In a similar way, reasons of antisymmetry in the flow field produced by S_U lead to the upwash generated at wing box n, m, 0 on S_L due to S_U:

$$w_{LU}(n, m, 0) = \sum_{\nu,\mu} (w_U)_{\nu,\mu} W_{\bar{\nu},\bar{\mu},(-d)}, \qquad (11\text{-}38\text{a})$$

or

$$\{w_{LU}\}_{\substack{\text{wing} \\ \text{only}}} = [W_{\bar{\nu},\bar{\mu},(-d)}]\{w_U\}. \qquad (11\text{-}38\text{b})$$

When writing the matrix formulas for φ_U and φ_L, w_{UL} and w_{LU} must be *subtracted* from the upwash that would be present at wing boxes on

S_U and S_L, respectively, in the absence of the interfering partner. Thus we obtain

$$\{\varphi_U\} = -\frac{b_1}{B}[\Phi_{\bar{\nu},\bar{\mu},0}](\{w_U\} - \{w_{UL}^*\}),\qquad(11\text{–}39)$$

where the last column covers wing and diaphragm boxes but zeros are inserted for the latter. Similarly,

$$\{\varphi_L\} = -\frac{b_1}{B}[\Phi_{\bar{\nu},\bar{\mu},0}](\{w_L\} - \{w_{LU}^*\}).\qquad(11\text{–}40)$$

Making appropriate substitutions for w_{UL}^* and w_{LU}^*, we get

$$\{\varphi_U\} = -\frac{b_1}{B}[\Phi_{\bar{\nu},\bar{\mu},0}](\{w_U\} - [W_{\bar{\nu},\bar{\mu},d}^*]\{w_L\}),\qquad(11\text{–}41\text{a})$$

$$\{\varphi_L\} = -\frac{b_1}{B}[\Phi_{\bar{\nu},\bar{\mu},0}](\{w_L\} - [W_{\bar{\nu},\bar{\mu},(-d)}^*]\{w_U\}),\qquad(11\text{–}41\text{b})$$

where the meaning of the notation for W^* is obvious in the light of the foregoing remarks.

We now have a set of coupled equations in φ and w. The values of φ are equated to zero at all diaphragm boxes, whereas w is known at all wing boxes, so the system is determinate. A solution procedure, at least in principle, is straightforward.

2. Intersecting Vertical and Horizontal Stabilizers.

Surfaces that interfere but are not parallel present no new conceptual difficulties. Once again, the source sheet representing the flow on one side of either surface is analyzed as if the other were not there, except that the mean-surface "normal-wash" distribution must be modified to account for interference. Thus, consider the empennage arrangement shown in Fig. 11–8. Diaphragms are shaded. The rightward sidewash at S_V due to S_U and its diaphragm is

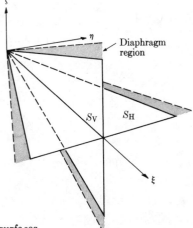

$$v_{\text{VH}}(n, 0, l) = \sum_{\nu,\mu}(w_\text{H})_{\nu,\mu}V_{\bar{\nu},\bar{\mu},l}.$$

$$(11\text{–}42\text{a})$$

Fig. 11–8. Intersecting supersonic surfaces.

Here $\bar{\mu} = m - \mu = 0 - \mu = -\mu$. The matrix abbreviation is

$$\{v_{VH}\} = [V_{\bar{\nu}, -\mu, l}]\{w_H\}. \tag{11-42b}$$

In a similar way, we find

$$w_{HV}(n, m, 0) = - \sum_{\nu, l} (v_V)_{\nu, l} V_{\bar{\nu}, -l, m} \tag{11-43a}$$

or

$$\{w_{HV}\} = - [V_{\nu, -l, m}]\{v_V\}. \tag{11-43b}$$

The potential formulas can then be written as follows and solved by substitutions like those discussed in the case of the biplane:

$$\{\varphi_H\} = - \frac{b_1}{B} [\Phi_{\bar{\nu}, m, 0}](\{w_H\} - \{w_{HV}\}), \tag{11-44a}$$

$$\{\varphi_V\} = - \frac{b_1}{B} [\Phi_{\bar{\nu}, 0, l}](\{v_V\} - \{v_{VH}\}). \tag{11-44b}$$

Other supersonic interference problems can be handled in a similar manner. In all cases, a computational scheme can be found to make the solutions for the lifting pressures a determinate problem. Together with some numerical results, a list of references reporting progress towards mechanization of the foregoing supersonic methods will be found in Ashley, Widnall, and Landahl (1965).

12

Transonic Small-Disturbance Flow

12–1 Introduction

A transonic flow is one in which local particle speeds both greater and
less than sonic speed are found mixed together. Thus in the lower transonic
range (ambient M slightly less than unity) there are one or more supersonic regions embedded in the subsonic flow and, similarly, in the upper
transonic range the supersonic flow encloses one or more subsonic flow
regions. Some typical transonic flow patterns are sketched in Fig. 12–1.
Since in a transonic flow the body travels at nearly the same speed as
the forward-going disturbances that it generates, one would expect that
the flow perturbations are generally greater near $M = 1$ than in purely
subsonic or supersonic flow. That this is indeed so is borne out by experimental results like those shown in Figs. 12–2 and 12–3, which show that
the drag and lift coefficients are maximum in the transonic range. In the
early days of high-speed flight, many doubted that supersonic aeroplanes
could ever be built because of the "sonic barrier," the sharp increase in
drag experienced near M equal to unity.

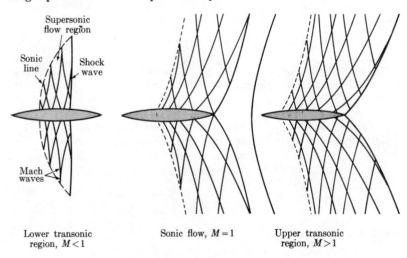

Fɪɢ. 12–1. Typical transonic flow patterns.

227

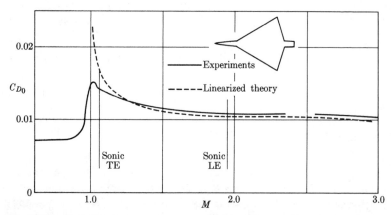

Fig. 12–2. Zero-lift drag coefficient for a delta-wing model. (Adapted from Holdaway and Mellenthin, 1960. Courtesy of National Aeronautics and Space Administration.)

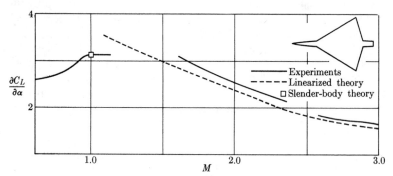

Fig. 12–3. Lift-curve slope for a delta-wing model. (Adapted from Holdaway and Mellenthin, 1960. Courtesy of National Aeronautics and Space Administration.)

Many of the special physical features, and the associated analytical difficulties, of a transonic flow may be qualitatively understood by considering the simplest case of one-dimensional fluid motion in a stream tube. Combination of the Euler equation

$$U \frac{dU}{dx} = - \frac{1}{\rho} \frac{dp}{dx}$$

with the definition of the speed of sound

$$\left(\frac{\partial p}{\partial \rho} \right)_s = a^2$$

and with the equation of continuity

$$\frac{d}{dx}(\rho US) = 0,$$

where $S(x)$ is the stream-tube area, yields after some manipulation

$$\frac{dp}{\rho U^2} = \frac{dS}{S[1 - (U^2/a^2)]}. \qquad (12\text{–}1)$$

This relation shows that for $U/a = 1$ the flow will resist with an infinite force any stream-tube area changes, i.e., it will effectively make the flow incompressible to gross changes in the stream-tube area (but not to curvature changes or lateral displacement of a stream-tube pattern). Therefore, the crossflow in planes normal to the free-stream direction will tend to be incompressible, as in the case of the flow near a slender body, so that much of the analysis of Chapter 6 applies in the transonic range to configurations that are not necessarily slender. This point will be discussed further below. It is evident that because of the stream-tube area constraint, there will be a tendency for a stronger cross flow within the stream tube and hence the effect of finite span will be maximum near $M = 1$. From (12–1) it also follows that in order to avoid large perturbation pressures and hence high drag one should avoid large (and sudden) cross-sectional area changes, which in essence is the statement of the transonic area rule discussed in Chapter 6. For the same reason one can see that the boundary layer can have a substantial influence on a transonic pressure distribution, since it provides a region of low-speed flow which is less "stiff" to area changes and hence can act as a "buffer" smoothing out area changes.

From such one-dimensional flow considerations, one practical difficulty also becomes apparent, namely that of wind tunnel testing at transonic speeds. Although a flow of $M = 1$ can be obtained in the minimum-area section of a nozzle with a moderate pressure ratio, the addition of a model, however small, will change the area distribution so that the flow no longer will correspond to an unbounded one of sonic free-stream speed. This problem was solved in the early 1950's with the development of slotted-wall wind tunnels in which the wall effects are eliminated or minimized by using partially open walls.

The main difficulty in the theoretical analysis of transonic flow is that the equations for small-disturbance flow are basically nonlinear, in contrast to those for subsonic and supersonic flow. This again may be surmised from equations like (12–1), because even a small velocity change caused by a pressure change will have a large effect on the pressure-area relation. So far, no satisfactory general method exists for solving the transonic small-perturbation equations. In the case of two-dimensional flow it is possible, through the interchange of dependent and independent variables, to transform the nonlinear equations into linear ones in the hodograph plane. However, solutions by the hodograph method have been obtained only for special simple airfoil shapes and, again, two-dimensional flow solutions are of rather limited practical usefulness for transonic speeds. For axisymmetric and three-dimensional flow, various approximate methods have been suggested, some of which will be discussed below.

12–2 Small-Perturbation Flow Equations

That the small-perturbation theory for sub- and supersonic flow breaks down at transonic speeds becomes evident from the linearized differential equations (5–29) and (6–21) for the perturbation potential, which in the limit of $M \rightarrow 1$ become

$$\varphi_{zz} = 0, \qquad \text{for two-dimensional flow,} \qquad (12\text{–}2)$$

$$\frac{1}{r}\, \varphi_r + \varphi_{rr} = 0, \qquad \text{for axisymmetric flow.} \qquad (12\text{–}3)$$

Thus, both the inner and outer flows will be described by the same differential equation, and it will in general not be possible to satisfy the boundary condition of vanishing perturbation velocities at infinity. For transonic flow it will hence be necessary to consider a different expansion that retains *at least one more* term in the equation for the first-order outer flow.

In searching for such an expansion we may be guided by experiments. By testing airfoils, or bodies of revolution, of the same shape but different thickness ratios (affine bodies) in a sonic flow one will find that, as the thickness ratio is decreased, not only will the flow disturbances decrease, as would be expected, but also the disturbance pattern will persist to larger distances (see Fig. 12–4).

This would suggest that the significant portion of the outer flow will recede farther and farther away from the body as its thickness tends towards zero. In order to preserve, in the limit of vanishing body thickness, those portions of the outer flow field in which the condition of vanishing flow perturbations is to be applied we must therefore "compress" this (in the mathematical sense). Taking first the case of a two-dimensional airfoil with thickness but no lift, we shall therefore consider an expansion of the following form:

$$\Phi^i = U_\infty[x + \epsilon \Phi_1^i(x, z) + \epsilon^2 \Phi_2^i(x, z) + \cdots] \qquad (12\text{–}4)$$

for the inner flow and

$$\Phi^o = U_\infty[x + \epsilon \Phi_1^o(x, \zeta) + \epsilon^2 \Phi_2^i(x, \zeta) + \cdots] \qquad (12\text{–}5)$$

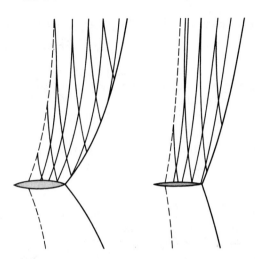

Fɪɢ. 12–4. Flow patterns at $M = 1$ (hypothetical) for affine bodies of two different thickness ratios.

for the outer flow. Here ϵ is a small nondimensional parameter measuring the perturbation level and

$$\zeta = \delta z, \tag{12–6}$$

with δ being a function of ϵ so that $\delta \to 0$ as $\epsilon \to 0$. Both ϵ and δ are related to the thickness ratio τ of the airfoil in a way that will be determined subsequently. For the first term in the inner expansion we obtain upon substituting (12–4) into the differential equation (1–74) for Φ, and assuming $1 - M^2 = 0(\epsilon)$, the same equation as (12–2):

$$\Phi^i_{1zz} = 0. \tag{12–7}$$

This result shows that the inner flow has a solution of the same form as in the previously considered sub- and supersonic flows, except that it is now valid in a larger region $z = 0(1)$. Next, substitution of the outer expansion (12–5) into (1–74) and (1–67) gives

$$[a^2 - U_\infty^2 - 2\epsilon U_\infty^2 \Phi^o_{1x} - \epsilon^2 U_\infty^2 (\Phi^o_{1x})^2]\Phi^o_{1xx}$$
$$+ \delta^2[a^2 - U_\infty^2 \epsilon^2 \, \delta^2 (\Phi^o_{1\zeta})^2]\Phi^o_{1\zeta\zeta}$$
$$+ 2U_\infty^2(1 + \epsilon\Phi^o_{1x})\epsilon \, \delta^2\Phi^o_{1x\zeta}\Phi^o_{1\zeta} + \text{higher order terms} = 0, \tag{12–8}$$

where

$$a^2 = a_\infty^2 - \frac{\gamma - 1}{2} U_\infty^2[2\epsilon\Phi^o_{1x} + \epsilon^2(\Phi^o_{1x})^2 + \epsilon^2 \, \delta^2(\Phi^o_{1\zeta})^2]. \tag{12–9}$$

By neglecting all terms of higher order in ϵ or δ we may simplify these to

$$[1 - M^2 - \epsilon M^2(\gamma + 1)\Phi^o_{1x}]\Phi^o_{1xx} + \delta^2\Phi^o_{1\zeta\zeta} = 0. \qquad (12\text{--}10)$$

In a transonic flow $1 - M^2 = 0(\epsilon)$ so that the first term is of order ϵ, and a nondegenerate equation is obtained by setting $\delta^2 \sim \epsilon$. For later purposes it is convenient to choose

$$\delta = M\sqrt{\epsilon(\gamma + 1)}, \qquad (12\text{--}11)$$

in which case the differential equation (12–10) takes the form

$$-\left[\frac{M^2 - 1}{\epsilon M^2(\gamma + 1)} + \Phi^o_{1x}\right]\Phi^o_{1xx} + \Phi^o_{1\zeta\zeta} = 0. \qquad (12\text{--}12)$$

Notice that (12–12) is basically nonlinear and that the Mach number enters only through the transonic parameter

$$K_1 = \frac{M^2 - 1}{\epsilon M^2(\gamma + 1)}. \qquad (12\text{--}13)$$

In order to relate ϵ to the thickness ratio τ of the airfoil, we have to match the outer and inner flows. The inner solution, which is determined so as to satisfy the tangency condition on the airfoil surface, gives that

$$\Phi^i_{1z} = \pm\frac{\tau}{\epsilon}\frac{d\bar{g}}{dx}, \qquad (12\text{--}14)$$

where the upper sign is to be used for the region above the airfoil, that is, for $z \geq \tau g$, and the lower sign for the region below the airfoil. Since $\partial/\partial z = \delta\partial/\partial\zeta$, application of the limit matching principle will thus give

$$\Phi^o_{1\zeta}(x, 0\pm) = \pm\frac{\tau}{\delta\epsilon}\frac{d\bar{g}}{dx}. \qquad (12\text{--}15)$$

For the expansion to be meaningful in the limit of $\epsilon \to 0$ we therefore set

$$\tau/\delta\epsilon = 1, \qquad (12\text{--}16)$$

which, upon use of (12–11), gives

$$\epsilon = (\tau/M)^{2/3}(\gamma + 1)^{-1/3}. \qquad (12\text{--}17)$$

The disturbance magnitude is thus of order $\tau^{2/3}$, which is to be compared with τ in the sub- and supersonic cases.

For the pressure on the airfoil we obtain, after series expansion and use of matching, that to first order

$$C_p = -2(\tau/M)^{2/3}(\gamma + 1)^{-1/3}\Phi^o_{1x}(x, 0). \qquad (12\text{--}18)$$

The expansion procedure having served its purpose, we may now, as usual, define a perturbation velocity potential $\varphi = \epsilon\Phi_1^o$, and return to the original coordinates. This gives the following formulation of the transonic small-disturbance airfoil problem:

$$[1 - M^2 - M^2(\gamma + 1)\varphi_x]\varphi_{xx} + \varphi_{zz} = 0, \qquad (12\text{–}19)$$

$$\varphi_z(x, 0\pm) = \pm\tau\frac{d\overline{g}}{dx}, \qquad (12\text{–}20)$$

with the pressure coefficient on the airfoil given by

$$C_p = -2\varphi_x(x, 0). \qquad (12\text{–}21)$$

It should be noted that (12–19) describes correctly small-perturbation sub- and supersonic flows, as well, because then the nonlinear term becomes negligible. The nonlinearity in the transonic speed range is essential to the problem since the equation must correctly account for the mixed subsonic-supersonic character of the flow. The sign of the coefficient of the first term determines, to first order, whether the local Mach number is less than one (the sign is positive and the flow governed by an elliptical differential equation) or greater than one (the sign is negative and the differential equation hyperbolic).

For axisymmetric flow we may proceed in a similar way* to seek series solutions of the forms

$$\Phi^i = U_\infty[x + \epsilon\Phi_1^i(x, r) + \epsilon^2\Phi_2^i(x, r) + \cdots] \qquad (12\text{–}22)$$

for the inner flow and

$$\Phi^o = U_\infty[x + \epsilon\Phi_1^o(x, \rho) + \epsilon^2\Phi_2^o(x, \rho) + \cdots], \qquad (12\text{–}23)$$

for the outer flow, where

$$\rho = r\delta. \qquad (12\text{–}24)$$

By substituting (12–23) into the differential equation for Φ we obtain, upon neglecting higher-order terms,

$$[1 - M^2 - \epsilon M^2(\gamma + 1)\Phi_{1x}^o]\Phi_{1xx}^o + \delta^2\left[\frac{1}{\rho}\Phi_{1\rho}^o + \Phi_{1\rho\rho}^o\right] = 0. \qquad (12\text{–}25)$$

This process again leads to the choice (12–11) for δ, and the differential equation thus becomes

$$-(K_1 + \Phi_{1x}^o)\Phi_{1xx}^o + \frac{1}{\rho}\Phi_{1\rho}^o + \Phi_{1\rho\rho}^o = 0, \qquad (12\text{–}26)$$

---

* The expansion procedure for axisymmetric flow was carried out in great detail by Cole and Messiter (1956).

with K_1 given by (12–13). The inner flow, as in the slender-body case, is this time found to be governed by

$$\frac{1}{r}\Phi^i_{1r} + \Phi^i_{1rr} = 0,$$

$$\epsilon\Phi^{(1)}_{1r}(x, R) = \frac{dR}{dx} = \tau\frac{d\overline{R}}{dx}, \qquad (12\text{–}27)$$

with the solution

$$\epsilon\Phi^i_1(x, r) = \frac{1}{2\pi}S'(x)\ln r + g_1(x) = \tau^2\left[\frac{1}{2\pi}\overline{S}'(x)\ln r + \overline{g}_1(x)\right]. \quad (12\text{–}28)$$

By applying the limit matching principle to the radial velocity component we find that, with the choice

$$\epsilon =' \tau^2, \qquad (12\text{–}29)$$

the boundary condition for the outer flow will read

$$\lim_{\rho\to 0}[\rho\Phi^o_{1\rho}] = \frac{1}{2\pi}\overline{S}'(x) \equiv \frac{1}{2\pi\tau^2}S'(x). \qquad (12\text{–}30)$$

The perturbation velocity potential for the case of a slender axisymmetric body is thus of order τ^2, as compared to $\tau^{2/3}$ for the two-dimensional case. Since the disturbances are of (at least) an order of magnitude smaller in the axisymmetric case for a given thickness ratio, one would expect the true transonic region to be correspondingly smaller than in the two-dimensional case. In the notation of the perturbation velocity potential the transonic flow around an axisymmetric body is governed by

$$[1 - M^2 - M^2(\gamma + 1)\varphi_x]\varphi_{xx} + \frac{1}{r}\varphi_r + \varphi_{rr} = 0, \qquad (12\text{–}31)$$

with

$$\lim_{r\to 0}(r\varphi_r) = \frac{1}{2\pi}S'(x), \qquad (12\text{–}32)$$

$$C_p = -2\varphi_x - \varphi_r^2. \qquad (12\text{–}33)$$

Next, we shall consider the flow around configurations that are not necessarily slender but which have the surface everywhere inclined at a small angle to the free stream appropriate to the small-disturbance assumption, so that the linearized boundary condition corresponding to (6–48) is applicable. Thus

$$\epsilon\frac{\partial\Phi^i_1}{\partial n} = \frac{dn}{dx}. \qquad (12\text{–}34)$$

Instead of (12–27) we obtain this time

$$\Phi^i_{1yy} + \Phi^i_{1zz} = 0 \qquad (12\text{–}35)$$

for the inner solution, as in the case of a general slender body, Section 6–5. Thus, the inner solution may similarly be written with the aid of (2–124) formally as

$$\Phi_1^i = \frac{1}{2\pi} \oint \left(\frac{\partial \Phi_1^i}{\partial n} - \Phi_1^i \frac{\partial}{\partial n} \right) \ln r_1 \, ds_1 + \bar{g}_1(x). \qquad (12\text{--}36)$$

Here $r_1 = \sqrt{(y - y_1)^2 + (z - z_1)^2}$, y_1, z_1 is the point on the contour, and ds_1 is the contour element of the cross section. The integration constant $\bar{g}_1(x)$ must be found by matching to the outer flow. The significant information provided by (12–36) is that, in the outer limit,

$$r = \sqrt{y^2 + z^2} \to \infty,$$

the inner flow approaches an axisymmetric one of the form

$$\Phi_1^{io} \sim \frac{1}{2\pi} \ln r \oint \frac{\partial \Phi_1^i}{\partial n} \, ds_1 + \bar{g}_1(x) = \frac{S'(x)}{2\pi\epsilon} \ln r + \bar{g}_1(x), \quad (12\text{--}37)$$

where $S(x)$ is the cross-sectional area. The outer flow must therefore be equal to that around the equivalent body of revolution as in the slender-body case, and we have thus demonstrated the validity of the transonic equivalence rule resulting from the form of the first-order term in an asymptotic series expansion as the disturbance level ϵ, and $|M^2 - 1|$, both tend to zero. The approach followed is essentially that taken by Messiter (1957). A similar derivation was given by Guderley (1957).

There is no requirement on aspect ratio except that it should be finite so that $\delta A \to 0$ as $\epsilon \to 0$, in order for the outer flow to be axisymmetric in the limit. A consequence of this is that slender-body theory should provide a valid first-order approximation to lifting transonic flows for wings of finite (and moderate) aspect ratios. In Fig. 12–3 the slender-body value for the lift coefficient is compared with experimental results for a delta wing of $A = 2$. It is seen that the agreement is indeed excellent at $M = 1$.

For a wing of high aspect ratio, the first-order theory will provide a poor approximation for thickness ratios of engineering interest. A different expansion is then called for, which does not lead to an axisymmetric outer flow. We therefore introduce in the outer expansion

$$\eta = \delta y, \qquad \zeta = \delta z, \qquad (12\text{--}38)$$

with δ chosen as before, (12–11). This then gives the following equation for the first-order outer term:

$$-(K_1 + \Phi_{1x}^o)\Phi_{1xx}^o + \Phi_{1\eta\eta}^o + \Phi_{1\zeta\zeta}^o = 0. \qquad (12\text{--}39)$$

The matching will prescribe the normal velocity on the wing projection on $\zeta = 0$, which in view of (12–38) will have all spanwise dimensions reduced by the factor δ. Thus, if the limit of $\epsilon \to 0$, and hence $\delta \to 0$, is taken with the aspect ratio A kept constant, the projection in the x,y-plane will have a reduced aspect ratio $A\delta$ that will shrink to zero in the limit, and the previous case with an axisymmetric outer flow is then recovered. In this case we therefore instead consider the limit of $\epsilon \to 0$ with $A \to \infty$ in such a manner that

$$A\,\delta = K_2 \qquad (12\text{–}40)$$

approaches a constant. The reduced aspect ratio will then be finite and equal to K_2. The matching procedure now parallels that for the two-dimensional flow and the choice (12–17) for ϵ gives the boundary condition

$$\Phi_{1\zeta}^{o}(x,\,\eta,\,0\pm) = \pm\,\frac{\partial\bar{g}}{\partial x} \qquad (12\text{–}41)$$

to be satisfied on the wing projection of reduced aspect ratio

$$K_2 = \tau^{1/3}AM^{2/3}(\gamma + 1)^{1/3} \qquad (12\text{–}42)$$

Thus, the solution in this case depends on two transonic parameters K_1 and K_2. The previous case, for which the transonic equivalence rule holds, may be considered the limiting solution when

$$K_2 \to 0.$$

As the approach to zero is made the solution defined by (12–39) and (12–41) becomes, in the limit, proportional to $K_2 \ln K_2$. The two-dimensional case described by (12–12)–(12–14), or (12–19)–(12–21), is obtained as the limit of

$$K_2 \to \infty.$$

The most general transonic small-perturbation equation is thus

$$[1 - M^2 - M^2(\gamma + 1)\varphi_x]\varphi_{xx} + \varphi_{yy} + \varphi_{zz} = 0, \qquad (12\text{–}43)$$

with the boundary condition in case of a thin wing

$$\varphi_z(x, y, 0\pm) = \pm\tau\,\frac{\partial\bar{g}}{\partial x} \qquad \text{on wing projection } S_w. \qquad (12\text{–}44)$$

The pressure is given by (12–21) for a thin wing and by

$$C_p = -2\varphi_x - \varphi_y^2 - \varphi_z^2 \qquad (12\text{–}45)$$

for a slender configuration.

12–3 Similarity Rules

The first-order terms in the series expansion considered above provide similarity rules to relate the flow around affine bodies.* Taking first the two-dimensional case, we see from (12–18) that the reduced pressure coefficient

$$\widetilde{C}_p = \frac{[M^2(\gamma + 1)]^{1/3}}{\tau^{2/3}} C_p \qquad (12\text{--}46)$$

must be a function of x/c ($c = $ chord) and the parameter

$$K_1 = \frac{M^2 - 1}{\epsilon M^2(\gamma + 1)} = \frac{M^2 - 1}{[M^2\tau(\gamma + 1)]^{2/3}} \qquad (12\text{--}47)$$

only. This conclusion follows because the solution must be independent of scale (see Section 1–4) and K_1 is the only parameter that enters the boundary value problem defined by (12–12) and (12–15).

The total drag is obtained by integrating the pressure times the airfoil slope, which leads in a similar way to the result that

$$\widetilde{C}_D = \frac{[M^2(\gamma + 1)]^{1/3}}{\tau^{5/3}} C_D \qquad (12\text{--}48)$$

must be a function of K_1 only. The additional factor of τ enters because the slope is proportional to τ. Of course, (12–48) holds only for the wave drag, so that in order to use it to correlate measurements, the friction drag must be subtracted out. Such an application to biconvex-airfoil drag measurements by Michel, Marchaud, and LeGallo (1953) is illustrated in Fig. 12–5. As may be seen, the drag coefficients for the various airfoil thicknesses, when reduced this way, fall essentially on one single curve, thus confirming the validity of (12–48), and hence the small-perturbation equations.

Transonic similarity rules for two-dimensional flow were first derived by von Kármán (1947b) and Oswatitsch (1947). These rules also included the lifting case.

Rules for a slender body of revolution were formulated by Oswatitsch and Berndt (1950). There is an additional difficulty in this case associated with the logarithmic singularity at the axis. It follows from the formulation (12–26)–(12–30) for the outer flow and the matching to the inner flow

---

* These are bodies of similar shape, but stretched differently in the x-direction, or y-direction, or both. Thin airfoils of different thickness ratios constitute one class of affine bodies.

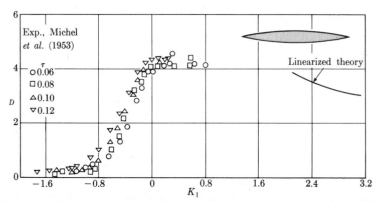

FIG. 12–5. Correlation of wave drag measurements on a circular-arc airfoil at zero lift by use of transonic similarity law (12–48). (Adapted from Spreiter and Alksne, 1958. Courtesy of National Aeronautics and Space Administration.)

as given by (12–28) that, near the body,

$$\Phi_1^o = \frac{1}{2\pi}\,\bar{S}'(x)\ln\rho + \bar{g}_1(x).\tag{12–49}$$

The transonic parameter

$$K_1 = \frac{M^2 - 1}{\epsilon M^2(\gamma + 1)} = \frac{M^2 - 1}{M^2\tau^2(\gamma + 1)}\tag{12–50}$$

thus enters into $\bar{g}_1$ only. By using (12–33) to calculate the pressure coefficient we find that, on the body,

$$C_p = -2\tau^2\left\{\frac{1}{2\pi}\,\bar{S}''(x)\ln\left[\tau^2 M\sqrt{\gamma + 1}\,\bar{R}\right] + \bar{g}'(x) + \tfrac{1}{2}(\bar{R}')^2\right\},\tag{12–51}$$

where $\bar{R}(x) = R(x)/\tau$. Thus,

$$\widetilde{C}_p = \left\{\frac{1}{\tau^2}\,C_p + \frac{1}{\pi}\,\bar{S}''(x)\ln\left[\tau^2 M\sqrt{\gamma + 1}\right]\right\}\tag{12–52}$$

is a function of K_1 and x/l only. An application of (12–52) to correlate the measured pressures on two bodies of different thickness ratios, carried out by Drougge (1959), is shown in Fig. 12–6. As may be seen, the correlation is almost perfect, except at the rearmost portions of the bodies where boundary layer separation occurs.

From (12–52) one can also construct an expression for the drag, as shown in the original paper by Oswatitsch and Berndt (1950). They found that

$$\widetilde{D} = \frac{D}{\frac{1}{2}\rho_\infty U_\infty^2\tau^4} + \frac{1}{2\pi}\,[\bar{S}'(l)]^2\ln\left[\tau^2 M\sqrt{\gamma + 1}\right]\tag{12–53}$$

must be a function of K_1 only.

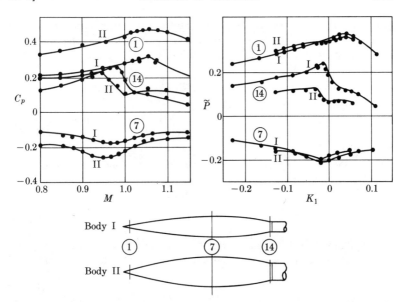

Fig. 12–6. Correlation of pressure measurements on two bodies of revolution using the transonic similarity law (12–52). $\widetilde{P} = \widetilde{C}_p + (1/\pi)\overline{S}'' \ln(\gamma + 1)$. (Adapted from Drougge, 1959. Courtesy of Aeronautical Research Institute of Sweden.)

It is a fairly straightforward matter to construct corresponding similarity rules for configurations of low-to-moderate aspect ratios. For wings of large aspect ratios one obtains results of the same form as for the two-dimensional case, except that now the reduced quantities depend on the second transonic parameter

$$K_2 = \tau^{1/3} A M^{2/3}(\gamma + 1)^{1/3}$$

as well as on K_1. The rules for three-dimensional wings were derived by Berndt (1950) and by Spreiter (1953).

12–4 Methods of Solution

Here we shall make a short review of some of the methods that have been proposed and used for solving the transonic nonlinear small-disturbance equations. The only case, so far, that has been found amenable to a mathematically exact treatment is the two-dimensional, for which it is possible to transform the nonlinear problem to a linear one by going to the hodograph plane. Let

$$\overline{u} = \Phi^o_{1x} + K_1, \qquad \overline{w} = \Phi^o_{1\zeta}, \qquad (12\text{–}54)$$

where K_1 is given by (12–47). The quantities $\bar{u}$ and $\bar{w}$ represent (to first order), with appropriate constants, the components of the difference between the local velocity and the speed of sound. Thus, on the sonic line

$$\bar{u} = 0, \tag{12–55}$$

with $\bar{w}$ taking on any value, and

$$\bar{u} = K_1, \qquad \bar{w} = 0 \tag{12–56}$$

at infinity.

We may then write (12–12) with the aid of the condition of irrotationality as follows:

$$-\bar{u}\bar{u}_x + \bar{w}_\zeta = 0, \tag{12–57}$$

$$\bar{u}_\zeta = \bar{w}_x. \tag{12–58}$$

We shall now transform this system so that $\bar{u}$ and $\bar{w}$ appear as the dependent variables. For this purpose we introduce a function $\psi(\bar{u}, \bar{w})$ such that

$$x = \psi_{\bar{u}}, \tag{12–59}$$

$$\zeta = \psi_{\bar{w}}. \tag{12–60}$$

It will be shown subsequently that the irrotationality condition (12–58) is thereby automatically satisfied. For from (12–59) and (12–60) it follows that

$$dx = \psi_{\bar{u}\bar{u}}\, d\bar{u} + \psi_{\bar{u}\bar{w}}\, d\bar{w}, \tag{12–61}$$

$$d\zeta = \psi_{\bar{w}\bar{u}}\, d\bar{u} + \psi_{\bar{w}\bar{w}}\, d\bar{w}, \tag{12–62}$$

which show that $dx = 0$ for

$$d\bar{w} = -(\psi_{\bar{u}\bar{u}}/\psi_{\bar{u}\bar{w}})\, d\bar{u}, \tag{12–63}$$

and that $d\zeta = 0$ for

$$d\bar{u} = -d\bar{w}(\psi_{\bar{w}\bar{w}}/\psi_{\bar{w}\bar{u}}). \tag{12–64}$$

Thus

$$\bar{u}_\zeta = \left(\frac{d\bar{u}}{d\zeta}\right)_{dx=0} = \frac{d\bar{u}}{[\psi_{\bar{w}\bar{u}} - \psi_{\bar{w}\bar{w}}(\psi_{\bar{u}\bar{u}}/\psi_{\bar{u}\bar{w}})]\, d\bar{u}}$$

$$= -\frac{\psi_{\bar{u}\bar{w}}}{D}, \tag{12–65}$$

where

$$D = \psi_{\bar{u}\bar{u}}\psi_{\bar{w}\bar{w}} - (\psi_{\bar{u}\bar{w}})^2$$

is the functional determinant (Jacobian).

Similarly, it is found by use of (12–64) that

$$\overline{w}_x = \left(\frac{d\overline{w}}{dx}\right)_{d\zeta=0} = -\frac{\psi_{\overline{u}\overline{w}}}{D},$$ (12–66)

which confirms (12–58).

Proceeding in this manner we find that

$$\overline{u}_x = \psi_{\overline{w}\overline{w}}/D$$ (12–67)

and

$$\overline{w}_\zeta = \psi_{\overline{u}\overline{u}}/D.$$ (12–68)

Hence (12–57) transforms to

$$-\overline{u}\psi_{\overline{w}\overline{w}} + \psi_{\overline{u}\overline{u}} = 0,$$ (12–69)

which is known as Tricomi's equation after Tricomi (1923), who first investigated its properties. It is seen that the linear equation (12–69) preserves the mixed subsonic-supersonic character of the original equation because it is hyperbolic for $\overline{u} > 0$ and elliptic for $\overline{u} < 0$. However, the linearization of the equation has not been bought without considerable sacrifice to the simplicity in the application of the boundary conditions. In fact, solutions have so far been obtained only for some very simple shapes.

To illustrate the difficulties involved, the boundary conditions for a simple wedge are formulated in Fig. 12–7. The case of a subsonic free stream for which $K_1 < 0$ is illustrated. In view of (12–54) the point $\overline{u} = K_1$, $\overline{w} = 0$ represents infinity in the physical flow field so that, following

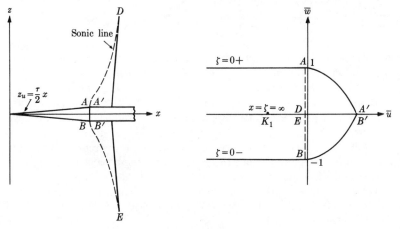

Fig. 12–7. The transonic flow around a nonlifting single wedge.

the definitions (12–59) and (12–60), the derivatives of ψ must be infinite at this point, i.e., the solution must have a singularity at $(K_1, 0)$. The form of this may be determined from the linearized subsonic solution, except for $M = 1$ which requires special treatment.

The transonic wedge solution has been worked out for M less than unity by Cole (1951) and by Yoshihara (1956), for $M = 1$ by Guderley and Yoshihara (1950), and for $M > 1$ by Vincenti and Wagoner (1952). The results for the drag coefficient are plotted in Fig. 12–8 together with experimental data obtained by Bryson (1952) and Liepmann and Bryson (1950). As seen, the agreement is excellent.

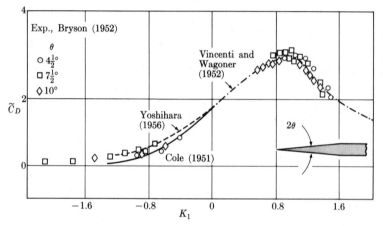

Fig. 12–8. Theoretical and experimental results for the drag of single-wedge airfoils. (From Spreiter and Alksne, 1958. Courtesy of the National Aeronautics and Space Administration.)

The hodograph method is not very useful for axisymmetric flow, since the factor $1/\rho$ in the second term of (12–26) makes the equation still non-linear after transformation to the hodograph plane. Because of such limitations, a considerable effort has been expended in finding methods that work directly in the physical plane. The solutions developed so far are all based on one or more approximations. In the method for two-dimensional flow proposed by Oswatitsch (1950) and developed in detail by Gullstrand (1951), the differential equation is rewritten as an integral equation by the aid of Green's theorem, and the nonlinear term is approximated under the implicit assumption that the value of an integral is less sensitive to errors in the approximations than is a derivative. Further improvements in this method have been introduced by Spreiter and Alksne (1955).

An approximation of a radically different kind was suggested by Oswatitsch and Keune (1955a) for treating the flow on the forward portion of a body of revolution at $M = 1$. In the differential equation (12–31) for $M = 1$,

$$-(\gamma + 1)\varphi_x\varphi_{xx} + \frac{1}{r}\,\varphi_r + \varphi_{rr} = 0, \qquad (12\text{–}70)$$

the nonlinear term was approximated by

$$(\gamma + 1)\varphi_x\varphi_{xx} = \lambda_p\varphi_x, \qquad (12\text{–}71)$$

where the constant λ_p is to be suitably chosen. The justification of this approximation is that on the forward portion of the body the flow is found to be everywhere accelerating at a fairly constant rate. Also, the resulting differential equation is parabolic, which intuitively is satisfying as an intermediate type between the elliptic and hyperbolic ones. The constant λ_p was chosen arbitrarily (but in a way consistent with the similarity law) so as to give good agreement with the measured pressure distribution in one case, and it was proposed to use this as a universal value in other cases.

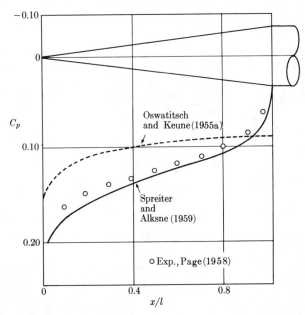

Fig. 12–9. Pressure distribution on a cone cylinder at $M = 1$. (From Spreiter and Alksne, 1959. Courtesy of National Aeronautics and Space Administration.)

Maeder and Thommen (1956) also used the approximation (12–71) for flows with M slightly different from unity and suggested a new, but still arbitrary, rule for determining λ_p.

An interesting extension of Oswatitsch's method, which removes the arbitrariness in selecting λ_p, has been presented by Spreiter and Alksne (1959). In this the parabolic equation resulting from the approximation (12–71) is first solved assuming λ_p constant, and the value of $u = \varphi_x$ is calculated on the body. Now $(\gamma + 1)u_x$ is restored in place of λ_p and a nonlinear differential equation of first order is obtained for u, which may be solved numerically. As an example, the pressure distribution on a slender cone-cylinder calculated this way is shown in Fig. 12–9 together with values obtained from the theory by Oswatitsch and Keune (1955) and measured values. As seen, the agreement with the improved theory is excellent and considerably better than with the original one.

Spreiter and Alksne (1958) also employed this technique with considerable success for two-dimensional flow, and for flows that have a Mach number slightly different from unity. In the latter case they replaced the nonlinear term

$$[1 - M^2 - M^2(\gamma + 1)\varphi_x]\varphi_{xx} \qquad (12\text{–}72)$$

by

$$\lambda\varphi_{xx}, \qquad (12\text{–}73)$$

and proceeded similarly to solve the resulting linear equation with λ constant. Thereupon (12–72) was resubstituted into the answer, producing a nonlinear first-order differential equation for $u = \varphi_x$ as before. They were able to show that in the two-dimensional supersonic case, this gave an answer that was identical to that given by simple-wave theory.

13

Unsteady Flow

13–1 Statement of the Problem

Chapters 4 through 12 have dealt with aerodynamic loading due to uniform flight of wings and bodies. Obviously no air vehicle remains indefinitely bathed in steady flow, but this idealization is justified on the basis that the time constants of unsteady motion are often very long compared to the interval required for transients in the fluid to die down to imperceptible levels. There exist important phenomena, however, where unsteadiness cannot be overlooked; rapid maneuvers, response to atmospheric turbulence, and flutter are familiar instances. We therefore end this book with a short review of some significant results on time-dependent loading of wings. These examples merely typify the extensive research that has lately been devoted to unsteady flow theory, both linearized and more exact. We hope that the reader will be able to construct parallels with steady-state counterparts and thus prepare himself to read the literature on oscillating nonplanar configurations, slender bodies, etc.

Let irrotationality be assumed, under the limitations set forth in Sections 1–1 and 1–7. The kinematics of the unsteady field are then fully described by a velocity potential Φ, governed by the differential equation (1–74), from which the speed of sound is formally eliminated using (1–67). (We let $\Omega = 0$ here.) Pressure distributions and generalized aerodynamic forces follow from (1–64).

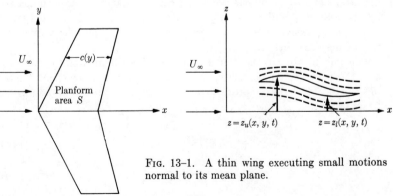

Fig. 13–1. A thin wing executing small motions normal to its mean plane.

Flow disturbances in a uniform stream

$$\Phi_0^o = U_\infty x \qquad (13\text{--}1)$$

are generated by a thin lifting surface (Fig. 13–1), which is performing rapid, small displacements in a direction generally normal to its x,y-plane projection. Thus the wing might be vibrating elastically, undergoing sudden roll or pitch aerobatics, or an encounter with gusty air might give rise to a situation mathematically and physically analogous to vibrations.

With z_u and z_l as given functions of position and time, we have no difficulty in reasoning that the boundary conditions which generalize the steady-state requirement of flow tangency at the surface (cf. 5–5) read as follows:

$$\Phi_z(x, y, z_u, t) = \frac{\partial z_u}{\partial x}\Phi_x(x, y, z_u, t) + \frac{\partial z_u}{\partial y}\Phi_y(x, y, z_u, t) + \frac{\partial z_u}{\partial t}$$
$$\Phi_z(x, y, z_l, t) = \frac{\partial z_l}{\partial x}\Phi_x(x, y, z_l, t) + \frac{\partial z_l}{\partial y}\Phi_y(x, y, z_l, t) + \frac{\partial z_l}{\partial t} \qquad (13\text{--}2)$$

for (x, y) on S. There is the usual auxiliary condition of vanishing disturbances at points remote from the wing and its wake, but for *compressible* fluid this must be refined to ensure that such disturbances behave like outward-propagating waves. The Kutta-Joukowsky hypothesis of continuous pressure at subsonic trailing edges is also applied, although we should observe that recent evidence (Ransleben and Abramson, 1962) has cast some doubt on its validity for cases of high-frequency oscillation.

Provided that there are no time-dependent variations of profile thickness, the upper and lower surface coordinates can be given by

$$z_u = \epsilon f_u(x, y, t) = \tau \overline{g}(x, y) + \theta \overline{h}(x, y, t)$$
$$z_l = \epsilon \overline{f}_l(x, y, t) = -\tau \overline{g}(x, y) + \theta \overline{h}(x, y, t). \qquad (13\text{--}3)$$

Here ϵ is a dimensionless small parameter measuring the maximum crosswise extension of the wing, including the space occupied by its unsteady displacement. Angle of attack α (5–1) can be thought of as encompassed by the θ-term; $\overline{g}$ and $\overline{h}$ are smooth functions as in steady motion; their x- and y-derivatives are everywhere of order unity; the t-derivative of $\overline{h}$ will be discussed below.

Recognizing that in the limit $\epsilon \to 0$ the wing collapses to the x,y-plane and the perturbation vanishes, we shall seek the leading terms by the method of matched asymptotic expansions. Let the inner and outer series be written

$$\Phi^o = U_\infty[x + \epsilon\Phi_1^o(x, y, z, t) + \cdots], \qquad (13\text{--}4)$$

$$\Phi^i = U_\infty[x + \epsilon\Phi_1^i(x, y, \overline{z}, t) + \cdots], \qquad (13\text{--}5)$$

where

$$\overline{z} = z/\epsilon, \qquad (13\text{--}6)$$

as in earlier developments. The presence of a uniform stream, which is clearly a solution of (1–74), has already been recognized in the zeroth-order terms in (13–4) and (13–5).

When we insert (13–3) into (13–2), a new question arises as to the size of $\partial \bar{f}_u / \partial t$, that is of $\partial \bar{h} / \partial t$. These derivatives may normally be expected to control the orders of magnitude of the time derivatives of Φ, hence of the terms that must be retained when (13–4)–(13–5) are substituted into (1–74). Here we shall avoid the complexities of this issue by requiring time and space rates of change to be of comparable magnitudes. For example, within the framework of linearized theory a sinusoidal oscillation can be represented by*

$$\bar{h}(x, y, t) = \hat{\bar{h}}(x, y)e^{i\omega t}. \tag{13–7}$$

The combination of (13–7) with (13–2), followed by a nondimensionalization of Φ_z and $\hat{\bar{h}}$ through division by U_∞ and typical length l, respectively, produces a term containing the factor

$$k \equiv \frac{\omega l}{U_\infty}. \tag{13–8}$$

Here k is known as the reduced frequency and our present intention is to specify that $k = O(1)$. For the rich variety of further reductions, even within the linearized framework, that result from other specifications on the magnitudes of k, M, etc., we cite Table I, Chapter 1, of the book by Miles (1959).

With the foregoing limitation on sizes of time derivatives, we find that the development of small-perturbation unsteady flow theory parallels the steps (5–6) through (5–31) quite closely. Thus the condition that vertical velocity W must vanish as $\epsilon \to 0$ shows, as in (5–12), that Φ_1^i must be independent of $\bar{z}$, say

$$\Phi_1^i = \bar{g}_1(x, y, t). \tag{13–9}$$

By combining (13–3) with (13–2), we conclude that Φ_2^i is the first term to possess a nonzero boundary condition,

$$\left. \begin{aligned} \Phi_{2\bar{z}}^i &= \frac{\partial \bar{f}_u}{\partial x} + \frac{1}{U_\infty} \frac{\partial \bar{f}_u}{\partial t} \quad \text{at} \quad \bar{z} = \bar{f}_u \\ \Phi_{2\bar{z}}^i &= \frac{\partial \bar{f}_l}{\partial x} + \frac{1}{U_\infty} \frac{\partial \bar{f}_l}{\partial t} \quad \text{at} \quad \bar{z} = \bar{f}_l \end{aligned} \right\} \quad \text{for } (x, y) \text{ on } S. \tag{13–10}$$

---

* This notation is commonly used in connection with mechanical or electrical vibrations and implies that the real (or imaginary) part of the right-hand side must be taken in order to recover the physical quantity of interest. Here $\hat{\bar{h}}$ is a complex function of position and allows for phase shifts between displacements of different points.

The differential equation

$$\Phi^i_{2\bar{z}\bar{z}} = 0 \qquad (13\text{--}11)$$

requires a solution linear in $\bar{z}$; thus

$$\Phi^i_2 = \bar{z}\left[\frac{\partial \bar{f}_u}{\partial x} + \frac{1}{U_\infty}\frac{\partial \bar{f}_u}{\partial t}\right] + \bar{g}_{2_u}(x, y, t), \qquad (13\text{--}12)$$

for $\bar{z} \geq \bar{f}_u$, with a similar form below the lower surface. As in steady flow, the z-velocities are seen to remain unchanged along vertical lines through the inner field, and it will be shown to serve as a "cushion" that transmits both W and pressure directly from the outer field to the wing.

It is an easy matter to extract the linear terms from (1–74) and derive the first-order outer differential equation

$$(1 - M^2)\Phi^o_{1_{xx}} + \Phi^o_{1_{yy}} + \Phi^o_{1_{zz}} - \frac{2M^2}{U_\infty}\Phi^o_{1_{xt}} - \frac{M^2}{U_\infty^2}\Phi^o_{1_{tt}} = 0. \qquad (13\text{--}13)$$

By matching with W derived from (13–12), we obtain indirectly the following boundary conditions:

$$\left.\begin{aligned}
\Phi^o_{1_z}(x, y, 0+, t) &= \frac{\partial \bar{f}_u}{\partial x} + \frac{1}{U_\infty}\frac{\partial \bar{f}_u}{\partial t} \\
\Phi^o_{1_z}(x, y, 0-, t) &= \frac{\partial \bar{f}_l}{\partial x} + \frac{1}{U_\infty}\frac{\partial \bar{f}_l}{\partial t}
\end{aligned}\right\} \quad \text{for } (x, y) \text{ on } S. \qquad (13\text{--}14)$$

Moreover, matching Φ itself identifies $\bar{g}_1$ with the potential Φ^o_1 at the inner limits $z = 0\pm$.

The linear dependence of Φ^o_1 on $\bar{f}_u$ and $\bar{f}_l$, evident from (13–14), suggests that, in a small-perturbation solution which does not proceed beyond first order in ϵ, we should deal separately with those portions of the flow that are symmetrical and antisymmetrical in z [cf. (5–32) or Sections 7–2 through 7–3]. Indeed, one may even isolate that part of $\bar{h}(x, y, t)$ from (13–3) that is both antisymmetrical and time-dependent. This we do, realizing that we may afterward superimpose both the thickness and lifting contributions of the steady field, but that neither has any *first-order* influence on the unsteady loading.

We again adopt a perturbation potential, given by

$$\epsilon\Phi^o_1 = \varphi(x, y, z, t) \qquad (13\text{--}15)$$

and satisfying, together with proper conditions at infinity, the following system:

$$(1 - M^2)\varphi_{xx} + \varphi_{yy} + \varphi_{zz} - \frac{2M^2}{U_\infty}\varphi_{xt} - \frac{M^2}{U_\infty^2}\varphi_{tt} = 0, \qquad (13\text{--}16)$$

$$\varphi_z = \theta\left[\frac{\partial \bar{h}}{\partial x} + \frac{1}{U_\infty}\frac{\partial \bar{h}}{\partial t}\right], \qquad \text{at } z = 0, \qquad (13\text{--}17)$$

for (x, y) on S. Corresponding to the pressure difference, φ has a discontinuity through S. As we shall see below, the Kutta-Joukowsky hypothesis also leads to unsteady discontinuities on the wake surface, which is approximated here by the part of the x,y-plane between the downstream wing-tip extensions.

Finally, a reduction of (1–64) and matching, to order ϵ, of Φ or Φ_x shows that

$$C_p = -2\varphi_x - \frac{2}{U_\infty} \varphi_t + O(\epsilon^2) \tag{13-18}$$

throughout the entire flow. (From Chapter 6, the reader will be able to reason that the small-perturbation Bernoulli equation again contains nonlinear terms when used in connection with unsteady motion of slender *bodies* rather than wings.)

13-2 Two-Dimensional, Constant-Density Flow

The best-known of the classical solutions for unsteady loading is the one, found almost simultaneously by five or six authors in the mid-1930's, for the oscillating thin airfoil at $M = 0$. In this case of nearly constant density, a key distinction disappears between the steady and unsteady problems because the flow must satisfy a two-dimensional Laplace equation

$$\varphi_{xx} + \varphi_{zz} = 0. \tag{13-19}$$

We may accordingly rely quite heavily on the results for a steadily lifting airfoil from Section 5–3, particularly on (5–58) and (5–73), which supply the needed inversion for the oscillatory integral equation while simultaneously enforcing Kutta's condition at the trailing edge.

With the lifting surface paralleling the x,y-plane between $x = 0$ and c, it can be assumed from (13–17) that $\varphi_z(x, 0, t)$ is known over that area and given by (dimensionless)

$$w_0(x, t) = \hat{w}_0(x)e^{i\omega t} \quad \text{for} \quad 0 \leq x \leq c. \tag{13-20}$$

The perturbation field has φ and u antisymmetric in z, and allowance must be made for φ-discontinuities through the x,y-plane for $x > 0$. Hence, (13–19) and all other conditions can be satisfied by a vortex sheet similar to the one described by (5–58) but extended downstream by replacing the upper limit with infinity. Equations (13–17) and (13–20) are introduced through

$$w_0(x, t) = -\frac{1}{2\pi} \oint_0^\infty \frac{\gamma(x_1, t)}{x - x_1} dx_1 \quad \text{for} \quad 0 \leq x \leq c, \tag{13-21}$$

where

$$\gamma = u(x, 0+, t) - u(x, 0-, t) = 2u(x, 0+, t). \tag{13-22}$$

For later convenience, we define an integrated vortex strength

$$\Gamma(x, t) = \int_0^x \gamma(x_1, t)\, dx_1 = 2\int_0^x \varphi_x(x_1, 0+, t)\, dx_1$$
$$= 2\varphi(x, 0+, t), \tag{13-23}$$

and note that $\Gamma(c, t)$ is the instantaneous circulation bound to the airfoil.

From (13–18) and the antisymmetry of C_p in z, we deduce that a (physically impossible) discontinuity of pressure through the wake is avoided only if

$$\varphi_x + \frac{1}{U_\infty} \varphi_t = 0, \tag{13-24}$$

for $x > c$ along $z = 0+$. Equation (13–24) is a partial differential expression for $\varphi(x, 0+, t)$, which is solved subject to continuity of φ at the trailing edge by

$$\varphi(x, 0+, t) = \varphi\left(c, 0+, t - \frac{x - c}{U_\infty}\right). \tag{13-25}$$

From (13–23) and (13–25) are derived the further relations for the wake:

$$\Gamma(x, t) = \Gamma\left(c, t - \frac{x - c}{U_\infty}\right), \tag{13-26}$$

$$\gamma(x, t) = -\frac{1}{U_\infty} \Gamma_t\left(c, t - \frac{x - c}{U_\infty}\right). \tag{13-27}$$

Equation (13–27) has the obvious interpretation that wake vortex elements are convected downstream approximately at the flight speed U_∞, after being shed as countervortices from the trailing edge at a rate equal to the variation of bound circulation.

We next introduce (13–27) into (13–21) and use the assumption that a linear, simple harmonic process has been going on indefinitely to replace all dependent variables with sinusoidal counterparts and cancel the common factor $e^{i\omega t}$:

$$\hat{w}_0(x) = -\frac{1}{2\pi} \oint_0^c \frac{\hat{\gamma}(x_1)\, dx_1}{x - x_1} + \hat{g}(x), \tag{13-28a}$$

where*

$$\hat{g}(x) = \frac{i\omega}{2\pi} \frac{\hat{\Gamma}(c)}{U_\infty} \int_c^\infty \frac{e^{-i(\omega/U_\infty)(x_1 - c)}}{x - x_1}\, dx_1. \tag{13-28b}$$

---

* If dimensionless x-variables are adopted in (13–28), based on reference length $l \equiv c/2$, it is clear how the aforementioned $k = \omega c/2U_\infty$ will arise as one parameter of the problem.

Multiplying both sides of (5–73) by 2 and using (5–56), we obtain an inversion of (5–58),

$$\gamma(x) = \frac{2}{\pi} \sqrt{\frac{c-x}{x}} \fint_0^c \frac{w_0(x_1)}{x-x_1} \sqrt{\frac{x_1}{c-x_1}}\, dx_1. \qquad (13\text{–}29)$$

If w_0 is replaced by $(\hat{w}_0 - \theta)$, (13–29) is suitable also for inverting (13–28), provided we have expressed $\hat{\Gamma}(c)$ in terms of known quantities. This final need is met by multiplying (13–28a) by $\sqrt{x/(c-x)}$ and integrating with respect to x along the chord. The integral on the left-hand side is easily evaluated for most continuous functions $\hat{w}_0(x)$. The two integrals appearing on the right are handled as follows:

(a)
$$I_1 = -\frac{1}{2\pi} \int_0^c \sqrt{\frac{x}{c-x}} \fint_0^c \frac{\hat{\gamma}(x_1)\, dx_1}{x-x_1}\, dx$$

$$= -\frac{1}{2\pi} \int_0^c \hat{\gamma}(x_1) \fint_0^c \sqrt{\frac{x}{c-x}} \frac{dx}{x-x_1}\, dx_1. \qquad (13\text{–}30)$$

The change of variable $x = (c/2)(1 - \cos\theta)$ converts the inner integral here into one of the familiar results arising in Glauert's solution for the lifting line,

$$\fint_0^c \sqrt{\frac{x}{c-x}} \frac{dx}{x-x_1} = -\fint_0^\pi \frac{(1-\cos\theta)\, d\theta}{\cos\theta - \cos\theta_1} = \pi. \qquad (13\text{–}31)$$

Whereupon,

(b)
$$I_1 = -\tfrac{1}{2}\int_0^c \hat{\gamma}(x_1)\, dx_1 = -\tfrac{1}{2}\hat{\Gamma}(c). \qquad (13\text{–}32)$$

$$I_2 = \frac{1}{2\pi} \int_0^c \sqrt{\frac{x}{c-x}} \int_c^\infty \frac{e^{-i(\omega/U_\infty)(x_1-c)}}{x-x_1}\, dx_1\, dx$$

$$= \frac{1}{2\pi} \int_c^\infty e^{-i(\omega/U_\infty)(x_1-c)} \int_0^c \sqrt{\frac{x}{c-x}} \frac{dx}{(x-x_1)}\, dx_1. \qquad (13\text{–}33)$$

For the inner integral here, $x_1 > c$ and (5–79) and (5–80) yield

$$\int_0^c \sqrt{\frac{x}{c-x}} \frac{dx}{(x-x_1)} = \pi\left[1 - \sqrt{\frac{x_1}{x_1-c}}\right]. \qquad (13\text{–}34)$$

The result

$$I_2 = \tfrac{1}{2}\int_c^\infty e^{-i(\omega/U_\infty)(x_1-c)}\left[1 - \sqrt{\frac{x_1}{x_1-c}}\right] dx_1 \qquad (13\text{–}35)$$

is seen to be properly convergent because the integrand vanishes as $x_1 \to \infty$. It is most easily evaluated by defining $\xi = (2x_1/c) - 1$ to give

$$I_2 = \frac{ce^{ik}}{4} \int_1^\infty e^{-ik\xi} \left[1 - \sqrt{\frac{\xi + 1}{\xi - 1}} \right] d\xi. \qquad (13\text{--}36)$$

The first term in the (13–36) brackets is straightforward, whereas the second may be identified with certain infinite integrals listed by Watson (1948):

$$\int_1^\infty \sqrt{\frac{\xi + 1}{\xi - 1}}\, e^{-ik\xi}\, d\xi = \int_1^\infty \frac{\xi e^{-ik\xi}}{\sqrt{\xi^2 - 1}}\, d\xi$$

$$+ \int_1^\infty \frac{e^{-ik\xi}}{\sqrt{\xi^2 - 1}}\, d\xi = -\frac{\pi}{2} H_1^{(2)}(k) - i\frac{\pi}{2} H_0^{(2)}(k), \qquad (13\text{--}37)$$

$H_n^{(2)}$ being the Hankel function of the second kind and order n. It should be noted that, for reasons of convergence, Watson limits the imaginary part of k in (13–37) to be *less* than zero. For use in (13–36), the result may be analytically continued to purely real values of k, however,

$$I_2 = \frac{c}{2} \left[\frac{\pi}{4} e^{ik}(H_1^{(2)}(k) + iH_0^{(2)}(k)) + \frac{1}{2ik} \right]. \qquad (13\text{--}38)$$

When I_1 and I_2 are substituted into the weighted integral of (13–28a), we obtain

$$\hat{\Gamma}(c) = \frac{4e^{-ik} \int_0^c \sqrt{x/(c - x)}\, \hat{w}_0(x)\, dx}{\pi ik[H_1^{(2)}(k) + iH_0^{(2)}(k)]}. \qquad (13\text{--}39)$$

(Incidentally, the appearance of $k \equiv \omega c/2U_\infty$ as argument in (13–39) and below explains the choice of $c/2$ as the "natural" reference length for subsonic unsteady problems of this kind.) Returning to (13–28a), we use (13–29) to express a formal solution analogous to (5–73):

$$\hat{\gamma}(x) = \frac{2}{\pi} \sqrt{\frac{c - x}{x}} \oint_0^c \frac{\hat{w}_0(x_1) - \hat{g}(x_1)}{x - x_1} \sqrt{\frac{x_1}{c - x_1}}\, dx_1. \qquad (13\text{--}40)$$

Through (13–39), (13–18), and (13–23) we are now able to work out the pressure distribution, lift, moment, etc. due to any motion of the airfoil. There is a somewhat more efficient way of presenting these results than has appeared in the literature, however, and we devote a few lines to describing it.

Let us define an auxiliary function, connected to $\hat{w}_0$ as is $\hat{\Gamma}$ to $\hat{\gamma}$:

$$\hat{p}(x) = \int_0^x \hat{w}_0(x_1)\, dx_1 = -\frac{1}{2\pi} \int_0^x \oint_0^\infty \frac{\hat{\gamma}(x_2)}{x_1 - x_2}\, dx_2\, dx_1. \qquad (13\text{--}41)$$

The last integral here, being improper, is evaluated by replacing the infinite limit with R, inverting order, and letting $R \to \infty$.

$$\int_0^x \oint_0^R \frac{\hat{\gamma}(x_2)\, dx_2\, dx_1}{x_1 - x_2} = \int_0^R \hat{\gamma}(x_2)[\ln|x - x_2| - \ln x_2]\, dx_2$$

$$= \left\{ \hat{\Gamma}(x_2) \ln\left|\frac{x - x_2}{x_2}\right| \right\}\Big|_0^R + \oint_0^R \frac{\hat{\Gamma}(x_2)}{x - x_2}\, dx_2$$

$$+ \int_0^R \frac{\hat{\Gamma}(x_2)}{x_2}\, dx_2 = \oint_0^\infty \frac{\hat{\Gamma}(x_1)}{x - x_1}\, dx_1 + 2\pi A, \quad (13\text{--}42)$$

where

$$A \equiv \frac{1}{2\pi} \int_0^\infty \frac{\hat{\Gamma}(x_1)\, dx_1}{x_1}. \quad (13\text{--}43)$$

The limit has, of course, been inserted before the last member of (13–42), and x_2 is replaced by x_1 as dummy variable. No principal value need be taken in the integral defining A, because $\hat{\Gamma}$ vanishes at the leading edge as $x_1^{1/2}$, leaving only an integrable singularity.

Thus we obtain

$$\hat{p}(x) = -\frac{1}{2\pi} \oint_0^\infty \frac{\hat{\Gamma}(x_1)\, dx_1}{x - x_1} - A. \quad (13\text{--}44)$$

The next step is to construct

$$\hat{w}_0 + i\frac{\omega}{U_\infty}\hat{p} = -\frac{1}{2\pi} \oint_0^\infty \frac{\hat{\gamma}(x_1) + [i\omega\hat{\Gamma}(x_1)/U_\infty]}{x - x_1}\, dx_1 - i\frac{\omega A}{U_\infty}, \quad (13\text{--}45)$$

and to notice from (13–18), (13–22), and (13–23) that the dimensionless pressure jump through the x-axis, positive in a sense to produce upward loading, is

$$\Delta C_p(x, t) = \Delta \hat{C}_p(x)e^{i\omega t} = \left[2\hat{\gamma}(x) + 2i\frac{\omega\hat{\Gamma}(x)}{U_\infty}\right]e^{i\omega t}. \quad (13\text{--}46)$$

In view of the vanishing of ΔC_p for $x > c$, (13–45) becomes

$$\hat{w}_0 + i\frac{\omega\hat{p}}{U_\infty} + i\frac{\omega A}{U_\infty} = -\frac{1}{4\pi} \oint_0^c \frac{\Delta\hat{C}_p(x_1)}{x - x_1}\, dx_1. \quad (13\text{--}47)$$

Immediate application of the inversion (13–29) yields

$$\Delta\hat{C}_p(x) = \frac{4}{\pi}\sqrt{\frac{c - x}{x}} \oint_0^c \frac{[\hat{w}_0 + i(\omega\hat{p}/U_\infty)]}{x - x_1}\sqrt{\frac{x_1}{c - x_1}}\, dx_1$$

$$- 4\frac{i\omega A}{U_\infty}\sqrt{\frac{c - x}{x}}. \quad (13\text{--}48)$$

An easy way of eliminating the constant A from (13–48) is to note that $\Delta \hat{C}_p$ and $2\hat{\gamma}$ approach one another at the leading edge $x = 0$, since $\hat{\Gamma}(0) = 0$. Multiplying through by $\sqrt{x}$ to cancel the singularity, we may therefore equate the following two limits. From (13–48):

$$\lim_{x \to 0} \left[\sqrt{\frac{x}{c}} \, \Delta \hat{C}_p(x) \right] = - \frac{4}{\pi} \int_0^c \frac{[\hat{w}_0 + i(\omega \hat{p}/U_\infty)] \, dx_1}{\sqrt{x_1(c - x_1)}} - 4 \frac{i\omega A}{U_\infty}, \qquad (13\text{–}49)$$

and from (13–40):

$$\lim_{x \to 0} \left[\sqrt{\frac{x}{c}} \, 2\hat{\gamma}(x) \right] = - \frac{4}{\pi} \int_0^c \frac{[\hat{w}_0 - \hat{g}]}{\sqrt{x_1(c - x_1)}} \, dx_1. \qquad (13\text{–}50)$$

Consequently,

$$i\frac{\omega A}{U_\infty} = - \frac{1}{\pi} \int_0^c i \frac{\omega \hat{p}(x_1)}{U_\infty} \frac{dx_1}{\sqrt{x_1(c - x_1)}}$$

$$- \frac{1}{\pi} \int_0^c \hat{g}(x_1) \frac{dx_1}{\sqrt{x_1(c - x_1)}} \cdot \qquad (13\text{–}51)$$

After replacing $\hat{g}(x_1)$ through (13–28b) and (13–39), we encounter the last of the two integrals evaluated in (13–37) and are ultimately led to

$$i\frac{\omega A}{U_\infty} = - \frac{1}{\pi} \int_0^c \frac{i\omega \hat{p}(x_1)}{U_\infty} \frac{dx_1}{\sqrt{x_1(c - x_1)}}$$

$$- \frac{2i}{\pi c} \frac{H_0^{(2)}(k)}{H_1^{(2)}(k) + iH_0^{(2)}(k)} \int_0^c \sqrt{\frac{x_1}{c - x_1}} \, \hat{w}_0(x_1) \, dx_1$$

$$= - \frac{1}{\pi} \int_0^c \frac{i\omega \hat{p}(x_1)}{U_\infty} \frac{dx_1}{\sqrt{x_1(c - x_1)}}$$

$$+ \frac{2}{\pi c} [C(k) - 1] \int_0^c \sqrt{\frac{x_1}{c - x_1}} \, \hat{w}_0(x_1) \, dx_1, \qquad (13\text{–}52)$$

where

$$C(k) = \frac{H_1^{(2)}(k)}{H_1^{(2)}(k) + iH_0^{(2)}(k)} \qquad (13\text{–}53)$$

is known as Theodorsen's function. We finally use (13–52) to eliminate A from (13–48) and obtain

$$\Delta \hat{C}_p(x) = \frac{4}{\pi} \sqrt{\frac{c - x}{x}} \oint_0^c \sqrt{\frac{x_1}{c - x_1}} \, \hat{w}_0(x_1) \, dx_1$$

$$+ \frac{4}{\pi} \sqrt{x(c - x)} \oint_0^c \frac{i\omega \hat{p}(x_1)}{U_\infty} \frac{dx_1}{(x - x_1)\sqrt{x_1(c - x_1)}}$$

$$+ \frac{8}{\pi c} [1 - C(k)] \sqrt{\frac{c - x}{x}} \int_0^c \sqrt{\frac{x_1}{c - x_1}} \, \hat{w}_0(x_1) \, dx_1. \qquad (13\text{–}54)$$

Working from (13-54) as a starting point, it is not difficult to prove such useful formulas as the following. The amplitude of the oscillatory lift force per unit span is

$$\frac{\hat{L}}{(\rho_\infty/2)U_\infty^2(c/2)} = -4C(k)\frac{2}{c}\int_0^c \sqrt{\frac{x_1}{c-x_1}}\,\hat{w}_0(x_1)\,dx_1$$

$$-\frac{4ik}{(c/2)^2}\int_0^c \sqrt{x_1(c-x_1)}\,\hat{w}_0(x_1)\,dx_1. \quad (13\text{-}55)$$

The amplitude of nose-up pitching moment per unit span, taken about an axis along the midchord line, is

$$\frac{\hat{M}_y}{(\rho_\infty/2)U_\infty^2(c^2/2)} = \frac{2}{c}\int_0^c \left[\sqrt{\frac{x_1}{c-x_1}} - \frac{4}{c}\sqrt{x_1(c-x_1)}\right]\hat{w}_0(x_1)\,dx_1$$

$$+ \int_0^c \frac{i\omega\hat{p}(x_1)}{U_\infty}\left[\frac{1}{\sqrt{x_1(c-x_1)}} - \frac{8}{c^2}\sqrt{x_1(c-x_1)}\right]dx_1$$

$$- C(k)\frac{2}{c}\int_0^c \sqrt{\frac{x_1}{c-x_1}}\,\hat{w}_0(x_1)\,dx_1. \quad (13\text{-}56)$$

More detail on the application and interpretation of these results will be found in such sources as Section 5-6 of Bisplinghoff, Ashley, and Halfman (1955). It is shown there, for instance, that $C(k)$ may be regarded as the lag in development of bound circulation due to the influence of the shed wake vortices. So-called "quasi-steady theory," which in one version corresponds to neglecting this wake effect, can be recovered by setting $C(k) = 1$ in (13-54)–(13-56).

13–3 Airfoils Oscillating at Supersonic and Sonic Speeds

The two-dimensional version of the problem (13-16)–(13-17) proves much more tractable when $M > 1$ than for $M = 0$. One particularly efficient approach begins by recognizing that any point along the chord can be affected only by conditions upstream, that the variable x has therefore the same unidirectional character as does time in linear-system transients, and that $\varphi = 0$ for $x < 0$. Hence Laplace transformation on x seems indicated. By writing the perturbation potential as $\hat{\varphi}e^{i\omega t}$, we modify (13-16), (13-17), and (13-20) to read

$$(M^2-1)\hat{\varphi}_{xx} - \hat{\varphi}_{zz} + \frac{2M^2 i\omega}{U_\infty}\hat{\varphi}_x - \frac{M^2\omega^2}{U_\infty^2}\hat{\varphi} = 0, \quad (13\text{-}57)$$

$$\hat{\varphi}_z(x,0) = \hat{w}_0(x), \quad \text{for} \quad x \geq 0. \quad (13\text{-}58)$$

We observe that the position of the trailing edge is immaterial when calculating the flow near the airfoil surface and that our solution should not depend on values of $\hat{w}_0(x)$ at $x > c$. This means also that Kutta's condition cannot be enforced, as is known also to be true (Chapter 8) in steady flow and for *finite* wings with supersonic trailing edges.

Let Laplace transformation be applied, according to the typical formula

$$\hat{\varphi}^*(z; s) = \int_0^\infty e^{-sx} \hat{\varphi}(x, z) \, dx. \tag{13-59}$$

The transformed equivalents of (13–57) and (13–58) are

$$\hat{\varphi}^*_{zz} = \mu^2 \hat{\varphi}^*, \tag{13-60}$$

$$\hat{\varphi}^*_z(0; s) = \hat{w}^*_0(s), \tag{13-61}$$

where

$$\mu = \left[s^2(M^2 - 1) + 2s \frac{M^2 i\omega}{U_\infty} - \frac{\omega^2 M^2}{U_\infty^2} \right]^{1/2}. \tag{13-62}$$

A general solution to (13–59) is

$$\hat{\varphi}^* = A e^{\mu z} + B e^{-\mu z}. \tag{13-63}$$

If the branch of the square root in (13–62) is chosen so as to keep the real part of μ positive, we must set $A = 0$ in the half-space $z \geq 0+$ to assure both the boundedness of disturbances at great distances and outward propagation. The solution satisfying (13–61) must therefore be

$$\hat{\varphi}^*(z; s) = -\frac{\hat{w}^*_0(s)}{\mu} e^{-\mu z}, \qquad \text{for} \qquad z \geq 0+. \tag{13-64}$$

According to the convolution theorem, the physical counterpart of (13–64) may be written

$$\hat{\varphi}(x, z) = -\int_0^x \hat{w}_0(x_1) \mathcal{L}^{-1} \left\{ \frac{e^{-\mu z}}{\mu} \right\} dx_1, \tag{13-65}$$

where $\mathcal{L}^{-1}$ is the inverse Laplace operator, and the result of this inversion is to be expressed as a function of $(x - x_1)$. After resort to a table of transforms, one derives the following:

$$\mathcal{L}^{-1}\left(\frac{e^{-\mu z}}{\mu} \right) = \begin{cases} \dfrac{1}{B} e^{-(i\omega M^2/U_\infty B^2)(x-x_1)} J_0\left(\dfrac{\omega M}{U_\infty B^2} \sqrt{(x - x_1)^2 - B^2 z^2} \right) \\ \qquad\qquad\qquad\qquad\qquad\qquad\qquad \text{for } x_1 < x - Bz \\ 0 \qquad\qquad \text{for } x_1 > x - Bz. \end{cases} \tag{13-66}$$

Here $B \equiv \sqrt{M^2 - 1}$ as in previous chapters. The potential, from (13–65) and (13–66), is for $z \geq 0+$

$$\bar{\varphi}(x, z) = -\frac{1}{B} \int_0^{x-Bz} \hat{w}_0(x_1) \exp\left[-i \frac{\omega M^2}{U_\infty B^2} (x - x_1)\right]$$

$$\times J_0\left[\frac{\omega M}{U_\infty B^2} \sqrt{(x - x_1)^2 - B^2 z^2}\right] dx_1. \qquad (13\text{–}67a)$$

For the lower half-space, the antisymmetry of $\bar{\varphi}$ yields

$$\bar{\varphi}(x, -z) = -\bar{\varphi}(x, z). \qquad (13\text{–}67b)$$

Upon introducing dimensionless coordinates, one sees that the "natural" reduced frequency for supersonic flow is some such combination as

$$(\omega c/U_\infty)(M^2/B^2);$$

this quantity tends to become very large as $M \to 1$, reflecting the short wavelength of some of the wave trains generated by an airfoil oscillating near sonic speed, which will be observed even when the frequency is quite low. For unswept wings, we may say that "unsteadiness" tends to be at a maximum in the transonic range.

More detail on the calculation of supersonic pressures from (13–18) and on other applications will be found in Chapter 6 of Bisplinghoff, Ashley, and Halfman (1955), in Miles (1959), and in references cited therein.

The essential nonlinearity of steady, two-dimensional transonic flow disappears in the oscillatory case when k is sufficiently large. The appropriate linear solution may be constructed by solving (13–57)–(13–58) for $M = 1$ in a way directly parallel to the above, or by taking the limit of (13–67) as $M \to 1$ with careful introduction of the asymptotic formula for J_0 of large argument. Either procedure yields, for $z \geq 0+$,

$$\bar{\varphi}(x, z; M = 1)$$

$$= -\int_0^x \hat{w}_0(x_1) \frac{\exp\left(-i(\omega/2U_\infty)[(x - x_1) + z^2/(x - x_1)]\right)}{\sqrt{2\pi i(\omega/U_\infty)(x - x_1)}} dx_1. \qquad (13\text{–}68)$$

Uses of this result and the circumstances under which it is valid are discussed in Chapters 1 and 2 of Landahl (1961), and references to further applications will be found there.

13–4 Indicial Motion in a Compressible Fluid

In the analysis of linear systems there exists a well-known duality between phenomena involving simple harmonic response and "indicial" phenomena—situations where an input or boundary undergoes a sudden step or impulsive change. Fourier's theorem enables problems of one type to be treated in terms of solutions of the other, and this is frequently the most useful avenue to follow in unsteady wing theory. There are cases, however, when a direct attack on the indicial motion is feasible.

As a particularly simple indicial problem, let us consider the initial development of flow near the upper surface of a wing (e.g., Fig. 7–1) when a step change occurs in the normal velocity of the surface. Such a specification demands that we reexamine the fundamental development of Section 13–1. Essentially what we are saying is that z_u in (13–3) is given by

$$\frac{\partial z_u}{\partial t} = \epsilon W_0 1(t), \qquad \text{for} \qquad (x, y) \quad \text{on} \quad S, \qquad (13\text{–}69)$$

where

$$1(t) = \begin{cases} 0, & t < 0 \\ 1, & t \geq 0 \end{cases} \qquad (13\text{–}70)$$

and $w_0 \equiv W_0/U_\infty$ is a constant of order unity. Clearly, in the vicinity of the time origin, there is now some interval where rates of change of flow properties are very large. It is useful to study this zone by defining

$$\bar{z} = \frac{z}{\epsilon} \qquad \bar{t} = \frac{t}{\epsilon} \qquad (13\text{–}71)$$

and replacing the inner series (13–5) by

$$\Phi^i = U_\infty[x + \epsilon\Phi_1^i(x, y, \bar{z}, \bar{t}) + \cdots]. \qquad (13\text{–}72)$$

Once again we are led to the conclusion that Φ_2^i carries the first significant disturbances, but now its differential equation and upper-surface boundary condition are

$$\Phi_{2_{\bar{z}\bar{z}}}^i = \frac{1}{a_\infty^2} \Phi_{2_{\bar{t}\bar{t}}}^i \qquad (13\text{–}73)$$

and

$$\Phi_{2_{\bar{z}}}^i = \frac{1}{U_\infty} \frac{\partial \bar{f}_u}{\partial t} = w_0 1(t), \qquad \text{at } \bar{z} = \bar{f}_u, \qquad \text{for } (x, y) \text{ on } S. \qquad (13\text{–}74)$$

(This statement is actually unchanged if w_0 depends on x and y.) Equations (13–73)–(13–74) describe the linearized field due to a one-dimensional piston moving impulsively into a gas at rest. The solution reads

$$\Phi_{2_{\bar{z}}}^i = \frac{1}{a_\infty} \Phi_{2_{\bar{t}}}^i = w_0 1\left(t - \frac{z}{a_\infty}\right), \qquad (13\text{–}75)$$

and it is easily shown that the overpressure on the wing surface (or piston face) is

$$p_u - p_\infty = \rho_\infty a_\infty \frac{\partial z_u}{\partial t}. \qquad (13\text{–}76)$$

All of these solutions are quite independent of flight Mach number M, so long as the disturbance velocities remain small compared with a_∞.

After a short time interval, the foregoing results make a continuous transition to solutions determined from (13-16)-(13-17). Moreover, for $t \gg c/U_\infty$, the indicial solution must settle down to the steady-state result for a wing at angle of attack $(-\epsilon w_0)$. This behavior can be demonstrated using the method of matched asymptotic expansions, but the details are much too complicated to deserve elaboration here.

Perhaps the most interesting aspect of (13-75) and (13-76) is their general applicability, when $M \gg 1$, for any small unsteady motion. At high Mach number, fluid particles pass the wing surface so rapidly that *all* of the disturbed fluid near this surface remains both in the inner z- and t-fields; except for large values of x far behind the trailing edge the outer field experiences no disturbance at all. Hence the piston formula, (13-76), yields for any instant the pressure distribution over the entire wing, a result which can also be extended into the nonlinear range (Lighthill, 1953).

13-5 Three-Dimensional Oscillating Wings

The general planar wing problem, (13-16)-(13-17), has stimulated some imaginative research in applied mathematics. For $M > 1$, there are many analytical solutions appropriate to particular wing planform shapes, such as rectangular or delta, and all details have been worked through for elementary modes of vibration like plunging and pitching. Miles (1959) constitutes a compendium of such supersonic information, as does Landahl (1961) for the vicinity of $M = 1$. In the range $0 < M < 1$ the only available exact linearized results pertain to the two-dimensional airfoil, whereas in constant-density fluid a complete and correct analysis has been published for a wing of circular planform.

Since the advent of high-speed computers, numerical methods have been elaborated to cover very general wing geometry and arbitrary continuous deflection shapes. The approach for subsonic speed has been through superposition of acceleration-potential doublets, culminating in complete lifting surface theories which generalize the steady-flow results of Section 7-6. The definitive works are those of Watkins *et al.* (1955, 1959).

The influence-coefficient methods mentioned in Chapters 8 and 11 have proved adaptable to supersonic wings, although there are some details of the treatment of singularities that have apparently been resolved only very recently. Nonplanar wings and interfering systems represent an extension that is likely to be mechanized successfully within a short period of time.

References and Author Index

The numbers in brackets show the page on which the references appear.

ABBOTT, I. H., VON DOENHOFF, A. E., and STIVERS, L. S., JR. (1945), *Summary of Airfoil Data*, NACA Report No. 824. Also issued as *Theory of Wing Sections*, McGraw-Hill, New York, 1949. [54]

ABRAMSON, H. N., *see* RANSLEBEN and ABRAMSON.

ACKERET, J. (1925), "Luftkräfte an Flügeln, die mit grösserer als Schallgeschwindigkeit bewegt werden," *Z. Flugtechn. Motorluftsch.* **XVI**, 72–74. [97, 155]

ADAMS, M. C., and SEARS, W. R. (1953), "Slender-Body Theory—Review and Extensions," *J. Aeron. Sci.* **20**, No. 2, 85–98. [102]

ALKSNE, A. Y., *see* SPREITER and ALKSNE.

ASHLEY, H., WIDNALL, S., and LANDAHL, M. T. (1965), "New Directions in Lifting Surface Theory," *AIAA Journal* **3**, No. 1, 3–16. [217, 226]

ASHLEY, H., *see also* BISPLINGHOFF *et al.*

BEANE, B. J. (1960), *Supersonic Characteristics of Partial Ring Wings*, Ph.D. Thesis, Department of Aeronautics and Astronautics, Massachusetts Institute of Technology. [207]

BEANE, B. J., DURGIN, F. H., and TILTON, E. L. (1963), *The Effect of Camber and Twist on Wing Pressure Distributions for Mach Numbers from 2 to 7.6*, Parts I and II, USAF Aeronautical Systems Division TDR 62–557. [220]

BEANE, B. J., *see also* GRAHAM, E. V. *et al.*, and BROWAND *et al.*

BERMAN, J. H., *see* WATKINS and BERMAN.

BERNDT, S. B. (1949), *Three Component Measurements and Flow Investigation of Plane Delta Wings at Low Speeds and Zero Yaw*, KTH AERO TN 4, Royal Institute of Technology, Stockholm. [122]

BERNDT, S. B. (1950), *Similarity Laws for Transonic Flow about Wings of Finite Aspect Ratio*, KTH AERO TN 14, Royal Institute of Technology, Stockholm. [239]

BERNDT, S. B., and ORLIK-RÜCKEMANN, K. (1949), *Comparison Between Lift Distributions of Plane Delta Wings and Low Speeds and Zero Yaw*, KTH AERO TN 10, Royal Institute of Technology, Stockholm. [151]

BERNDT, S. B., *see also* OSWATITSCH and BERNDT.

BETZ, A., *see* PRANDTL *et al.*

BISPLINGHOFF, R. L., ASHLEY, H., and HALFMAN, R. L. (1955), *Aeroelasticity*, Addison-Wesley, Reading, Massachusetts. [255, 257]

BOLLAY, W. (1937), "A Theory for Rectangular Wings of Small Aspect Ratio," *J. Aeron. Sci.* **4**, 294–296. [113]

BOYD, J. W., *see* FRICK and BOYD.

BRETHERTON, F. P. (1962), "Slow Viscous Motion Round a Cylinder in Simple Shear," *J. Fluid Mech.* **12**, 591–613. [62]

BROWAND, F. K., BEANE, B. J., and NOWLAN, D. T. (1962), *The Design and Test at Mach Number 2.5 of Two Low-Wave-Drag Ring-Wing Configurations of Aspect Ratio 1.3 and 2.6*, Rand Corporation, Memorandum RM-2933-PR. [207]

BRYSON, A. E. (1952), *On Experimental Investigation of Transonic Flow Past Two-Dimensional Wedge and Circular-Arc Sections Using a Mach-Zehnder Interferometer*, NACA Report 1094. [242]

BRYSON, A. E., *see also* LIEPMANN and BRYSON.

CLARKE, J., *see* FERRI et al.

COHEN, D., *see* JONES and COHEN.

COLE, J. D. (1951), "Drag of a Finite Wedge at High Subsonic Speeds," *J. Math. Phys.* **30**, No. 2, 79–93. [242]

COLE, J. D., and MESSITER, A. F. (1956), *Expansion Procedures and Similarity Laws for Transonic Flow, Part 1. Slender Bodies at Zero Incidence*, Calif. Inst. of Tech., USAF OSR Technical Note 56–1. (See also *Z. Angew. Math. Phys.* **8**, 1–25, 1957.) [233]

COLE, J. D., *see also* LAGERSTROM and COLE.

CONE, C. D., JR. (1962), *The Theory of Induced Lift and Minimum Induced Drag of Nonplanar Lifting Systems*, NASA Technical Report R–139. [220]

CUNNINGHAM, H. J., and WOOLSTON, D. S. (1958), *Developments in the Flutter Analysis of General Planform Wings Using Unsteady Air Forces from the Kernel Function Procedure*, Proceedings of the National Specialists Meeting on Dynamics and Aeroelasticity, Ft. Worth Division, Institute of the Aeronautical Sciences. [151]

VON DOENHOFF, A. E., *see* ABBOT et al.

DROUGGE, G. (1959), *An Experimental Investigation of the Interference between Bodies of Revolution at Transonic Speeds with Special Reference to the Sonic and Supersonic Area Rules*, Aeronautical Research Institute of Sweden, Report 83. [106, 169, 238, 239]

DUGUNDJI, J., *see* PINES et al.

DURGIN, F. H., *see* BEANE et al.

ELLIS, M. C., JR., and HASEL, L. E. (1947), *Preliminary Tests at Supersonic Speeds of Triangular and Swept-Back Wings*, NACA R & M 26L17. [169]

ERDELYI, A. (1961), "Expansion Procedure for Singular Perturbations," *Atti Accad. Sci. Torino Classe Sci. Fis. Mat. Nat.* **95**, 651–672. [62]

ETKIN, B. E. (1955), *Numerical Integration Methods for Supersonic Wings in Steady and Oscillatory Motion*, University of Toronto, Institute of Aerophysics, Report 36. [220]

ETKIN, B. E., and WOODWARD, F. A. (1954), *Lift Distributions on Supersonic Wings with Subsonic Leading Edges and Arbitrary Angle of Attack Distribution*, Proceedings of the Second Canadian Symposium on Aerodynamics, University of Toronto. See also *J. Aeron. Sci.* **21**, No. 11, 783–784 (1954), Readers' Forum. [161]

EVVARD, J. C. (1950), *Use of Source Distributions for Evaluating Theoretical Aerodynamics of Thin Finite Wings at Supersonic Speeds*, NACA Report 951. [157, 162, 171]

FALKNER, V. M. (1948), *The Solution of Lifting Plane Problems by Vortex Lattice Theory*, British A.R.C., R & M 2591. [148]

FELDMAN, F. K. (1948), *Untersuchung von symmetrischen Tragflügelprofilen bei hohen Unterschallgeschwindigkeiten in einem geschlossenen Windkanal*, Mitt. Inst. für Aerodynamik, A. G. Gebr. Leeman and Col, Zurich. [126]

FERRARI, C. (1957), "Interaction Problems," Section C of *Aerodynamic Components of Aircraft at High Speeds*, Volume VII of Princeton Series in High Speed Aerodynamic and Jet Propulsion, A. F. Donovan and H. R. Lawrence, editors, Princeton Univ. Press, Princeton. [208, 209, 210]

FERRI, A. (1940), *Experimental Results with Airfoils Tested in the High-Speed Tunnel at Guidona*, NACA TM 946. [98]

FERRI, A., CLARKE, J., and TING, L. (1957), "Favorable Interference in Lifting Systems in Supersonic Flow," *J. Aeron. Sci.* **24**, No. 11, 791–804. [191]

FRICK, C. W., and BOYD, J. W. (1948), *Investigation at Supersonic Speed* (M = 1.53) *of the Pressure Distribution over a 63° Swept Airfoil of Biconvex Section at Zero Lift*, NACA RM A8C22. [156]

FRIEDRICHS, K. O. (1953), *Special Topics in Fluid Dynamics*, 120–130. New York Univ. [62]

FRIEDRICHS, K. O. (1954), *Special Topics in Analysis*, New York Univ. [62]

GARRICK, I. E. (1957), "Nonsteady Wing Characteristics," Section F of *Aerodynamic Components of Aircraft at High Speeds*, Volume VII of Princeton Series in High Speed Aerodynamics and Jet Propulsion, A. F. Donovan and H. R. Lawrence, editors, Princeton Univ. Press, Princeton. [19]

GARRICK, I. E., *see also* THEODORSEN and GARRICK.

GERMAIN, P. (1957a), "Sur le minimum de trainée d'une aile de form en plan donnée," *Compt. Rend.* **244**, 2691–2693. [193]

GERMAIN, P. (1957b), "Aile symetrique à portance nulle et de volume donnée réalisant le minimum de trainée en encoulement supersonique," *Compt. Rend.* **244**, 2691–2693. [193]

GLAUERT, H. (1924), *A Theory of Thin Airfoils*, British A.R.C., R & M 910. See also Glauert, H., *The Elements of Aerofoil and Airscrew Theory*, second edition, Cambridge Univ. Press, New York, 1947. [91, 96, 124, 126, 178, 209, 251]

GÖTHERT, B. (1940), *Ebene und räumliche Strömungen bei hohen Unterschalls-geschwindigkeit*, Lilienthal Gesellschaft für Luftfahrtforschung, Bericht 127, translated as NACA TM 1105. [124, 126, 209]

GRAHAM, D. J., NITZBERG, G. E., and OLSON, R. N. (1945), *A Systematic Investigation of Pressure Distributions at High Speeds over Five Representative NACA Low-Drag and Conventional Airfoil Sections*, NACA Report 832. [96]

GRAHAM, E. W. (1956), *The Calculation of Minimum Supersonic Drag by Solution of an Equivalent Two-Dimensional Potential Problem*, Douglas Aircraft Report SM–22666. [193, 204, 206, 207, 220]

GRAHAM, E. W., LAGERSTROM, P. A., LICHER, R. M., and BEANE, B. J. (1957), *A Theoretical Investigation of the Drag of Generalized Aircraft Configurations in Supersonic Flow*, NACA TM 1421. [183, 198]

GUDERLEY, K. G. (1957), *Theorie shallnaher Strömungen*, Springer Verlag, Berlin, Göttingen/Heidelberg. Also published in English translation, *The Theory of Transonic Flow*, Pergamon Press, London, 1962. [109, 235]

GUDERLEY, K. G., and YOSHIHARA, H. (1950), "The Flow over a Wedge Profile at Mach Number 1," *J. Aeron. Sci.* **17**, No. 11, 723–735. [242]

GULLSTRAND, T. (1951), *The Flow over Symmetrical Aerofoils without Incidence in the Lower Transonic Range*, Royal Institute of Technology, Stockholm, KTH AERO TN 20. [242]

HAACK, W. (1947), *Geschussformen kleinsten Wellenwiederstandes*, Lilienthal-Gesellschaft, Bericht 139. [180, 181, 188, 203]

HALFMAN, R. L., see BISPLINGHOFF et al.

HARDER, K. C., and KLUNKER, E. B. (1957), *On Slender-Body Theory and the Area Rule at Transonic Speeds*, NACA Report 1315. [109]

HARDER, K. C., and RENNEMANN, C., JR. (1957), *On Boattail Bodies of Revolution Having Minimum Wave Drag*, NACA Report 1271. [181]

HASEL, L. E., see ELLIS and HASEL.

HAYES, W. D. (1947), *Linearized Supersonic Flow*, North American Aviation, Rept. No. A.L. 222, Los Angeles, California. [181, 183, 184, 192, 193, 196]

HAYES, W. D., and PROBSTEIN, R. F. (1959), *Hypersonic Flow Theory*, Academic Press, New York and London. [7]

HAYES, W. E., *Topology and Fluid Mechanics*, to be published. [42, 43]

HEASLET, M. A., see LOMAX and HEASLET.

HOLDAWAY, G. H. (1953), *Comparison of Theoretical and Experimental Zero-Lift Drag-Rise Characteristics of Wing-Body-Tail Combinations Near the Speed of Sound*, NACA RM A53H17. [187, 188]

HOLDAWAY, G. H. (1954), *An Experimental Investigation of Reduction in Transonic Drag Rise at Zero Lift by the Addition of Volume to the Fuselage of a Wing-Body-Tail Configuration and a Comparison with Theory*, NACA RM A54F22. [189]

HOLDAWAY, G. H., and MELLENTHIN, J. A. (1960), *Investigations at Mach Numbers of 0.20 to 3.50 of Blended Wing-Body Combinations of Sonic Design with Diamond, Delta, and Arrow Plan Forms*, NASA TM X-372. [228]

HSU, P. T. (1958), *Some Recent Developments in the Flutter Analysis of Low Aspect Ratio Wings*, Proceedings of National Specialists Meeting on Dynamics of Elasticity, Ft. Worth Div., Institute of the Aeronautical Sciences. [151]

HSU, P. T., see also ZARTARIAN and HSU.

JONES, R. (1925), *The Distribution of Normal Pressures on a Prolate Spheroid*, British A.R.C., R & M 1061. [41]

JONES, R. T. (1946), *Properties of Low-Aspect Ratio Pointed Wings at Speeds Below and Above the Speed of Sound*, NACA Report 835. [123]

JONES, R. T. (1951), "The Minimum Drag of Thin Wings in Frictionless Flow," *J. Aeron. Sci.* **18**, No. 2, 75–81. [198]

JONES, R. T. (1952), "Theoretical Determination of the Minimum Drag of Airfoils at Supersonic Speeds," *J. Aeron. Sci.* **19**, No. 12, 813–822. [190, 198]

JONES, R. T., and COHEN, D. (1960), *High Speed Wing Theory*, Princeton Aeronautical Paperback No. 6, Princeton Univ. Press, Princeton. [58, 126, 156, 169]

KAPLUN, S. (1954), "The Role of Coordinate Systems in Boundary-Layer Theory," *Z. Angew. Math. Phys.* **5**, No. 2, 111–135. [62]

KAPLUN, S., and LAGERSTROM, P. (1957), "Asymptotic Expansions of Navier-Stokes Solutions for Small Reynolds Numbers," *J. Math. Mech.* **5**, 585–593. See also note by P. Lagerstrom in the same issue. [62, 69]

VON KÁRMÁN, T. (1927), *Berechnung der Druckverteilung an Luftschiffkörpern*, Abhandlung aus dem Aerodynamischen Institut an der Technischen Hochschule, Aachen, No. 6, 3–17. See also *Collected Works of Theodore von Kármán*, Butterworths, London, 1956. [40].

VON KÁRMÁN, T. (1936), *The Problems of Resistance in Compressible Fluids*, Proceedings of the Fifth Volta Congress, R. Accad. D'Italia (Rome), 222–283. [178, 180, 181, 202]

VON KÁRMÁN, T. (1947a), "Supersonic Aerodynamics—Principles and Applications," *J. Aeron. Sci.* **14**, No. 7, 373–409. [192]

VON KÁRMÁN, T. (1947b), "The Similarity Law of Transonic Flow," *J. Math. Phys.* **XXVI**, 182–190. [238]

VON KÁRMÁN, T., and TREFFTZ, E. (1918), "Potentialströmung um gegebene Tragflächenquerschnitte," *Z. Flugtechn. Motorluftsch.* **9**, 111–116. [52, 53]

KEUNE, F., *see* OSWATITSCH and KEUNE.

KLEBANOFF, P. S., TIDSTROM, K. D., and SARGENT, L. M. (1962), "The Three-Dimensional Nature of Boundary Layer Instability," *J. Fluid Mech.* **12**, Part 1, 1–12. [72]

KLUNKER, E. B., *see* HARDER and KLUNKER.

KOGAN, M. N. (1957), "On Bodies of Minimum Drag in a Supersonic Gas Stream," *Prikl. Mat. Mekhan.* **XXI**, No. 2, 207–212. [193]

KRASILSHCHIKOVA, E. A. (1951), "Zapiski 154," *Mekhanika* **4**, 181–239; also translated as NACA TM 1386 (1956). [157, 162, 171]

LAGERSTROM, P. A., and COLE, J. D. (1955), "Examples Illustrating Expansion Procedures for the Navier-Stokes Equations," *J. Rat. Mech. Anal.* **4**, 817–882. [62]

LAGERSTROM, P. A., *see also* KAPLUN *et al.* and GRAHAM, E. W. *et al.*

LAMB, SIR HORACE (1945), *Hydrodynamics*, sixth edition, Dover Publications, New York. [1, 21, 22, 27, 31, 33, 35, 37, 41]

LANDAHL, M. T. (1961), *Unsteady Transonic Flow*, Pergamon Press, London. [257, 259]

LANDAHL, M. T., *see also* ASHLEY *et al.*

LANGE and WACKE (1948), *Test Reports on Three- and Six-Component Measurements on a Series of Tapered Wings of Small Aspect Ratio* (partial report: *Triangular Wing*) NACA TM No. 1176. [122]

LAWRENCE, H. R. (1951), "The Lift Distribution of Low Aspect Ratio Wings at Subsonic Speeds," *J. Aeron. Sci.* **18**, No. 10, 683–695. [148]

LE GALLO, J., *see* MICHEL *et al.*

LEGENDRE, R. (1953), "Ecoulement au voisinage de la pointe avant d'une aile à forte flèche aux incidences moyennes," *Rech. Aeron.* No. 31, 3–8, Novembre-Décembre (1952) and No. 32, 3–6, Janvier-Fevrier. [113]

LICHER, R. M., *see* GRAHAM, E. W. *et al.*

LIEPMANN, H. W., and BRYSON, A. E. (1950), "Transonic Flow Past Wedge Sections," *J. Aeron. Sci.* **17**, No. 12, 745–755. [242]

LIEPMANN, H. W., and ROSHKO, A. (1957), *Elements of Gasdynamics*, John Wiley, New York. [126]

LIGHTHILL, M. J. (1948), "Supersonic Flow Past Slender Pointed Bodies the Slope of Whose Meridian Section is Discontinuous," *Quart. J. Mech. Appl. Math.* **1**, 90–102. [107]

LIGHTHILL, M. J. (1953), "Oscillating Airfoils at High Mach Number," *J. Aeron. Sci.* **20**, No. 6, 402–406. [259]

LINNALOUTO, V. (1951), *A Numerical Integration Method for Calculating the Pressure Distribution at Supersonic Speeds for Wings with Subsonic Leading Edge at Symmetric Flow Conditions*, SAAB TN 6, Sweden. [170]

LOMAX, H., and HEASLET, M. A. (1956), "Recent Developments in the Theory of Wing-Body Wave Drag," *J. Aeron. Sci.* **23**, No. 12, 1061–1074. [186, 188, 191]

MAEDER, P. R., and THOMMEN, H. U. (1956), "Some Results of Linearized Transonic Flow about Slender Airfoils and Bodies of Revolution," *J. Aeron. Sci.* **23**, No. 2, 187–188. [244]

MANGLER, K. W. (1951), *Improper Integrals in Theoretical Aerodynamics*, British A.R.C., R & M 2424. [132, 133, 150, 212]

MANGLER, K. W. (1955), *Calculation of the Load Distribution over a Wing with Arbitrary Camber and Twist at Sonic Speed*, British A.R.C., R & M 3102. [123]

MANGLER, K. W., and SMITH, J. H. B. (1956), *Flow Past Slender Delta Wings with Leading Edge Separation*, IX International Congress in Applied Mechanics, Brussels, Proceedings, Vol. 2, 137–145. [113]

MARCHAUD, F., *see* MICHEL *et al.*

MELLENTHIN, J. A., *see* HOLDAWAY and MELLENTHIN.

MESSITER, A. F. (1957), *Expansion Procedures and Similarity Laws for Transonic Flow*, USAF OSR TN 57-626. [235]

MESSITER, A. F., *see also* COLE and MESSITER.

MICHEL, R., MARCHAUD, F., and LE GALLO, J. (1953), *Etude des écoulements transoniques autour des profils lenticulaires, à incidence nulle*, O.N.E.R.A. Pub. No. 65. [237, 238]

MILES, J. W. (1959), *Potential Theory of Unsteady Supersonic Flow*, Cambridge Univ. Press, London. [247, 257, 259]

MILNE-THOMPSON, L. M. (1960), *Theoretical Hydrodynamics*, fourth edition, Macmillan, New York. [1, 4, 9, 21, 46, 47, 48, 57, 58]

MULTHOPP, H. (1938), *The Calculation of the Lift Distribution of Aerofoils*, British Ministry of Aircraft Production, R.T.P. Translation 2392 (from Luftfahrtforschung, Bd. 15, No. 4). [145, 209]

MULTHOPP, H. (1941), "Zur Aerodynamik des Flugzeugrumpfes." *Luftfahrtforsch.* **18**, 52–66. (Translated as "Aerodynamics of the Fuselage," NACA TM 1036.) [209]

MULTHOPP, H. (1950), *Methods for Calculating the Lift Distribution of Wings* (Subsonic Lifting Surface Theory), British A.R.C., R & M 2884. [148]

MUNK, M. M. (1924), *The Aerodynamic Forces on Airship Hulls*, NACA Report 184. [122]

MUNK, M. M. (1950), "The Reversal Theorem of Linearized Supersonic Airfoil Theory," *J. Appl. Phys.* **21**, 159–161. [192, 198]

MUSKHELISHVILI, N. J. (1953), *Singular Integral Equations*, second edition, translated by J. R. M. Radok, P. Noordhoff Ltd., Groningen, The Netherlands. [91]

NEURINGER, J., see PINES et al.

NIKOLSKY, A. A. (1956), *On the Theory of Axially Symmetric Supersonic Flow and Flows with Axisymmetric Hodographs*, IX Int. Congress Appl. Mech., Brussels. [192]

NIKURADSE, J. (1942), *Laminare Reibungschichten an der längsangerströmten Platte*, Monograph Zentrale Wiss. Berichtswesen, Berlin. [79]

NITZBERG, G. E., see GRAHAM, E. W.

NOWLAN, T. D., see BROWAND et al.

OLSON, R. N., see GRAHAM, E. W. et al.

ORLIK-RÜCKEMANN, K., see BERNDT and ORLIK-RÜCKEMANN.

OSWATITSCH, K. (1947), *A New Law of Similarity for Profiles Valid in the Transonic Region*, Roy. Aer. Establ. TN 1902. [238]

OSWATITSCH, K. (1950), *Die Geschwindigheitsverteilung bei lokalen Überschallgebieten an flachen Profielen*, *Z. Angew. Math. Mech.* **XXX**/1, 2, 17–24. [238]

OSWATITSCH, K., and BERNDT, S. B. (1950), *Aerodynamic Similarity of Axisymmetric Transonic Flow around Slender Bodies*, KTH AERO TN 15, Royal Institute of Technology, Stockholm. [242]

OSWATITSCH, K., and KEUNE, F. (1955), "Ein Äquivalenzsatz für nichtangestellte Flügel kleiner Spannweite in Schallnaher Strömung," *Z. Flugwiss.* **3**, No. 2, 29–46. [102, 109, 110, 237, 244]

OSWATITSCH, K., and KEUNE, F. (1955a), *The Flow around Bodies of Revolution at Mach Number 1*, Proc. Conf. on High-Speed Aeronautics, Polytechnic Institute of Brooklyn, N.Y., June 20–22, 113–131. [243, 244]

PAGE, W. A. (1958), *Experimental Study of the Equivalence of Transonic Flow about Slender Cone-Cylinders or Circular and Elliptic Cross Sections*, NACA TN 4233. [243]

PINES, S., DUGUNDJI, J., and NEURINGER, J. (1955), "Aerodynamic Flutter Derivatives for a Flexible Wing with Supersonic and Subsonic Edges," *J. Aeron. Sci.* **22**, No. 10, 693–700. [170, 220]

PINKERTON, R. M. (1936), *Calculated and Measured Pressure Distributions over the Midspan Section of the NACA 4412 Airfoil*, NACA Report 563. [59]

PRANDTL, L., WIESELSBERGER, C., and BETZ, A. (1921), *Ergebnisse der Aerodynamischen Versuchsanstalt zu Göttingen*, I. Lieferung, R. Oldenbourg, München und Berlin. [141, 142]

PROBSTEIN, R. E., *see* HAYES and PROBSTEIN.

PUCKETT, A. E. (1946), "Supersonic Wave Drag of Thin Airfoils," *J. Aeron. Sci.* **13**, No. 9, 475–484. [155]

RANSLEBEN, G. E., JR., and ABRAMSON, H. N. (1962), *Experimental Determination of Oscillatory Lift and Moment Distributions on Fully Submerged Flexible Hydrofoils*, Southwest Research Institute, Report No. 2, Contract No. Nonr-3335(00). [246]

REISSNER, E. (1949), *Note on the Theory of Lifting Surfaces*, Proc. Nat. Acad. Sci. U.S. **35**, No. 4, 208–215. [143, 146, 148]

RENNEMANN, C., *see* HARDER and RENNEMANN.

RIEGELS, F. (1961), *Aerofoil Sections*, English translation by D. Randall, Butterworths, London. [52]

ROSHKO, A., *see* LIEPMANN and ROSHKO.

RUNYAN, H. L., *see* WATKINS *et al.*

SACKS, A. H., *see* SPREITER and SACKS.

SARGENT, L. M., *see* KLEBANOFF *et al.*

SAUNDERS, G. H. (1963), *Aerodynamic Characteristics of Wings in Ground Proximity*, M. Sc. Thesis, Department of Aeronautics and Astronautics, Massachusetts Institute of Technology. [217, 218]

SCHLICHTING, H. (1960), *Boundary Layer Theory*, fourth edition, McGraw-Hill, New York. [63, 71, 77, 79, 80]

SEARS, W. R. (1947), "On Projectiles of Minimum Drag," *Quart. Appl. Math.* **4**, No. 4, 361–366. [180, 181, 188, 203]

SEARS, W. R., *see also* ADAMS and SEARS.

SHAPIRO, A. H. (1953), *The Dynamics and Thermodynamics of Compressible Fluid Flow*, I and II, Ronald Press, New York. [1, 15]

SMITH, J. H. B., *see* MANGLER and SMITH.

SPENCE, D. A. (1954), "Prediction of the Characteristics of Two-Dimensional Airfoils, *J. Aer. Sci.* **21**, 577–587. [59]

SPREITER, J. R. (1950), *The Aerodynamic Forces on Slender Plane- and Cruciform-Wing and Body Combination*, NACA Report 962. [112]

SPREITER, J. R. (1953), *On the Application of Transonic Similarity Rules to Wings of Finite Span*, NACA Report 1153. [239]

SPREITER, J. R., and ALKSNE, A. (1955), *Theoretical Prediction of Pressure Distributions on Nonlifting Airfoils at High Subsonic Speeds*, NACA Report 1217. [242]

SPREITER, J. R., and ALKSNE, A. Y. (1958), *Thin Airfoil Theory Based on Approximate Solution of the Transonic Flow Equation*, NACA Report 1359. [238, 244]

SPREITER, J. R., and ALKSNE, A. Y. (1959), *Slender-body Theory Based on Approximate Solution of the Transonic Flow Equation*, NASA Technical Report R-2. [243, 244]

SPREITER, J. R., and SACKS, A. H. (1951), "The Rolling Up of the Trailing Vortex Sheet and Its Effect on the Downwash Behind Wings," *J. Aeron. Sci.* **18**, No. 1, 21–32. [135]

STARK, V. J. E. (1964), *Calculation of Aerodynamic Forces on Two Oscillating Finite Wings at Low Supersonic Mach Numbers*, SAAB TN 53, Sweden. [170]

STEWART, A. J. (1946), "The Lift of the Delta Wing at Supersonic Speeds," *Quart. Appl. Math.* **IV**, 3, 246–254. [162]

STIVERS, L. S., see ABBOT et al.

THEODORSEN, T. (1931), *Theory of Wing Sections of Arbitrary Shape*, NACA Report 411. [54, 55, 56, 57, 59]

THEODORSEN, T. (1935), *General Theory of Aerodynamic Instability and the Mechanism of Flutter*, NACA Report 496. [254]

THEODORSEN, T., and GARRICK, I. E. (1933), *General Potential Theory of Arbitrary Wing Sections*, NACA Report 452. [54]

THOMMEN, H. U., see MAEDER and THOMMEN.

THWAITES, B., editor (1960), *Incompressible Aerodynamics*, Oxford Univ. Press, Oxford. [21, 59]

TIDSTROM, K. D., see KLEBANOFF et al.

TILTON, E. L., see BEANE et al.

TING, L., see FERRI et al.

TREFFTZ, E., see VON KÁRMÁN and TREFFTZ.

TRICOMI, F. (1923), "Sulle equazioni lineari alle derivate parziali di 2° ordine di tipo misto," *Atti Accad. Naz. Lincei, Ser. Quinta Cl. Sci. Fis. Mat. Nat.* **XIV**, 134–247. [241]

URSELL, F., and WARD, G. N. (1950), "On Some General Theorems in the Linearized Theory of Compressible Flow," *Quart. J. Mech. Appl. Math.* **3**, Part 3, 326–348. [192]

VANDREY, F. (1937), "Zur theoretischen Behandlung des gegenseitigen Einflusses von Tragflügel und Rumpf." *Luftfahrtforsch.* **14**, 347–355. [209]

VAN DYKE, M. D. (1956), *Second-Order Subsonic Airfoil Theory Including Edge Effects*, NACA Report 1274. [97]

VAN DYKE, M. D. (1963), *Lifting-Line Theory as a Singular-Perturbation Problem*, Stanford University Report SUDAER No. 165. [142]

VAN DYKE, M. D. (1964), *Perturbation Methods in Fluid Mechanics*, Academic Press, New York. [62, 69, 80, 85, 93]

VINCENTI, W. G., and WAGONER, C. B. (1952), *Transonic Flow Past a Wedge Profile with Detached Bow Wave*, NACA Report 1095. [242]

WACKE, see LANGE and WACKE.

WAGONER, C. B., see VINCENTI and WAGONER.

WARD, G. N. (1949), "Supersonic Flow Past Slender Pointed Bodies," *Quart. J. Mech. Appl. Math.* **2**, Part I, 75–97. [109, 113]

WARD, G. N. (1956), *On the Minimum Drag of Thin Lifting Bodies in Steady Supersonic Flows*, British A.R.C. Report 18, 711, FM 2459. [193]

WARD, G. N., see also URSELL and WARD.

WATKINS, C. E., and BERMAN, J. H. (1956), *On the Kernel Function of the Integral Equation Relating Lift and Downwash Distributions of Oscillating Wings in Supersonic Flow*, NACA Report 1257. [220]

WATKINS, C. E., RUNYAN, H. L. and WOOLSTON, D. S. (1955), *On the Kernel Function of the Integral Equation Relating the Lift and Downwash Distribution of Oscillating Finite Wings in Subsonic Flow*, NACA Report 1234. [148, 259]

WATKINS, C. E., RUNYAN, H. L. and WOOLSTON, D. S. (1959), *A Systematic Kernel Function Procedure for Determining Aerodynamic Forces on Oscillating or Steady Finite Wings at Subsonic Speeds*, NASA Technical Report R-48. [148, 149, 150, 151, 213, 214]

WATSON, G. N. (1948), *A Treatise on the Theory of Bessel Functions*, second edition, The Macmillan Company, London. [252]

WEISSINGER, J.(1947), *The Lift Distribution of Swept-Back Wings*, NACA TM 1120. [145]

WHITCOMB, R. T. (1956), *A Study of the Zero-Lift Drag-Rise Characteristics of Wing-Body Combinations Near the Speed of Sound*, NACA Report 1273. [117, 118]

WIDNALL, S. E., see ASHLEY et al.

WIESELSBERGER, C., see PRANDTL et al.

WOODWARD, F. A., see ETKIN and WOODWARD.

WOOLSTON, D. A., see WATKINS et al.

YOSHIHARA, H. (1956), *On the Flow Over a Finite Wedge in the Lower Transonic Region*, USAF Wright Aeronautical Development Center, T.R. 56–268. [242]

YOSHIHARA, H., see also GUDERLEY and YOSHIHARA.

ZARTARIAN, G., and HSU, P. T. (1955), *Theoretical Studies on the Prediction of Unsteady Supersonic Airloads on Elastic Wings*, Part I, *Investigations on the Use of Oscillatory Supersonic Aerodynamic Influence Coefficients*, USAF Wright Aeronautical Development Center, T.R. 56–97 (Confidential, Title Unclassified). [170, 221, 223]

ZARTARIAN, G. (1956), *Theoretical Studies on the Prediction of Unsteady Supersonic Airloads on Elastic Wings*, Part II, *Rules for Application of Oscillatory Supersonic Aerodynamic Influence Coefficients*, USAF Wright Aeronautical Development Center, T.R. 56–97. [223]

ZHILIN, YU. L. (1957), "Wings of Minimum Drag," *Prikl. Mat. Mekhan.* **XXI**, 213–220. [193]

List of Symbols

Major Notation for General Use

a Speed of sound

$\mathbf{a}$ Fluid particle acceleration

$A(\equiv b^2/S)$ Aspect ratio of wing

b Wingspan; width of plate

$B(\equiv \sqrt{M^2 - 1})$ Supersonic parameter

$B(x, y, z, t) = 0$ Equation describing the surface of a wing or body

c Chord length of wing; radius of circular cylinder

$C_D(\equiv D/\frac{1}{2}\rho_\infty U_\infty^2 S)$ Drag coefficient

$C_L(\equiv L/\frac{1}{2}\rho_\infty U_\infty^2 S)$ Lift coefficient

$C_p(\equiv (p - p_\infty)/\frac{1}{2}\rho_\infty U_\infty^2 S)$ Pressure coefficient

D Drag, or streamwise force experienced by wing or body

g Acceleration due to gravity

$i(\equiv \sqrt{-1})$ Imaginary unit

$\mathbf{i}, \mathbf{j}, \mathbf{k}$ Unit vectors in directions of x, y, z, respectively

$k(\equiv \omega l/U_\infty)$ Reduced frequency

l Reference length of object in flow (usually chord length or other streamwise dimension); lift per unit span; real constant identifying singular points of Joukowsky-Kutta transformation

L Lift, or force normal to flight direction in plane of symmetry, experienced by wing or body

$M(\equiv U_\infty/a_\infty)$ Flight Mach number

n Magnitude of $\mathbf{n}$

$\mathbf{n}$ Unit vector normal to surface (usually directed outward from volume V bounded by surface S)

$O(\ldots)$ Identifies a quantity of the same order as or smaller than $(\ldots)$

p Static pressure in fluid

q Magnitude of $\mathbf{q}$; "complex velocity" $(u - iw)$ in the x, z-plane

$\mathbf{q}$ Disturbance velocity vector of fluid particles $(\mathbf{Q} = U_\infty \mathbf{i} + \mathbf{q})$

$\mathbf{Q}$ Total velocity vector of fluid particles (dimensional)

$\mathbf{r}$ Position vector

S Denotes any surface; examples are the closed boundary of volume V, the projected mean surface of a wing, or the cross-sectional area of a slender body

t Time coordinate (dimensional)

T Absolute temperature; kinetic energy; total forward force theoretically experienced at a subsonic leading edge of a thin wing

u, v, w x, y, z-components of $\mathbf{q}$ (made dimensionless by division with U_∞)

U, V, W x, y, z-components of $\mathbf{Q}$ (dimensional)

U_∞ Flight speed or free-stream speed

v_n Disturbance velocity component normal to a surface (made dimensionless by division with U_∞)

V Volume of body or portion of fluid

x, y, z Rectangular Cartesian coordinates with x parallel to free stream (either dimensional or made dimensionless by division with l)

α Angle of attack of wing, measured from some convenient reference attitude

$\beta(\equiv \sqrt{1 - M^2})$ Prandtl-Glauert parameter

γ Ratio of specific heats of gas; dimensionless spanwise circulation component, per unit chordwise distance, bound to wing

$\Gamma\left(\equiv \oint \mathbf{Q} \cdot d\mathbf{s}\right)$ Circulation bound to airfoil or wing (dimensional)

ϵ Small parameter in asymptotic expansion procedure (various physical meanings); conformal angular distortion in Theodorsen procedure

ρ Density of fluid

σ Small spherical surface surrounding field point P

Σ Summation sign; surface forming the outer boundary of a fluid mass

φ Disturbance velocity potential (dimensions of length, or made dimensionless by division with l) ($\Phi = U_\infty(x + \varphi)$)

Φ Total velocity potential; complex potential function of constant-density flow perturbation ($\mathbf{Q} = \nabla\Phi$)

ψ Disturbance stream function

Ψ Total stream function; acceleration potential

ω Circular frequency of simple harmonic motion

$d\mathbf{s}$ Vector differential element of arc

$\dfrac{D}{Dt}\left(\equiv \dfrac{\partial}{\partial t} + (\mathbf{Q} \cdot \nabla)\right)$ Substantial or material derivative

$(\ldots)_\infty$ Identifies ambient properties of the uniform stream

$(\ldots)^i$ Identifies properties of "inner" solution in method of matched asymptotic expansions

$(\ldots)_n^i$ nth term (coefficient of ϵ^n) in series for "inner" solution

$(\ldots)^o$ Identifies properties of "outer" solution in method of matched asymptotic expansions

$(\ldots)_n^o$ nth term (coefficient of ϵ^n) in series for "outer" solution

Specialized Notation

$a_n(n = 1, 2, 3, \ldots)$	Complex constants in conformal transformation series
$a, b, c, \ldots$	Constants associated with corners of a polygon in Schwarz-Christoffel transformation
A_n	Fourier coefficients used in wing and slender-body load and area distributions
$A, B, C, P, Q, R, \ldots$	Inertia coefficients of a body moving through liquid
c_p, c_v	Specific heats at constant pressure and volume, respectively
$C_n(\equiv A_n + iB_n)$	Complex constants in Theodorsen transformation series
D_i	Induced drag of wing or body
$E(k)$	Complete elliptic integral of the second kind and modulus k
$\tilde{f}$	Source strength (dimensionless).
f_u, f_l	Functions describing the shape of upper and lower surfaces of wing
$\mathbf{F}$	Vector body force per unit mass; resultant force vector
F_x, F_y	Components of force experienced by a two-dimensional profile
$g(x)$	Function of x occurring in slender-body theory
$\bar{g}(x)$	[also $\bar{g}(x, y)$] Normalized semithickness distribution of wing or airfoil
h	Specific enthalpy
$\bar{h}(x)$	[also $\bar{h}(x, y)$, $\bar{h}(x, y, t)$] Normalized camber distribution of wing or airfoil
K	Kernel function of lifting surface integral equation
$\bar{l}$	Lifting element (dimensionless)
$m(\equiv \beta/\tan \Lambda)$	Parameter measuring subsonic ($m < 1$) or supersonic ($m > 1$) character of wing leading edge
m, n	Integers identifying terms in spanwise and chordwise series, respectively, describing circulation distribution on a wing
$(m, n); (v, \mu, l)$	Numbers identifying centers of "receiving" and "sending" area elements in supersonic wing theory
m'	Parameter related to m by Eq. (8–39)
$\mathbf{M}$	Resultant moment vector
M_0	Moment about axis through origin experienced by a two-dimensional profile
r	Distance from boundary point to field point
r, θ	Polar coordinates
$\mathrm{Re}(\equiv U_\infty l/\nu)$	Reynolds number
$\mathbf{Re}\{\}$	Real part of a complex number
$R(x)$	Radius of body of revolution
$\bar{R}_h$	Hyperbolic radius between points in supersonic flow field

s	Specific entropy
$\tilde{s}$	Side force element
$s(x)$	Semispan at chordwise station x on a low aspect-ratio wing
$\mathbf{u}(= \mathbf{i}u + \mathbf{j}v + \mathbf{k}w)$	Linear velocity of rigid solid
$\hat{U}, \hat{V}, \hat{W}$	Analytic functions of ξ in conical-flow theory, such that $\hat{U} = u + i\tilde{u}$, etc., u being the x-velocity component
U_r, U_θ	Velocity components in polar coordinates (dimensional)
U_ξ, U_η	Velocity components in ξ, η-plane
$V_{\bar{\nu},\bar{\mu},l}; W_{\bar{\nu},\bar{\mu},l}$	Aerodynamic influence coefficients for sidewash and upwash, respectively
$\mathcal{W}(\equiv \Phi + i\Psi)$	Complex potential function for two-dimensional flow
$X(\equiv y + iz)$	Complex variable in y, z-plane
$Y(\equiv x + iz)$	Complex variable in x, z-plane
$z_0(y)$	Coordinate describing location of mean surface of non-planar wing
$Z(\equiv x + iy)$	Complex variable in x, y-plane
Z'	Complex variable for intermediate plane used in Theodorsen's procedure
$\alpha, \beta, \gamma, \ldots$	Interior angles between sides of polygon in Schwarz-Christoffel transformation
$\alpha_{Z.L.}$	Angle of attack, measured from attitude at which $L = 0$
$\zeta(\equiv \zeta + i\eta)$	Complex variable in plane to be transformed into physical plane
$\boldsymbol{\zeta}(\equiv \nabla \times \mathbf{Q})$	Fluid vorticity vector
η, ζ	Transformed conical-flow variables
θ	Fractional camber of airfoil or wing
θ, η	Dimensionless chordwise and spanwise variables replacing x, y in subsonic wing theory
$\boldsymbol{\lambda}(= \mathbf{i}\lambda_1 + \mathbf{j}\lambda_2 + \mathbf{k}\lambda_3)$	Vector of angular momentum of liquid mass
Λ	Leading-edge sweep angle of wing
μ	Dynamic coefficient of viscosity
$\nu(\equiv \mu/\rho)$	Kinematic coefficient of viscosity
ξ	Radial conical coordinate
$\boldsymbol{\xi}$	Used generally for linear momentum of liquid or gaseous mass (components $\xi_1, \xi_2,$ and ξ_3).
τ	Semithickness or thickness ratio of airfoil or wing; dimensionless time coordinate
φ, ψ, θ	Dimensionless coordinates use in Theodorsen's procedure
$\boldsymbol{\varphi}(\equiv \mathbf{i}\varphi_1 + \mathbf{j}\varphi_2 + \mathbf{k}\varphi_3)$ $\boldsymbol{\chi}(\equiv \mathbf{i}x_1 + \mathbf{j}x_2 + \mathbf{k}x_3)$	Velocity potentials associated with unit linear and angular motion, respectively, of a solid through a liquid
$\Phi_{\bar{\nu},\bar{\mu},l}$	Aerodynamic influence coefficient for velocity potential
$\psi(y)$	Angle between y-axis and mean surface of nonplanar wing
$\boldsymbol{\omega}(\equiv \mathbf{i}p + \mathbf{j}q + \mathbf{k}r)$	Angular velocity of rigid solid
Ω	Potential of conservative body force field ($F \equiv \nabla\Omega$)

$(\cdots)_B$ Identifies base of slender body

$(\cdots)_F, (\cdots)_R$ Properties of a given (forward) flow and a second flow obtained from it by operations that include reversing the free-stream direction

$(\cdots)_l, (\cdots)_u$ Associated with lower and upper surfaces, respectively, of airfoil or wing

$(\cdots)_{LE}, (\cdots)_{TE}$ Associated with leading and trailing edges, respectively, of wing

$(\cdots)_{NP}$ Identifies a nonplanar lifting surface

$(\cdots)_1$ Dummy variable of integration, replacing the same quantity without subscript

Index

A CATALOG OF SELECTED
DOVER BOOKS
IN ALL FIELDS OF INTEREST

A CATALOG OF SELECTED DOVER
BOOKS IN ALL FIELDS OF INTEREST

DRAWINGS OF REMBRANDT, edited by Seymour Slive. Updated Lippmann, Hofstede de Groot edition, with definitive scholarly apparatus. All portraits, biblical sketches, landscapes, nudes. Oriental figures, classical studies, together with selection of work by followers. 550 illustrations. Total of 630pp. 9⅛ × 12¼.
21485-0, 21486-9 Pa., Two-vol. set $25.00

GHOST AND HORROR STORIES OF AMBROSE BIERCE, Ambrose Bierce. 24 tales vividly imagined, strangely prophetic, and decades ahead of their time in technical skill: "The Damned Thing," "An Inhabitant of Carcosa," "The Eyes of the Panther," "Moxon's Master," and 20 more. 199pp. 5⅜ × 8½. 20767-6 Pa. $3.95

ETHICAL WRITINGS OF MAIMONIDES, Maimonides. Most significant ethical works of great medieval sage, newly translated for utmost precision, readability. Laws Concerning Character Traits, Eight Chapters, more. 192pp. 5⅜ × 8½.
24522-5 Pa. $4.50

THE EXPLORATION OF THE COLORADO RIVER AND ITS CANYONS, J. W. Powell. Full text of Powell's 1,000-mile expedition down the fabled Colorado in 1869. Superb account of terrain, geology, vegetation, Indians, famine, mutiny, treacherous rapids, mighty canyons, during exploration of last unknown part of continental U.S. 400pp. 5⅜ × 8½. 20094-9 Pa. $6.95

HISTORY OF PHILOSOPHY, Julián Marías. Clearest one-volume history on the market. Every major philosopher and dozens of others, to Existentialism and later. 505pp. 5⅜ × 8½. 21739-6 Pa. $8.50

ALL ABOUT LIGHTNING, Martin A. Uman. Highly readable non-technical survey of nature and causes of lightning, thunderstorms, ball lightning, St. Elmo's Fire, much more. Illustrated. 192pp. 5⅜ × 8½. 25237-X Pa. $5.95

SAILING ALONE AROUND THE WORLD, Captain Joshua Slocum. First man to sail around the world, alone, in small boat. One of great feats of seamanship told in delightful manner. 67 illustrations. 294pp. 5⅜ × 8½. 20326-3 Pa. $4.50

LETTERS AND NOTES ON THE MANNERS, CUSTOMS AND CONDITIONS OF THE NORTH AMERICAN INDIANS, George Catlin. Classic account of life among Plains Indians: ceremonies, hunt, warfare, etc. 312 plates. 572pp. of text. 6⅛ × 9¼. 22118-0, 22119-9 Pa. Two-vol. set $15.90

ALASKA: The Harriman Expedition, 1899, John Burroughs, John Muir, et al. Informative, engrossing accounts of two-month, 9,000-mile expedition. Native peoples, wildlife, forests, geography, salmon industry, glaciers, more. Profusely illustrated. 240 black-and-white line drawings. 124 black-and-white photographs. 3 maps. Index. 576pp. 5⅜ × 8½. 25109-8 Pa. $11.95

THE BOOK OF BEASTS: Being a Translation from a Latin Bestiary of the Twelfth Century, T. H. White. Wonderful catalog real and fanciful beasts: manticore, griffin, phoenix, amphivius, jaculus, many more. White's witty erudite commentary on scientific, historical aspects. Fascinating glimpse of medieval mind. Illustrated. 296pp. 5⅜ × 8¼. (Available in U.S. only) 24609-4 Pa. $5.95

FRANK LLOYD WRIGHT: ARCHITECTURE AND NATURE With 160 Illustrations, Donald Hoffmann. Profusely illustrated study of influence of nature—especially prairie—on Wright's designs for Fallingwater, Robie House, Guggenheim Museum, other masterpieces. 96pp. 9¼ × 10¾. 25098-9 Pa. $7.95

FRANK LLOYD WRIGHT'S FALLINGWATER, Donald Hoffmann. Wright's famous waterfall house: planning and construction of organic idea. History of site, owners, Wright's personal involvement. Photographs of various stages of building. Preface by Edgar Kaufmann, Jr. 100 illustrations. 112pp. 9¼ × 10.
23671-4 Pa. $7.95

YEARS WITH FRANK LLOYD WRIGHT: Apprentice to Genius, Edgar Tafel. Insightful memoir by a former apprentice presents a revealing portrait of Wright the man, the inspired teacher, the greatest American architect. 372 black-and-white illustrations. Preface. Index. vi + 228pp. 8¼ × 11. 24801-1 Pa. $9.95

THE STORY OF KING ARTHUR AND HIS KNIGHTS, Howard Pyle. Enchanting version of King Arthur fable has delighted generations with imaginative narratives of exciting adventures and unforgettable illustrations by the author. 41 illustrations. xviii + 313pp. 6⅛ × 9¼. 21445-1 Pa. $5.95

THE GODS OF THE EGYPTIANS, E. A. Wallis Budge. Thorough coverage of numerous gods of ancient Egypt by foremost Egyptologist. Information on evolution of cults, rites and gods; the cult of Osiris; the Book of the Dead and its rites; the sacred animals and birds; Heaven and Hell; and more. 956pp. 6⅛ × 9¼.
22055-9, 22056-7 Pa., Two-vol. set $20.00

A THEOLOGICO-POLITICAL TREATISE, Benedict Spinoza. Also contains unfinished *Political Treatise.* Great classic on religious liberty, theory of government on common consent. R. Elwes translation. Total of 421pp. 5⅜ × 8½.
20249-6 Pa. $6.95

INCIDENTS OF TRAVEL IN CENTRAL AMERICA, CHIAPAS, AND YU-CATAN, John L. Stephens. Almost single-handed discovery of Maya culture; exploration of ruined cities, monuments, temples; customs of Indians. 115 drawings. 892pp. 5⅜ × 8½. 22404-X, 22405-8 Pa., Two-vol. set $15.90

LOS CAPRICHOS, Francisco Goya. 80 plates of wild, grotesque monsters and caricatures. Prado manuscript included. 183pp. 6⅜ × 9⅜. 22384-1 Pa. $4.95

AUTOBIOGRAPHY: The Story of My Experiments with Truth, Mohandas K. Gandhi. Not hagiography, but Gandhi in his own words. Boyhood, legal studies, purification, the growth of the Satyagraha (nonviolent protest) movement. Critical, inspiring work of the man who freed India. 480pp. 5⅜ × 8½. (Available in U.S. only)
24593-4 Pa. $6.95

ILLUSTRATED DICTIONARY OF HISTORIC ARCHITECTURE, edited by Cyril M. Harris. Extraordinary compendium of clear, concise definitions for over 5,000 important architectural terms complemented by over 2,000 line drawings. Covers full spectrum of architecture from ancient ruins to 20th-century Modernism. Preface. 592pp. 7½ × 9⅝. 24444-X Pa. $14.95

THE NIGHT BEFORE CHRISTMAS, Clement Moore. Full text, and woodcuts from original 1848 book. Also critical, historical material. 19 illustrations. 40pp. 4⅝ × 6. 22797-9 Pa. $2.25

THE LESSON OF JAPANESE ARCHITECTURE: 165 Photographs, Jiro Harada. Memorable gallery of 165 photographs taken in the 1930's of exquisite Japanese homes of the well-to-do and historic buildings. 13 line diagrams. 192pp. 8⅜ × 11¼. 24778-3 Pa. $8.95

THE AUTOBIOGRAPHY OF CHARLES DARWIN AND SELECTED LETTERS, edited by Francis Darwin. The fascinating life of eccentric genius composed of an intimate memoir by Darwin (intended for his children); commentary by his son, Francis; hundreds of fragments from notebooks, journals, papers; and letters to and from Lyell, Hooker, Huxley, Wallace and Henslow. xi + 365pp. 5⅜ × 8. 20479-0 Pa. $5.95

WONDERS OF THE SKY: Observing Rainbows, Comets, Eclipses, the Stars and Other Phenomena, Fred Schaaf. Charming, easy-to-read poetic guide to all manner of celestial events visible to the naked eye. Mock suns, glories, Belt of Venus, more. Illustrated. 299pp. 5¼ × 8¼. 24402-4 Pa. $7.95

BURNHAM'S CELESTIAL HANDBOOK, Robert Burnham, Jr. Thorough guide to the stars beyond our solar system. Exhaustive treatment. Alphabetical by constellation: Andromeda to Cetus in Vol. 1; Chamaeleon to Orion in Vol. 2; and Pavo to Vulpecula in Vol. 3. Hundreds of illustrations. Index in Vol. 3. 2,000pp. 6⅛ × 9¼. 23567-X, 23568-8, 23673-0 Pa., Three-vol. set $36.85

STAR NAMES: Their Lore and Meaning, Richard Hinckley Allen. Fascinating history of names various cultures have given to constellations and literary and folkloristic uses that have been made of stars. Indexes to subjects. Arabic and Greek names. Biblical references. Bibliography. 563pp. 5⅜ × 8½. 21079-0 Pa. $7.95

THIRTY YEARS THAT SHOOK PHYSICS: The Story of Quantum Theory, George Gamow. Lucid, accessible introduction to influential theory of energy and matter. Careful explanations of Dirac's anti-particles, Bohr's model of the atom, much more. 12 plates. Numerous drawings. 240pp. 5⅜ × 8½. 24895-X Pa. $4.95

CHINESE DOMESTIC FURNITURE IN PHOTOGRAPHS AND MEASURED DRAWINGS, Gustav Ecke. A rare volume, now affordably priced for antique collectors, furniture buffs and art historians. Detailed review of styles ranging from early Shang to late Ming. Unabridged republication. 161 black-and-white drawings, photos. Total of 224pp. 8⅜ × 11¼. (Available in U.S. only) 25171-3 Pa. $12.95

VINCENT VAN GOGH: A Biography, Julius Meier-Graefe. Dynamic, penetrating study of artist's life, relationship with brother, Theo, painting techniques, travels, more. Readable, engrossing. 160pp. 5⅜ × 8½. (Available in U.S. only) 25253-1 Pa. $3.95

HOW TO WRITE, Gertrude Stein. Gertrude Stein claimed anyone could understand her unconventional writing—here are clues to help. Fascinating improvisations, language experiments, explanations illuminate Stein's craft and the art of writing. Total of 414pp. 4⅝ × 6⅜. 23144-5 Pa. $5.95

ADVENTURES AT SEA IN THE GREAT AGE OF SAIL: Five Firsthand Narratives, edited by Elliot Snow. Rare true accounts of exploration, whaling, shipwreck, fierce natives, trade, shipboard life, more. 33 illustrations. Introduction. 353pp. 5⅜ × 8½. 25177-2 Pa. $7.95

THE HERBAL OR GENERAL HISTORY OF PLANTS, John Gerard. Classic descriptions of about 2,850 plants—with over 2,700 illustrations—includes Latin and English names, physical descriptions, varieties, time and place of growth, more. 2,706 illustrations. xlv + 1,678pp. 8½ × 12¼. 23147-X Cloth. $75.00

DOROTHY AND THE WIZARD IN OZ, L. Frank Baum. Dorothy and the Wizard visit the center of the Earth, where people are vegetables, glass houses grow and Oz characters reappear. Classic sequel to *Wizard of Oz*. 256pp. 5⅜ × 8. 24714-7 Pa. $4.95

SONGS OF EXPERIENCE: Facsimile Reproduction with 26 Plates in Full Color, William Blake. This facsimile of Blake's original "Illuminated Book" reproduces 26 full-color plates from a rare 1826 edition. Includes "The Tyger," "London," "Holy Thursday," and other immortal poems. 26 color plates. Printed text of poems. 48pp. 5¼ × 7. 24636-1 Pa. $3.50

SONGS OF INNOCENCE, William Blake. The first and most popular of Blake's famous "Illuminated Books," in a facsimile edition reproducing all 31 brightly colored plates. Additional printed text of each poem. 64pp. 5¼ × 7. 22764-2 Pa. $3.50

PRECIOUS STONES, Max Bauer. Classic, thorough study of diamonds, rubies, emeralds, garnets, etc.: physical character, occurrence, properties, use, similar topics. 20 plates, 8 in color. 94 figures. 659pp. 6⅛ × 9¼. 21910-0, 21911-9 Pa., Two-vol. set $14.90

ENCYCLOPEDIA OF VICTORIAN NEEDLEWORK, S. F. A. Caulfeild and Blanche Saward. Full, precise descriptions of stitches, techniques for dozens of needlecrafts—most exhaustive reference of its kind. Over 800 figures. Total of 679pp. 8⅛ × 11. Two volumes. Vol. 1 22800-2 Pa. $10.95
Vol. 2 22801-0 Pa. $10.95

THE MARVELOUS LAND OF OZ, L. Frank Baum. Second Oz book, the Scarecrow and Tin Woodman are back with hero named Tip, Oz magic. 136 illustrations. 287pp. 5⅜ × 8½. 20692-0 Pa. $5.95

WILD FOWL DECOYS, Joel Barber. Basic book on the subject, by foremost authority and collector. Reveals history of decoy making and rigging, place in American culture, different kinds of decoys, how to make them, and how to use them. 140 plates. 156pp. 7⅞ × 10¾. 20011-6 Pa. $7.95

HISTORY OF LACE, Mrs. Bury Palliser. Definitive, profusely illustrated chronicle of lace from earliest times to late 19th century. Laces of Italy, Greece, England, France, Belgium, etc. Landmark of needlework scholarship. 266 illustrations. 672pp. 6⅛ × 9¼. 24742-2 Pa. $14.95

ILLUSTRATED GUIDE TO SHAKER FURNITURE, Robert Meader. All furniture and appurtenances, with much on unknown local styles. 235 photos. 146pp. 9 × 12. 22819-3 Pa. $7.95

WHALE SHIPS AND WHALING: A Pictorial Survey, George Francis Dow. Over 200 vintage engravings, drawings, photographs of barks, brigs, cutters, other vessels. Also harpoons, lances, whaling guns, many other artifacts. Comprehensive text by foremost authority. 207 black-and-white illustrations. 288pp. 6 × 9. 24808-9 Pa. $8.95

THE BERTRAMS, Anthony Trollope. Powerful portrayal of blind self-will and thwarted ambition includes one of Trollope's most heartrending love stories. 497pp. 5⅜ × 8½. 25119-5 Pa. $8.95

ADVENTURES WITH A HAND LENS, Richard Headstrom. Clearly written guide to observing and studying flowers and grasses, fish scales, moth and insect wings, egg cases, buds, feathers, seeds, leaf scars, moss, molds, ferns, common crystals, etc.—all with an ordinary, inexpensive magnifying glass. 209 exact line drawings aid in your discoveries. 220pp. 5⅜ × 8½. 23330-8 Pa. $3.95

RODIN ON ART AND ARTISTS, Auguste Rodin. Great sculptor's candid, wide-ranging comments on meaning of art; great artists; relation of sculpture to poetry, painting, music; philosophy of life, more. 76 superb black-and-white illustrations of Rodin's sculpture, drawings and prints. 119pp. 8⅜ × 11¼. 24487-3 Pa. $6.95

FIFTY CLASSIC FRENCH FILMS, 1912–1982: A Pictorial Record, Anthony Slide. Memorable stills from Grand Illusion, Beauty and the Beast, Hiroshima, Mon Amour, many more. Credits, plot synopses, reviews, etc. 160pp. 8¼ × 11. 25256-6 Pa. $11.95

THE PRINCIPLES OF PSYCHOLOGY, William James. Famous long course complete, unabridged. Stream of thought, time perception, memory, experimental methods; great work decades ahead of its time. 94 figures. 1,391pp. 5⅜ × 8½. 20381-6, 20382-4 Pa., Two-vol. set $19.90

BODIES IN A BOOKSHOP, R. T. Campbell. Challenging mystery of blackmail and murder with ingenious plot and superbly drawn characters. In the best tradition of British suspense fiction. 192pp. 5⅜ × 8½. 24720-1 Pa. $3.95

CALLAS: PORTRAIT OF A PRIMA DONNA, George Jellinek. Renowned commentator on the musical scene chronicles incredible career and life of the most controversial, fascinating, influential operatic personality of our time. 64 black-and-white photographs. 416pp. 5⅜ × 8¼. 25047-4 Pa. $7.95

GEOMETRY, RELATIVITY AND THE FOURTH DIMENSION, Rudolph Rucker. Exposition of fourth dimension, concepts of relativity as Flatland characters continue adventures. Popular, easily followed yet accurate, profound. 141 illustrations. 133pp. 5⅜ × 8½. 23400-2 Pa. $3.50

HOUSEHOLD STORIES BY THE BROTHERS GRIMM, with pictures by Walter Crane. 53 classic stories—Rumpelstiltskin, Rapunzel, Hansel and Gretel, the Fisherman and his Wife, Snow White, Tom Thumb, Sleeping Beauty, Cinderella, and so much more—lavishly illustrated with original 19th century drawings. 114 illustrations. x + 269pp. 5⅜ × 8½. 21080-4 Pa. $4.50

SUNDIALS, Albert Waugh. Far and away the best, most thorough coverage of ideas, mathematics concerned, types, construction, adjusting anywhere. Over 100 illustrations. 230pp. 5⅜ × 8½. 22947-5 Pa. $4.00

PICTURE HISTORY OF THE NORMANDIE: With 190 Illustrations, Frank O. Braynard. Full story of legendary French ocean liner: Art Deco interiors, design innovations, furnishings, celebrities, maiden voyage, tragic fire, much more. Extensive text. 144pp. 8⅞ × 11¾. 25257-4 Pa. $9.95

THE FIRST AMERICAN COOKBOOK: A Facsimile of "American Cookery," 1796, Amelia Simmons. Facsimile of the first American-written cookbook published in the United States contains authentic recipes for colonial favorites—pumpkin pudding, winter squash pudding, spruce beer, Indian slapjacks, and more. Introductory Essay and Glossary of colonial cooking terms. 80pp. 5⅜ × 8½. 24710-4 Pa. $3.50

101 PUZZLES IN THOUGHT AND LOGIC, C. R. Wylie, Jr. Solve murders and robberies, find out which fishermen are liars, how a blind man could possibly identify a color—purely by your own reasoning! 107pp. 5⅜ × 8½. 20367-0 Pa. $2.00

THE BOOK OF WORLD-FAMOUS MUSIC—CLASSICAL, POPULAR AND FOLK, James J. Fuld. Revised and enlarged republication of landmark work in musico-bibliography. Full information about nearly 1,000 songs and compositions including first lines of music and lyrics. New supplement. Index. 800pp. 5⅜ × 8¼. 24857-7 Pa. $14.95

ANTHROPOLOGY AND MODERN LIFE, Franz Boas. Great anthropologist's classic treatise on race and culture. Introduction by Ruth Bunzel. Only inexpensive paperback edition. 255pp. 5⅜ × 8½. 25245-0 Pa. $5.95

THE TALE OF PETER RABBIT, Beatrix Potter. The inimitable Peter's terrifying adventure in Mr. McGregor's garden, with all 27 wonderful, full-color Potter illustrations. 55pp. 4¼ × 5½. (Available in U.S. only) 22827-4 Pa. $1.75

THREE PROPHETIC SCIENCE FICTION NOVELS, H. G. Wells. *When the Sleeper Wakes, A Story of the Days to Come* and *The Time Machine* (full version). 335pp. 5⅜ × 8½. (Available in U.S. only) 20605-X Pa. $5.95

APICIUS COOKERY AND DINING IN IMPERIAL ROME, edited and translated by Joseph Dommers Vehling. Oldest known cookbook in existence offers readers a clear picture of what foods Romans ate, how they prepared them, etc. 49 illustrations. 301pp. 6⅛ × 9¼. 23563-7 Pa. $6.00

SHAKESPEARE LEXICON AND QUOTATION DICTIONARY, Alexander Schmidt. Full definitions, locations, shades of meaning of every word in plays and poems. More than 50,000 exact quotations. 1,485pp. 6½ × 9¼. 22726-X, 22727-8 Pa., Two-vol. set $27.90

THE WORLD'S GREAT SPEECHES, edited by Lewis Copeland and Lawrence W. Lamm. Vast collection of 278 speeches from Greeks to 1970. Powerful and effective models; unique look at history. 842pp. 5⅜ × 8½. 20468-5 Pa. $10.95

CATALOG OF DOVER BOOKS

THE BLUE FAIRY BOOK, Andrew Lang. The first, most famous collection, with many familiar tales: Little Red Riding Hood, Aladdin and the Wonderful Lamp, Puss in Boots, Sleeping Beauty, Hansel and Gretel, Rumpelstiltskin; 37 in all. 138 illustrations. 390pp. 5⅜ × 8½.   21437-0 Pa. $5.95

THE STORY OF THE CHAMPIONS OF THE ROUND TABLE, Howard Pyle. Sir Launcelot, Sir Tristram and Sir Percival in spirited adventures of love and triumph retold in Pyle's inimitable style. 50 drawings, 31 full-page. xviii + 329pp. 6½ × 9¼.   21883-X Pa. $6.95

AUDUBON AND HIS JOURNALS, Maria Audubon. Unmatched two-volume portrait of the great artist, naturalist and author contains his journals, an excellent biography by his granddaughter, expert annotations by the noted ornithologist, Dr. Elliott Coues, and 37 superb illustrations. Total of 1,200pp. 5⅜ × 8.
Vol. I 25143-8 Pa. $8.95
Vol. II 25144-6 Pa. $8.95

GREAT DINOSAUR HUNTERS AND THEIR DISCOVERIES, Edwin H. Colbert. Fascinating, lavishly illustrated chronicle of dinosaur research, 1820's to 1960. Achievements of Cope, Marsh, Brown, Buckland, Mantell, Huxley, many others. 384pp. 5¼ × 8¼.   24701-5 Pa. $6.95

THE TASTEMAKERS, Russell Lynes. Informal, illustrated social history of American taste 1850's–1950's. First popularized categories Highbrow, Lowbrow, Middlebrow. 129 illustrations. New (1979) afterword. 384pp. 6 × 9.
23993-4 Pa. $6.95

DOUBLE CROSS PURPOSES, Ronald A. Knox. A treasure hunt in the Scottish Highlands, an old map, unidentified corpse, surprise discoveries keep reader guessing in this cleverly intricate tale of financial skullduggery. 2 black-and-white maps. 320pp. 5⅜ × 8½. (Available in U.S. only)   25032-6 Pa. $5.95

AUTHENTIC VICTORIAN DECORATION AND ORNAMENTATION IN FULL COLOR: 46 Plates from "Studies in Design," Christopher Dresser. Superb full-color lithographs reproduced from rare original portfolio of a major Victorian designer. 48pp. 9¼ × 12¼.   25083-0 Pa. $7.95

PRIMITIVE ART, Franz Boas. Remains the best text ever prepared on subject, thoroughly discussing Indian, African, Asian, Australian, and, especially, Northern American primitive art. Over 950 illustrations show ceramics, masks, totem poles, weapons, textiles, paintings, much more. 376pp. 5⅜ × 8.   20025-6 Pa. $6.95

SIDELIGHTS ON RELATIVITY, Albert Einstein. Unabridged republication of two lectures delivered by the great physicist in 1920–21. *Ether and Relativity* and *Geometry and Experience.* Elegant ideas in non-mathematical form, accessible to intelligent layman. vi + 56pp. 5⅜ × 8½.   24511-X Pa. $2.95

THE WIT AND HUMOR OF OSCAR WILDE, edited by Alvin Redman. More than 1,000 ripostes, paradoxes, wisecracks: Work is the curse of the drinking classes, I can resist everything except temptation, etc. 258pp. 5⅜ × 8½.   20602-5 Pa. $3.95

ADVENTURES WITH A MICROSCOPE, Richard Headstrom. 59 adventures with clothing fibers, protozoa, ferns and lichens, roots and leaves, much more. 142 illustrations. 232pp. 5⅜ × 8½.   23471-1 Pa. $3.95

PLANTS OF THE BIBLE, Harold N. Moldenke and Alma L. Moldenke. Standard reference to all 230 plants mentioned in Scriptures. Latin name, biblical reference, uses, modern identity, much more. Unsurpassed encyclopedic resource for scholars, botanists, nature lovers, students of Bible. Bibliography. Indexes. 123 black-and-white illustrations. 384pp. 6 × 9. 25069-5 Pa. $8.95

FAMOUS AMERICAN WOMEN: A Biographical Dictionary from Colonial Times to the Present, Robert McHenry, ed. From Pocahontas to Rosa Parks, 1,035 distinguished American women documented in separate biographical entries. Accurate, up-to-date data, numerous categories, spans 400 years. Indices. 493pp. 6½ × 9¼. 24523-3 Pa. $9.95

THE FABULOUS INTERIORS OF THE GREAT OCEAN LINERS IN HISTORIC PHOTOGRAPHS, William H. Miller, Jr. Some 200 superb photographs capture exquisite interiors of world's great "floating palaces"—1890's to 1980's: *Titanic, Ile de France, Queen Elizabeth, United States, Europa*, more. Approx. 200 black-and-white photographs. Captions. Text. Introduction. 160pp. 8⅜ × 11¼. 24756-2 Pa. $9.95

THE GREAT LUXURY LINERS, 1927–1954: A Photographic Record, William H. Miller, Jr. Nostalgic tribute to heyday of ocean liners. 186 photos of Ile de France, Normandie, Leviathan, Queen Elizabeth, United States, many others. Interior and exterior views. Introduction. Captions. 160pp. 9 × 12. 24056-8 Pa. $9.95

A NATURAL HISTORY OF THE DUCKS, John Charles Phillips. Great landmark of ornithology offers complete detailed coverage of nearly 200 species and subspecies of ducks: gadwall, sheldrake, merganser, pintail, many more. 74 full-color plates, 102 black-and-white. Bibliography. Total of 1,920pp. 8⅜ × 11¼. 25141-1, 25142-X Cloth. Two-vol. set $100.00

THE SEAWEED HANDBOOK: An Illustrated Guide to Seaweeds from North Carolina to Canada, Thomas F. Lee. Concise reference covers 78 species. Scientific and common names, habitat, distribution, more. Finding keys for easy identification. 224pp. 5⅜ × 8½. 25215-9 Pa. $5.95

THE TEN BOOKS OF ARCHITECTURE: The 1755 Leoni Edition, Leon Battista Alberti. Rare classic helped introduce the glories of ancient architecture to the Renaissance. 68 black-and-white plates. 336pp. 8⅜ × 11¼. 25239-6 Pa. $14.95

MISS MACKENZIE, Anthony Trollope. Minor masterpieces by Victorian master unmasks many truths about life in 19th-century England. First inexpensive edition in years. 392pp. 5⅜ × 8½. 25201-9 Pa. $7.95

THE RIME OF THE ANCIENT MARINER, Gustave Doré, Samuel Taylor Coleridge. Dramatic engravings considered by many to be his greatest work. The terrifying space of the open sea, the storms and whirlpools of an unknown ocean, the ice of Antarctica, more—all rendered in a powerful, chilling manner. Full text. 38 plates. 77pp. 9¼ × 12. 22305-1 Pa. $4.95

THE EXPEDITIONS OF ZEBULON MONTGOMERY PIKE, Zebulon Montgomery Pike. Fascinating first-hand accounts (1805–6) of exploration of Mississippi River, Indian wars, capture by Spanish dragoons, much more. 1,088pp. 5⅜ × 8½. 25254-X, 25255-8 Pa. Two-vol. set $23.90

A CONCISE HISTORY OF PHOTOGRAPHY: Third Revised Edition, Helmut Gernsheim. Best one-volume history—camera obscura, photochemistry, daguerreotypes, evolution of cameras, film, more. Also artistic aspects—landscape, portraits, fine art, etc. 281 black-and-white photographs. 26 in color. 176pp. 8⅜ × 11¼. 25128-4 Pa. $12.95

THE DORÉ BIBLE ILLUSTRATIONS, Gustave Doré. 241 detailed plates from the Bible: the Creation scenes, Adam and Eve, Flood, Babylon, battle sequences, life of Jesus, etc. Each plate is accompanied by the verses from the King James version of the Bible. 241pp. 9 × 12. 23004-X Pa. $8.95

HUGGER-MUGGER IN THE LOUVRE, Elliot Paul. Second Homer Evans mystery-comedy. Theft at the Louvre involves sleuth in hilarious, madcap caper. "A knockout."—Books. 336pp. 5⅜ × 8½. 25185-3 Pa. $5.95

FLATLAND, E. A. Abbott. Intriguing and enormously popular science-fiction classic explores the complexities of trying to survive as a two-dimensional being in a three-dimensional world. Amusingly illustrated by the author. 16 illustrations. 103pp. 5⅜ × 8½. 20001-9 Pa. $2.00

THE HISTORY OF THE LEWIS AND CLARK EXPEDITION, Meriwether Lewis and William Clark, edited by Elliott Coues. Classic edition of Lewis and Clark's day-by-day journals that later became the basis for U.S. claims to Oregon and the West. Accurate and invaluable geographical, botanical, biological, meteorological and anthropological material. Total of 1,508pp. 5⅜ × 8½.
21268-8, 21269-6, 21270-X Pa. Three-vol. set $25.50

LANGUAGE, TRUTH AND LOGIC, Alfred J. Ayer. Famous, clear introduction to Vienna, Cambridge schools of Logical Positivism. Role of philosophy, elimination of metaphysics, nature of analysis, etc. 160pp. 5⅜ × 8½. (Available in U.S. and Canada only) 20010-8 Pa. $2.95

MATHEMATICS FOR THE NONMATHEMATICIAN, Morris Kline. Detailed, college-level treatment of mathematics in cultural and historical context, with numerous exercises. For liberal arts students. Preface. Recommended Reading Lists. Tables. Index. Numerous black-and-white figures. xvi + 641pp. 5⅜ × 8½.
24823-2 Pa. $11.95

28 SCIENCE FICTION STORIES, H. G. Wells. Novels, *Star Begotten* and *Men Like Gods,* plus 26 short stories: "Empire of the Ants," "A Story of the Stone Age," "The Stolen Bacillus," "In the Abyss," etc. 915pp. 5⅜ × 8½. (Available in U.S. only)
20265-8 Cloth. $10.95

HANDBOOK OF PICTORIAL SYMBOLS, Rudolph Modley. 3,250 signs and symbols, many systems in full; official or heavy commercial use. Arranged by subject. Most in Pictorial Archive series. 143pp. 8⅛ × 11. 23357-X Pa. $5.95

INCIDENTS OF TRAVEL IN YUCATAN, John L. Stephens. Classic (1843) exploration of jungles of Yucatan, looking for evidences of Maya civilization. Travel adventures, Mexican and Indian culture, etc. Total of 669pp. 5⅜ × 8½.
20926-1, 20927-X Pa., Two-vol. set $9.90

DEGAS: An Intimate Portrait, Ambroise Vollard. Charming, anecdotal memoir by famous art dealer of one of the greatest 19th-century French painters. 14 black-and-white illustrations. Introduction by Harold L. Van Doren. 96pp. 5⅜ × 8½.
25131-4 Pa. $3.95

PERSONAL NARRATIVE OF A PILGRIMAGE TO ALMANDINAH AND MECCAH, Richard Burton. Great travel classic by remarkably colorful personality. Burton, disguised as a Moroccan, visited sacred shrines of Islam, narrowly escaping death. 47 illustrations. 959pp. 5⅜ × 8½. 21217-3, 21218-1 Pa., Two-vol. set $17.90

PHRASE AND WORD ORIGINS, A. H. Holt. Entertaining, reliable, modern study of more than 1,200 colorful words, phrases, origins and histories. Much unexpected information. 254pp. 5⅜ × 8½. 20758-7 Pa. $4.95

THE RED THUMB MARK, R. Austin Freeman. In this first Dr. Thorndyke case, the great scientific detective draws fascinating conclusions from the nature of a single fingerprint. Exciting story, authentic science. 320pp. 5⅜ × 8½. (Available in U.S. only) 25210-8 Pa. $5.95

AN EGYPTIAN HIEROGLYPHIC DICTIONARY, E. A. Wallis Budge. Monumental work containing about 25,000 words or terms that occur in texts ranging from 3000 B.C. to 600 A.D. Each entry consists of a transliteration of the word, the word in hieroglyphs, and the meaning in English. 1,314pp. 6⅜ × 10.
23615-3, 23616-1 Pa., Two-vol. set $27.90

THE COMPLEAT STRATEGYST: Being a Primer on the Theory of Games of Strategy, J. D. Williams. Highly entertaining classic describes, with many illustrated examples, how to select best strategies in conflict situations. Prefaces. Appendices. xvi + 268pp. 5⅜ × 8½. 25101-2 Pa. $5.95

THE ROAD TO OZ, L. Frank Baum. Dorothy meets the Shaggy Man, little Button-Bright and the Rainbow's beautiful daughter in this delightful trip to the magical Land of Oz. 272pp. 5⅜ × 8. 25208-6 Pa. $4.95

POINT AND LINE TO PLANE, Wassily Kandinsky. Seminal exposition of role of point, line, other elements in non-objective painting. Essential to understanding 20th-century art. 127 illustrations. 192pp. 6½ × 9¼. 23808-3 Pa. $4.50

LADY ANNA, Anthony Trollope. Moving chronicle of Countess Lovel's bitter struggle to win for herself and daughter Anna their rightful rank and fortune—perhaps at cost of sanity itself. 384pp. 5⅜ × 8½. 24669-8 Pa. $6.95

EGYPTIAN MAGIC, E. A. Wallis Budge. Sums up all that is known about magic in Ancient Egypt: the role of magic in controlling the gods, powerful amulets that warded off evil spirits, scarabs of immortality, use of wax images, formulas and spells, the secret name, much more. 253pp. 5⅜ × 8½. 22681-6 Pa. $4.00

THE DANCE OF SIVA, Ananda Coomaraswamy. Preeminent authority unfolds the vast metaphysic of India: the revelation of her art, conception of the universe, social organization, etc. 27 reproductions of art masterpieces. 192pp. 5⅜ × 8½.
24817-8 Pa. $5.95

CHRISTMAS CUSTOMS AND TRADITIONS, Clement A. Miles. Origin, evolution, significance of religious, secular practices. Caroling, gifts, yule logs, much more. Full, scholarly yet fascinating; non-sectarian. 400pp. 5⅜ × 8½.
23354-5 Pa. $6.50

THE HUMAN FIGURE IN MOTION, Eadweard Muybridge. More than 4,500 stopped-action photos, in action series, showing undraped men, women, children jumping, lying down, throwing, sitting, wrestling, carrying, etc. 390pp. 7⅞ × 10⅝.
20204-6 Cloth. $19.95

THE MAN WHO WAS THURSDAY, Gilbert Keith Chesterton. Witty, fast-paced novel about a club of anarchists in turn-of-the-century London. Brilliant social, religious, philosophical speculations. 128pp. 5⅜ × 8½. 25121-7 Pa. $3.95

A CEZANNE SKETCHBOOK: Figures, Portraits, Landscapes and Still Lifes, Paul Cezanne. Great artist experiments with tonal effects, light, mass, other qualities in over 100 drawings. A revealing view of developing master painter, precursor of Cubism. 102 black-and-white illustrations. 144pp. 8¾ × 6⅜. 24790-2 Pa. $5.95

AN ENCYCLOPEDIA OF BATTLES: Accounts of Over 1,560 Battles from 1479 B.C. to the Present, David Eggenberger. Presents essential details of every major battle in recorded history, from the first battle of Megiddo in 1479 B.C. to Grenada in 1984. List of Battle Maps. New Appendix covering the years 1967–1984. Index. 99 illustrations. 544pp. 6½ × 9¼. 24913-1 Pa. $14.95

AN ETYMOLOGICAL DICTIONARY OF MODERN ENGLISH, Ernest Weekley. Richest, fullest work, by foremost British lexicographer. Detailed word histories. Inexhaustible. Total of 856pp. 6½ × 9¼.
21873-2, 21874-0 Pa., Two-vol. set $17.00

WEBSTER'S AMERICAN MILITARY BIOGRAPHIES, edited by Robert McHenry. Over 1,000 figures who shaped 3 centuries of American military history. Detailed biographies of Nathan Hale, Douglas MacArthur, Mary Hallaren, others. Chronologies of engagements, more. Introduction. Addenda. 1,033 entries in alphabetical order. xi + 548pp. 6½ × 9¼. (Available in U.S. only)
24758-9 Pa. $11.95

LIFE IN ANCIENT EGYPT, Adolf Erman. Detailed older account, with much not in more recent books: domestic life, religion, magic, medicine, commerce, and whatever else needed for complete picture. Many illustrations. 597pp. 5⅜ × 8½.
22632-8 Pa. $8.50

HISTORIC COSTUME IN PICTURES, Braun & Schneider. Over 1,450 costumed figures shown, covering a wide variety of peoples: kings, emperors, nobles, priests, servants, soldiers, scholars, townsfolk, peasants, merchants, courtiers, cavaliers, and more. 256pp. 8⅜ × 11¼. 23150-X Pa. $7.95

THE NOTEBOOKS OF LEONARDO DA VINCI, edited by J. P. Richter. Extracts from manuscripts reveal great genius; on painting, sculpture, anatomy, sciences, geography, etc. Both Italian and English. 186 ms. pages reproduced, plus 500 additional drawings, including studies for *Last Supper, Sforza* monument, etc. 860pp. 7⅞ × 10¾. (Available in U.S. only) 22572-0, 22573-9 Pa., Two-vol. set $25.90

CATALOG OF DOVER BOOKS

THE ART NOUVEAU STYLE BOOK OF ALPHONSE MUCHA: All 72 Plates from "Documents Decoratifs" in Original Color, Alphonse Mucha. Rare copyright-free design portfolio by high priest of Art Nouveau. Jewelry, wallpaper, stained glass, furniture, figure studies, plant and animal motifs, etc. Only complete one-volume edition. 80pp. 9⅜ × 12¼. 24044-4 Pa. $8.95

ANIMALS: 1,419 COPYRIGHT-FREE ILLUSTRATIONS OF MAMMALS, BIRDS, FISH, INSECTS, ETC., edited by Jim Harter. Clear wood engravings present, in extremely lifelike poses, over 1,000 species of animals. One of the most extensive pictorial sourcebooks of its kind. Captions. Index. 284pp. 9 × 12. 23766-4 Pa. $9.95

OBELISTS FLY HIGH, C. Daly King. Masterpiece of American detective fiction, long out of print, involves murder on a 1935 transcontinental flight—"a very thrilling story"—NY Times. Unabridged and unaltered republication of the edition published by William Collins Sons & Co. Ltd., London, 1935. 288pp. 5⅜ × 8½. (Available in U.S. only) 25036-9 Pa. $4.95

VICTORIAN AND EDWARDIAN FASHION: A Photographic Survey, Alison Gernsheim. First fashion history completely illustrated by contemporary photographs. Full text plus 235 photos, 1840–1914, in which many celebrities appear. 240pp. 6½ × 9¼. 24205-6 Pa. $6.00

THE ART OF THE FRENCH ILLUSTRATED BOOK, 1700–1914, Gordon N. Ray. Over 630 superb book illustrations by Fragonard, Delacroix, Daumier, Doré, Grandville, Manet, Mucha, Steinlen, Toulouse-Lautrec and many others. Preface. Introduction. 633 halftones. Indices of artists, authors & titles, binders and provenances. Appendices. Bibliography. 608pp. 8⅜ × 11¼. 25086-5 Pa. $24.95

THE WONDERFUL WIZARD OF OZ, L. Frank Baum. Facsimile in full color of America's finest children's classic. 143 illustrations by W. W. Denslow. 267pp. 5⅜ × 8½. 20691-2 Pa. $5.95

FRONTIERS OF MODERN PHYSICS: New Perspectives on Cosmology, Relativity, Black Holes and Extraterrestrial Intelligence, Tony Rothman, et al. For the intelligent layman. Subjects include: cosmological models of the universe; black holes; the neutrino; the search for extraterrestrial intelligence. Introduction. 46 black-and-white illustrations. 192pp. 5⅜ × 8½. 24587-X Pa. $6.95

THE FRIENDLY STARS, Martha Evans Martin & Donald Howard Menzel. Classic text marshalls the stars together in an engaging, non-technical survey, presenting them as sources of beauty in night sky. 23 illustrations. Foreword. 2 star charts. Index. 147pp. 5⅜ × 8½. 21099-5 Pa. $3.50

FADS AND FALLACIES IN THE NAME OF SCIENCE, Martin Gardner. Fair, witty appraisal of cranks, quacks, and quackeries of science and pseudoscience: hollow earth, Velikovsky, orgone energy, Dianetics, flying saucers, Bridey Murphy, food and medical fads, etc. Revised, expanded In the Name of Science. "A very able and even-tempered presentation."—The New Yorker. 363pp. 5⅜ × 8. 20394-8 Pa. $5.95

ANCIENT EGYPT: ITS CULTURE AND HISTORY, J. E Manchip White. From pre-dynastics through Ptolemies: society, history, political structure, religion, daily life, literature, cultural heritage. 48 plates. 217pp. 5⅜ × 8½. 22548-8 Pa. $4.95

SIR HARRY HOTSPUR OF HUMBLETHWAITE, Anthony Trollope. Incisive, unconventional psychological study of a conflict between a wealthy baronet, his idealistic daughter, and their scapegrace cousin. The 1870 novel in its first inexpensive edition in years. 250pp. 5⅜ × 8½. 24953-0 Pa. $4.95

LASERS AND HOLOGRAPHY, Winston E. Kock. Sound introduction to burgeoning field, expanded (1981) for second edition. Wave patterns, coherence, lasers, diffraction, zone plates, properties of holograms, recent advances. 84 illustrations. 160pp. 5⅜ × 8¼. (Except in United Kingdom) 24041-X Pa. $3.50

INTRODUCTION TO ARTIFICIAL INTELLIGENCE: SECOND, EN-LARGED EDITION, Philip C. Jackson, Jr. Comprehensive survey of artificial intelligence—the study of how machines (computers) can be made to act intelligently. Includes introductory and advanced material. Extensive notes updating the main text. 132 black-and-white illustrations. 512pp. 5⅜ × 8½. 24864-X Pa. $8.95

HISTORY OF INDIAN AND INDONESIAN ART, Ananda K. Coomaraswamy. Over 400 illustrations illuminate classic study of Indian art from earliest Harappa finds to early 20th century. Provides philosophical, religious and social insights. 304pp. 6⅜ × 9⅜. 25005-9 Pa. $8.95

THE GOLEM, Gustav Meyrink. Most famous supernatural novel in modern European literature, set in Ghetto of Old Prague around 1890. Compelling story of mystical experiences, strange transformations, profound terror. 13 black-and-white illustrations. 224pp. 5⅜ × 8½. (Available in U.S. only) 25025-3 Pa. $5.95

ARMADALE, Wilkie Collins. Third great mystery novel by the author of *The Woman in White* and *The Moonstone.* Original magazine version with 40 illustrations. 597pp. 5⅜ × 8½. 23429-0 Pa. $7.95

PICTORIAL ENCYCLOPEDIA OF HISTORIC ARCHITECTURAL PLANS, DETAILS AND ELEMENTS: With 1,880 Line Drawings of Arches, Domes, Doorways, Facades, Gables, Windows, etc., John Theodore Haneman. Sourcebook of inspiration for architects, designers, others. Bibliography. Captions. 141pp. 9 × 12. 24605-1 Pa. $6.95

BENCHLEY LOST AND FOUND, Robert Benchley. Finest humor from early 30's, about pet peeves, child psychologists, post office and others. Mostly unavailable elsewhere. 73 illustrations by Peter Arno and others. 183pp. 5⅜ × 8½. 22410-4 Pa. $3.95

ERTÉ GRAPHICS, Erté. Collection of striking color graphics: *Seasons, Alphabet, Numerals, Aces* and *Precious Stones.* 50 plates, including 4 on covers. 48pp. 9⅜ × 12¼. 23580-7 Pa. $6.95

THE JOURNAL OF HENRY D. THOREAU, edited by Bradford Torrey, F. H. Allen. Complete reprinting of 14 volumes, 1837–61, over two million words; the sourcebooks for *Walden,* etc. Definitive. All original sketches, plus 75 photographs. 1,804pp. 8½ × 12¼. 20312-3, 20313-1 Cloth., Two-vol. set $80.00

CASTLES: THEIR CONSTRUCTION AND HISTORY, Sidney Toy. Traces castle development from ancient roots. Nearly 200 photographs and drawings illustrate moats, keeps, baileys, many other features. Caernarvon, Dover Castles, Hadrian's Wall, Tower of London, dozens more. 256pp. 5⅜ × 8¼. 24898-4 Pa. $5.95

AMERICAN CLIPPER SHIPS: 1833–1858, Octavius T. Howe & Frederick C. Matthews. Fully-illustrated, encyclopedic review of 352 clipper ships from the period of America's greatest maritime supremacy. Introduction. 109 halftones. 5 black-and-white line illustrations. Index. Total of 928pp. 5⅜ × 8½.
25115-2, 25116-0 Pa., Two-vol. set $17.90

TOWARDS A NEW ARCHITECTURE, Le Corbusier. Pioneering manifesto by great architect, near legendary founder of "International School." Technical and aesthetic theories, views on industry, economics, relation of form to function, "mass-production spirit," much more. Profusely illustrated. Unabridged translation of 13th French edition. Introduction by Frederick Etchells. 320pp. 6⅛ × 9¼. (Available in U.S. only) 25023-7 Pa. $8.95

THE BOOK OF KELLS, edited by Blanche Cirker. Inexpensive collection of 32 full-color, full-page plates from the greatest illuminated manuscript of the Middle Ages, painstakingly reproduced from rare facsimile edition. Publisher's Note. Captions. 32pp. 9⅜ × 12¼. 24345-1 Pa. $4.50

BEST SCIENCE FICTION STORIES OF H. G. WELLS, H. G. Wells. Full novel *The Invisible Man*, plus 17 short stories: "The Crystal Egg," "Aepyornis Island," "The Strange Orchid," etc. 303pp. 5⅜ × 8½. (Available in U.S. only)
21531-8 Pa. $4.95

AMERICAN SAILING SHIPS: Their Plans and History, Charles G. Davis. Photos, construction details of schooners, frigates, clippers, other sailcraft of 18th to early 20th centuries—plus entertaining discourse on design, rigging, nautical lore, much more. 137 black-and-white illustrations. 240pp. 6⅛ × 9¼.
24658-2 Pa. $5.95

ENTERTAINING MATHEMATICAL PUZZLES, Martin Gardner. Selection of author's favorite conundrums involving arithmetic, money, speed, etc., with lively commentary. Complete solutions. 112pp. 5⅜ × 8½. 25211-6 Pa. $2.95

THE WILL TO BELIEVE, HUMAN IMMORTALITY, William James. Two books bound together. Effect of irrational on logical, and arguments for human immortality. 402pp. 5⅜ × 8½. 20291-7 Pa. $7.50

THE HAUNTED MONASTERY and THE CHINESE MAZE MURDERS, Robert Van Gulik. 2 full novels by Van Gulik continue adventures of Judge Dee and his companions. An evil Taoist monastery, seemingly supernatural events; overgrown topiary maze that hides strange crimes. Set in 7th-century China. 27 illustrations. 328pp. 5⅜ × 8½. 23502-5 Pa. $5.00

CELEBRATED CASES OF JUDGE DEE (DEE GOONG AN), translated by Robert Van Gulik. Authentic 18th-century Chinese detective novel; Dee and associates solve three interlocked cases. Led to Van Gulik's own stories with same characters. Extensive introduction. 9 illustrations. 237pp. 5⅜ × 8½.
23337-5 Pa. $4.95

Prices subject to change without notice.

Available at your book dealer or write for free catalog to Dept. GI, Dover Publications, Inc., 31 East 2nd St., Mineola, N.Y. 11501. Dover publishes more than 175 books each year on science, elementary and advanced mathematics, biology, music, art, literary history, social sciences and other areas.